Radio Recombination Lines

Astrophysics and Space Science Library

For other titles published in this series, go to
http://www.springer.com/5664

Radio Recombination Lines

Their Physics and Astronomical Applications

by

M.A. Gordon

and

R.L. Sorochenko

 Springer

M.A. Gordon
National Radio Astronomy Observatory
Tucson
NRAO
949 North Cherry Avenue
Tucson AZ 85721-0655
USA
mgordon@nrao.edu

R.L. Sorochenko
Russian Academy of Sciences
P.N. Lebedev Physical Institute
Laninsky Prospect 53
Moskva
Russia 119991

ISSN: 0067-0057
ISBN: 978-0-387-09604-9   e-ISBN: 978-0-387-09691-9
DOI: 10.1007/978-0-387-09691-9

Library of Congress Control Number: 2008931183

Printed on acid-free paper

springer.com

To our children and grandchildren, to their children, and to all of the next generations, who may benefit from a deeper understanding of the universe in which they live.

Preface

Recombination lines at radio wavelengths have been – and still are – a powerful tool for modern astronomy. For more than 30 years, they have allowed astronomers to probe the gases from which stars form. They have even been detected in the Sun.

In addition, observations of these spectral lines facilitate basic research into the atom, in forms and environments that can only exist in the huge dimensions and extreme conditions of cosmic laboratories.

We intend this book to serve as a tourist's guide to the world of Radio Recombination Lines. It contains three divisions: a history of their discovery, the physics of how they form and how their voyage to us influences their spectral profiles, and a description of their many astronomical contributions to date. The appendix includes supplementary calculations that may be useful to some astronomers. This material also includes tables of line frequencies from 12 MHz to 30 THz ($\lambda = 10\,\mu$m) as well as FORTRAN computer code to calculate the fine-structure components of the lines, to evaluate radial matrix integrals, and to calculate the departure coefficients of hydrogen in a cosmic environment. It also describes how to convert observational to astrophysical units. The text includes extensive references to the literature to assist readers who want more details.

We appreciate the help of L.W. Avery, D.S. Balser, T.M. Bania, T. Bastian, J.H. Bieging, H.J.J. Blom, N.G. Bochkarev, R.L. Brown, L.A. Bureeva (Minaeva), W.B. Burton, T. Alan Clark, Z.F. Dravskikh, W.C. Erickson, P.A. Feldman, H.C. Goldwire, W.M. Goss, H.R. Griem, C. Heiles, D. Hoang-Binh, D. Hollenbach, S. Japenga, V.S. Lisitsa, F.J. Lockman, J.M. Moran, P. Palmer, Y.N. Parijskij, H.E. Payne, P.A. Pinto, V.O. Ponomarev, B. Robinson, A. Roshi, E.R. Seaquist, M. Simon, P.J. Smiley, G.T. Smirnov, I.I. Sobelman, V.S. Strelnitski, Y. Terzian, R.I. Thompson, J.K.G. Watson, A. Wettstein, and T.L. Wilson.

We are grateful to N.S. Kardashev and P.A. Vanden Bout for providing the support that enabled us to write this book.

This book was written mainly by exchanging Email; we authors did not meet during this collaboration. We thank V.R. Sorochenko (RLS's physicist son) for his great help with these communications as well as for his draft translation of the Russian part of the text.

Contents

List of Figures

List of Tables

Chapter 1
Introduction

Abstract This chapter describes the early theory and initial detection of radio recombination lines from astronomical objects. The focus is historical.

1.1 The Cosmos as a Laboratory

The history of science shows many close connections between physics and astronomy. It is well known that a number of physical laws evolved from a base of astronomical observations. For example, Kepler observed and, later Newton derived, the laws of gravitation while studying the motion of planets and their satellites. The existence of thermonuclear energy was solidly established when it explained the energy balance of the Sun and stars. The anomalous shift of Mercury's perihelion showed us that Newton's gravitational theory was incomplete; this observation helped lead Einstein to the more comprehensive theory of General Relativity.

The cosmos is a wonderful laboratory. There, physicists find that matter can have very high and very low temperatures. It can have ultrahigh and ultralow densities. It can occupy huge volumes. It can exist in states impossible to duplicate in a terrestrial laboratory – states that are not always in dynamical or thermal equilibrium. This extreme diversity of matter in the cosmos is one of the reasons that astronomical observations and astrophysical studies are so valuable.

1.2 Spectral Lines in Astronomy

Low-density cosmic matter gives us a unique opportunity to study elementary processes in atoms and molecules by means of the phenomenon of spectral lines. This is important for physics. It is worthwhile to remember that the

M.A. Gordon, R.L. Sorochenko, *Radio Recombination Lines*, Astrophysics and Space Science Library 282, DOI: 10.1007/978-0-387-09691-9_1, first spectral lines – the Fraunhofer lines – were first detected in astronomic objects, in the spectra of the Sun and the stars. These observations stimulated the development of laboratory spectroscopy.

Emission of spectral lines from cosmic objects became an essential tool in astronomy. The frequency of each line is unique and identifies the atom, ion, or molecule emitting that radiation. Knowing the line frequency through laboratory measurements or through calculations, astronomers can determine the velocity shift of the line and, by local kinematics and by the Hubble law, estimate the distance of the emitting region. The line intensities are related to the number of atoms along the line of sight within the telescope's field of view. The line widths are produced by a combination of the motion of the emitting atoms, of perturbations to the radiation induced by magnetic fields, and by the difficulty that the photons experienced passing through the medium. In this way, the line shapes are the record of what the photons experienced when they were created and in their voyage to us.

The opportunities to investigate spectral lines in astronomy broadened considerably with the extension of astronomical observations into the radio regime, now known as "radio astronomy." One enormous advantage of the radio regime relative to the optical was that the spectral window could be shifted from high to low frequencies, thereby obtaining high spectral resolution at easily managed frequencies. Called "superheterodyne" conversion, this process was developed in the early 1900s to enhance radio receivers for communications. Implementing this technique in the optical regime involves solving difficult physics problems. At present, only limited applications exist.

Spectral lines from a great number of cosmic atoms and molecules are now available throughout the electromagnetic spectrum. In this book, we consider a special class of these spectral lines, namely, spectral lines resulting from transitions between highly excited atomic levels. Conceptually, these lines appear after the recombination of ions and electrons to form atoms, leaving the electrons in levels with high principal quantum numbers n. These newly bound electrons jump downward from level to level much like going down a flight of stairs, losing energy in each jump by radiating it away in the form of a spectral line. When these lines appear in radio regime, they are called "radio recombination lines" (RRLs).

The study of RRLs has revealed a number of surprising new concepts for physics and astronomy. For example, in ultralow-density regions of the interstellar medium (ISM), an atom can exist with electrons in very high quantum levels – up to $n \approx 1,000$ and, correspondingly, with huge diameters approaching 0.1 mm. We can observe the spectral lines from these giant atoms over a wide range of radio waves, from millimeter to decameter wavelengths. Because interstellar atoms are sensitive to variations in gas densities and temperatures in any region, their RRL emission sends us information about the structure of their cosmic environments. And, as we shall see, the basic physics underlying these atomic lines are easy to understand.

1.3 The Bohr Atom

To understand the early searches for RRLs, we first need to discuss a basic physics model known today as the "Bohr atom." This model explains atomic emission lines in a simple way.

Line radiation caused by transitions between atomic levels was detected about 100 years ago. These lines were grouped into series such as the then well-known Lyman, Balmer, and Paschen line series emitted by hydrogen in the ultraviolet (UV), visible, and infrared (IR) wavelength ranges. Physicists soon found empirically that the frequencies ν of the lines in these series could be represented by a simple formula:

$$\nu_{n_2 \to n_1} = Rc \left(\frac{1}{n_1{}^2} - \frac{1}{n_2{}^2} \right), \quad n_2 > n_1 > 0, \tag{1.1}$$

where n_1 and n_2 are positive integers, c is the speed of light, and the constant R was called the Rydberg constant. Each line series could be fitted by choosing a value for n_1 and then sequentially entering values for n_2. For example, $n_1 = 1$ would give the Lyman series; $n_1 = 2$, the Balmer series; $n_1 = 3$, the Paschen series; and so on. Examination of (1.1) shows that the lines of each series become closer together as n_2 increases, forming a "series limit" when $n_2 \Rightarrow \infty$ of

$$\nu_{n_2 = \infty \to n_1} = \frac{Rc}{n_1{}^2} \tag{1.2}$$

beyond which the lines become a continuum clearly visible on the spectral plates.

What are the physics behind these empirical formulas? From these observations, Bohr (1913) developed his quantum theory of the atom – a mathematically simple theory that explained most of the series of atomic lines known at that time.

In this theory, Bohr postulated that atoms have discrete stationary energy levels; in other words, these energy levels are "quantized" rather than continuous. One can imagine a set of orbits of electrons circulating around the nucleus at quantized radii. Introducing discrete quantum numbers for angular momentum, Bohr assumed that only those orbits can exist for which the angular momentum L is a multiple of $h/2\pi$, i.e., described by following expression:

$$L = n\frac{h}{2\pi}, \quad n > 0 \tag{1.3}$$

$$= 1.0545919 \times 10^{-27} n \quad \text{erg sec,} \tag{1.4}$$

where h is a Planck's constant and n is any positive integer. Bohr's formulation allowed orbits of discrete diameters $2a$ given by

$$2a = \frac{n^2 h^2}{2\pi^2 m Z e^2} \tag{1.5}$$

$$= 1.05835 \times 10^{-8} n^2 \quad \text{cm}, \tag{1.6}$$

where m is the mass of the electron, e is the electronic charge in ESU, and Ze is the charge of the nucleus. Equation (1.6) indicates that the sizes of orbits as well as atom's sizes increase as n^2. Setting $n=1$ and $Z=1$ produces the radius of the first orbit of hydrogen, known as the "Bohr radius,"

$$a_0 = \frac{h^2}{4\pi^2 m e^2} \tag{1.7}$$

often used as a parameter in equations involving atomic physics.

Classical electrodynamics predicted orbital diameters by equating the electrical attraction between each electron and the nucleus to the centripetal acceleration:

$$\frac{Ze^2}{a^2} = \frac{mv^2}{a}, \quad \text{or} \tag{1.8}$$

$$2a = \frac{2Ze^2}{mv^2}, \tag{1.9}$$

so that every orbital diameter would be allowed depending upon the orbital speed v or the kinematic energy mv^2 of the electron.

The total energy E of an electron in a circular orbit is the sum of the electrical potential and the kinetic energy:

$$E = -\frac{Ze^2}{a} + \frac{1}{2}mv^2 \tag{1.10}$$

$$= -\frac{Ze^2}{a} + \frac{Ze^2}{2a} \tag{1.11}$$

$$= -\frac{Ze^2}{2a} \tag{1.12}$$

after substitution of (1.9). Using the quantization of orbits described by (1.6), Bohr calculated the energy E_n associated with each electronic orbit[1] n:

$$E_n = -\frac{2\pi^2 m e^4}{h^2} \frac{Z^2}{n^2} \tag{1.13}$$

$$= -2.17989724 \times 10^{-11} \frac{Z^2}{n^2} \quad \text{ergs.} \tag{1.14}$$

---

[1] A fundamental difference between the Bohr theory and classical electrodynamics is that in the Bohr theory, electrons do not radiate even though they are technically accelerating by changing direction.

Note that the energy of bound electrons must be negative. Because the energy of a photon is $h\nu$, the frequency of each atomic line would then be

$$\nu_{n_2 \to n_1} = \frac{E_{n_2} - E_{n_1}}{h}, \quad \text{or} \tag{1.15}$$

$$= \frac{2\pi^2 m Z^2 e^4}{h^3} \left(\frac{1}{n_1{}^2} - \frac{1}{n_2{}^2} \right) \tag{1.16}$$

$$= RcZ^2 \left(\frac{1}{n_1{}^2} - \frac{1}{n_2{}^2} \right), \tag{1.17}$$

which is identical with the empirical formula given by (1.1) if the effective nuclear charge $Z = 1$ and the Rydberg constant is

$$R = \frac{2\pi^2 m e^4}{h^3 c}. \tag{1.18}$$

Substituting into (1.18) the values of the physical constants listed in Table A.1, we derive $R = 109,737.35\,\text{cm}^{-1}$ which is close to the value of $109,675\,\text{cm}^{-1}$ obtained by Rydberg (1890) from measurements of hydrogen spectral lines. Such close agreement leaves no doubt regarding the validity of (1.14).

Later, Bohr (1914) did even better. The original theory assumed an infinitely small electronic mass orbiting the nucleus. He refined his earlier equations to use the center of mass as the centroid of the orbit and the reduced mass m_R in place of the orbiting electronic mass, so that

$$R = R_\infty \left(\frac{M}{M + m} \right), \tag{1.19}$$

where the coefficient

$$R_\infty = \frac{2\pi^2 m e^4}{c h^3} \tag{1.20}$$

and is now called the Rydberg constant for infinite mass. Section A.2.1 gives details. With this correction, the Rydberg constant for hydrogen $R_H = 109,677.57\,\text{cm}^{-1}$. The calculated and measured values of R_H now agreed within 0.002%.

There are additional refinements to the Bohr model that improve generality. These include consideration of elliptical orbits and the quantization of angular momentum. Section C.2 describes these calculations in detail.

Although our discussion has so far concentrated on hydrogen, these equations can also describe RRL spectra of multielectron atoms *and* ions. This "hydrogenic" model assumes that only one electron is in an excited level; the $Z - 1$ other electrons lie in or near ground levels. For neutral atoms, the net negative charge of the inner electrons would screen the positive charge of the nucleus, so that a lone outer electron would see only a single nuclear charge and $Z = 1$. For ions, a similar situation would obtain but with $Z > 1$. Table A.2 gives Rydberg constants for a few atoms[2] common to the cosmos.

1.3.1 Bohr Lines at Radio Wavelengths

The theory did not restrict the number of atomic levels nor the number of the line series. Bohr (1914) showed that, for large quantum numbers and for transitions from $n_2 = n + 1 \rightarrow n_1 = n$, (A.6) gives a series of line frequencies for Bohr lines of neutral hydrogen:[3]

$$\nu_H \approx \frac{2R_H c Z^2}{n^3}, \quad n \gg 1$$

$$= 6.58 \times 10^{15} \frac{1}{n^3} \quad \text{Hz} \tag{1.22}$$

with an accuracy of about 2–3% depending upon the frequency. Although unrealized at the time, substituting values of, say, $100 < n < 200$ into (1.22) will yield approximate frequencies for lines throughout the radio range.

1.3.2 Other Line Series

Bohr's model was a brilliant success. It not only explained the hydrogen line series observed up to the year 1913 but predicted new lines as well. However, research into spectral lines toward longer wavelengths proceeded slowly. The fourth atomic series for hydrogen with $n_1 = 4$ and the first line $\lambda_{5 \rightarrow 4} = 4.05\,\mu\text{m}$ was detected by Brackett (1922) 9 years after Bohr's theory had appeared; the fifth series with $n_1 = 5$ and the first line $\lambda_{6 \rightarrow 5} = 7.46\,\mu\text{m}$, by Pfund (1924) 11 years after; and the sixth series with $n_1 = 6$ and the first

---

[2] A property of the Bohr line series expressed by (1.17) and (1.19) is that the entire line series can be shifted in frequency by changing the Rydberg constant R. Radial velocities will cause similar shifts. This means that identification of the atomic species emitting these lines in a cosmic environment, in principle, cannot be made simply on the basis of the observed frequencies – as can be done for molecular emission lines with their less regularly spaced frequencies. In practice, radial velocities for observed optical and molecular lines along the same sight lines help identification of the atomic species of Bohr lines from cosmic gas.

[3] The approximation comes from the first term of the binomial expansion of (1.17) when $n_2 \equiv n_1 + \Delta n$:

$$\nu_H \approx 2R_H c Z^2 \frac{\Delta n}{n^3} \left[1 - \frac{3}{2}\left(\frac{\Delta n}{n}\right) + 2\left(\frac{\Delta n}{n}\right)^2 - \frac{5}{2}\left(\frac{\Delta n}{n}\right)^3 + \cdots \right], \quad \Delta n \ll n.$$

$$\tag{1.21}$$

line $\lambda_{7\to6} = 12.3\,\mu m$, by Humphreys (1953) 40 years later as a result of a very fine measurements of spectra in a gas discharge.

With these studies, classical laboratory spectroscopy ran out of ability. Only new techniques could find new series and answer the question of how far the theory could go. In this quest, the frequencies of the lines moved from the optical through the infrared into the radio regime. Here, astronomy and, more exactly, its rapidly developing branch of radio astronomy came to the aid of physics.

1.4 Spectral Lines in Radio Astronomy

1.4.1 Theoretical Studies

The Dutch astronomer, van de Hulst (1945), was the first to consider the possibility of radio line radiation from transitions between highly excited levels of atoms in the ISM. In the same classical paper that predicted the $\lambda = 21$ cm line, van de Hulst also considered radiation from ionized hydrogen for both free–free and bound–bound transitions.

While calculation of the total emission in these lines is straightforward, the detectability of such RRLs would depend upon the distribution of this emission above the underlying continuum emission or, in other words, upon the shape of the emission lines. Although thermal conditions determine the amount of emission in the RRLs relative to the continuum emission, other effects like Stark broadening can widen the lines, spreading out the line emission in frequency, thereby reducing their peak intensities and, in turn, their detectability. van de Hulst derived an expression for the ratio of the peak intensity of the line I_L to the continuum intensity I_C:

$$\frac{I_L}{I_C} = 0.1 \frac{\nu}{\Delta\nu} \frac{h\nu}{kT}, \tag{1.23}$$

where $\Delta\nu$ is the full frequency width of the line, T is the temperature of medium, and k is a Boltzmann's constant. Conceptually, (1.23) describes the total emission in the line to be the product $I_L\Delta\nu$.

To estimate the Stark broadening, van de Hulst drew from an analysis by Inglis and Teller (1939) of optical and infrared line series in stellar spectra. Within the hydrogenic line series of stellar spectra (see Sect. 1.3), there is a wavelength, short of the series limit, at which distinct lines can no longer be seen. In frequency units, (1.17) models these series (for any given n_1) as $n_2 \to \infty$. At a critical value of n_2, the lines of that series merge into a continuum that continues until the series limit is reached.

The explanation for this line merging is simple. At this critical frequency (or wavelength), Stark broadening within the stellar atmosphere broadens the line to match the gap between it and the adjacent line at $n_2 + 1$. By counting lines, an astronomer determines this critical value of n_2 and calculates the wavelength (or frequency) separation to the next line at $n_2 + 1$. This separation must equal the amount of the line broadening and, consequently, is a measure of the electron density necessary to produce it. In this way, Inglis and Teller provided a method of determining gas densities in stellar atmospheres.

Using results from Inglis and Teller and an estimate for the density of the ISM, van de Hulst estimated the magnitude of Stark broadening to be

$$\frac{\Delta\nu}{\nu} = \left(\frac{\lambda}{100\,\mathrm{m}}\right)^{3/5}. \tag{1.24}$$

Although the original paper gives few details regarding this formula, the quantity $\Delta\nu/\nu \approx 0.02$ if $\lambda = 20\,\mathrm{cm}$, a typical wavelength considered for radio astronomy in 1944, e.g., the $\lambda = 21\,\mathrm{cm}$ line. At this wavelength, (1.24) shows the Stark broadening to be dramatically bigger than the thermal broadening that he correctly estimated as $\Delta\nu/\nu = 10^{-4}$. In the meter wavelength range where radio astronomers (Reber, 1944) were actually observing at that time, the Stark broadening would be even larger. Consequently, van de Hulst concluded that hydrogen lines caused by transitions between highly excited levels would be too broad and, therefore, too weak to be observed.[4]

Other astronomers were also pessimistic. Reber and Greenstein (1947) had considered hydrogen radio lines in their examination of the astronomical possibilities of radio wavelengths but had excluded them, "these [lines] have small intensity." Wild (1952) also considered RRLs but dismissed them because "these lines are so numerous that, without the presence of some selection mechanism they may be regarded merely as contributing toward a continuous spectrum."

Kardashev (1959) reached just the opposite conclusion. Although he was aware that Wild (1952) had dismissed the possibility of detecting lines, he was unaware of the very pessimistic van de Hulst (1945) study.[5] Kardashev made

---

[4] Many years later, Sullivan (1982) analyzed the working notes of van de Hulst while studying the history of radio astronomy. He found a place in these notes where van de Hulst appeared to have inverted the exponent in (1.24), i.e., the Stark broadening should vary as $(\lambda/100\mathrm{m})^{5/3}$. In fact, combining expressions from the two relevant papers (van de Hulst, 1945; Inglis and Teller, 1939) show that the exponent indeed should have been 5/3. The correct formula – not (1.24) – would predict smaller Stark broadening at radio wavelengths and, correspondingly, more intense line intensities. Sullivan did not comment on the choice of electron density used to derive (1.24). Probably, van de Hulst assumed N_e to be $1\,\mathrm{cm}^{-3}$. See Appendix C.1 for more details.

[5] The van de Hulst paper was published in a very rare edition of "Nederlands Tijdschrift voor Natuurkunde" that Soviet libraries did not have. The disruption of scientific contact during the second world war and, later, during the cold war years also played a role. According to Shklovsky (1956b; 1960), Soviet scientists had learned about the $\lambda = 21\,\mathrm{cm}$ line only from references and comments that appeared much later in journals that were

detailed calculations of the expected line widths and intensities of excited hydrogen RRLs in ionized nebulae (H II regions).

The earlier papers by highly respected astronomers created a difficult climate for optimism with respect to detections of RRLs. Parijskij (2002) recalls an ad hoc meeting at the IAU General Assembly in Moscow in August 1958, where well-known radio astronomers discussed with the then young Nicolay Kardashev the validity of his new, encouraging calculations (Kardashev, 1959). These probably included W.L. Erickson, G.B. Field, L. Goldberg, F.T. Haddock, J.P. Hagen, D.S. Heeschen, T.K. Menon, C.A. Muller, H.F. Weaver, and G.L. Westerhout (Kardashev, 2002). The discussion took place in a small room in a new building of Moscow University. Parijskij acted as interpreter. He recalls that the discussion was interesting but quite intense – one "of the deepest I have heard in my life" – with the experienced astronomers examining every calculation made by Kardashev. At the end of the 2-h meeting, they took some kind of a vote and decided that Kardashev might well be correct. This must have been a challenging experience for the young astronomer.

The principal difference between Kardashev and van de Hulst in these calculations lies in their approach to Stark broadening. Kardashev also used the Inglis–Teller relationship, but only for a rough estimate. Independently, he calculated Stark broadening from collisions of excited atoms with electrons as well as from quasistatic broadening. From this analysis, he concluded that, in H II regions with typical values of electron temperature $T_e = 10^4 \, K$ and density $N_e = 10^2 \, \mathrm{cm}^{-3}$, pressure broadening would have no significant influence on the line broadening at frequencies greater than 7,000 MHz. In other words, he concluded that line widths would be determined solely from thermal effects, i.e., from the frequency redistribution of emission from a Maxwellian gas according to Doppler effects giving rise to a Gaussian line shape.

After calculating an oscillator strength to determine the line intensities, Kardashev predicted that excited hydrogen radio lines would be observable by radio astronomical techniques in the range from the FIR to decimeter waves. He also showed that the $n \to n - 1$ transition would have highest intensity and, in addition to the hydrogen lines, the radio lines of helium would be detectable. The frequencies of the helium lines would be shifted relative to hydrogen because of the difference in the Rydberg constant (see Table A.2) due to its greater mass.

Subsequent calculations made it possible to define the intensities of expected radio lines more accurately and, thereby, to plan a search optimized in both frequency and in target sources (Sorochenko, 1965). To re-estimate the line intensities, the attention was again focused on Stark broadening. This time, the calculations used the theory of line broadening in a plasma as developed in early 1960s (Griem, 1960).

---

available in the USSR. In fact, the van de Hulst paper only appeared in the USSR after its translation (Sullivan, 1982) into English.

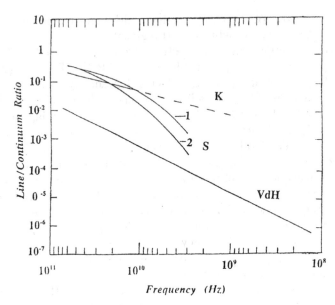

Fig. 1.1 Predicted line-to-continuum ratios for radio recombination lines as a function of frequency. *VdH* van de Hulst (1945) from (1.23) and (1.24), *K* Kardashev (1959) who calculated that Stark broadening may be neglected for $\nu > 7,000$ MHz, *S* Sorochenko (1965) who considered both thermal and Stark broadening for the two values of electron density $N_e = 100\,\mathrm{cm}^{-3}$ (1) and $N_e = 1,000\,\mathrm{cm}^{-3}$ (2). All calculations assume an electron temperature of $T_e = 10^4\,K$

Figure 1.1 summarizes the line-to-continuum ratios (I_L/I_C) from the papers mentioned. All calculations refer to the $n \to n-1$ transitions. One can see that van de Hulst (1945) strongly underestimated I_L/I_C, especially taking into account the probable adopted density $N_e = 1\,\mathrm{cm}^{-3}$. For Doppler broadening alone as calculated by Kardashev (1959) for the centimeter wavelength range, the L/C ratio is a few percent, at values of $N_e = 10^2\,\mathrm{cm}^{-3}$ appropriate for H II regions. If Stark broadening is taken into account, at $N_e \geq 10^2\,\mathrm{cm}^{-3}$ the L/C ratio decreases noticeably at frequencies $\nu < 10\,\mathrm{GHz}\,(\lambda = 3\,\mathrm{cm})$.

To estimate realistic circumstances for the detection of the radio lines, Sorochenko (1965) calculated their intensities in the units of brightness temperature T_b customary in radio astronomy. Figure 1.2 shows these expected values at line center $T_{b,l.c}$ as a function of wavelength λ and of N_e. These data are normalized to the value of the emission measure (EM) of the H II region. EM is a physical parameter calculated from observations of the continuum emission of an H II region and defined as $N_e^2 L\,\mathrm{cm}^{-6}\,\mathrm{pc}$, where L is the depth of an H II region in parsecs[6] (pc) along the line of sight through the H II region.

---

[6] Abbreviation for "parallactic arcsecond," the distance at which the average Earth–Sun distance subtends an angle of $1''$. One pc $= 3.0856 \times 10^{18}$ cm.

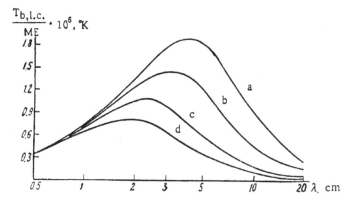

Fig. 1.2 Brightness temperature at the line center as a function of wavelength: (*a*) $N_e = 100 \, \text{cm}^{-3}$, (*b*) $N_e = 200 \, \text{cm}^{-3}$, (*c*) $N_e = 500 \, \text{cm}^{-3}$, and (*d*) $N_e = 1,000 \, \text{cm}^{-3}$

In the millimeter range, the effective size of the atoms is small, there is less collisional interaction with the ambient H II gas, and Stark (pressure) broadening is insignificant as a result. Here, only Doppler broadening determines the line widths and Fig. 1.2 shows the brightness temperature of the lines to increase with wavelength. For an electron density of $N_e > 100 \, \text{cm}^{-3}$, Stark broadening of the lines begins to manifest itself at centimeter wavelengths, spreading the line emission over a broader wavelength (frequency) range and reducing the peak intensity of the lines. As the wavelength increases further, the line intensities decline sharply. There is a peak or "turnover" in each of the curves, with the maximum of the brightness temperature shifting toward shorter wavelengths with larger densities.

Simple Bohr atom physics easily explains this effect. Because the longer wavelength lines are generated by atoms whose electrons are in larger orbits, the effective size of these atoms is larger, and their larger sizes render them more likely to interact or collide with the charged particles of the ambient H II gas. These collisions strip the atoms of the outer electrons, thereby removing their ability to radiate and, correspondingly, reducing the aggregate line intensity emitted by the H II region. The wavelength of the turnover is directly related to the probability of these collisions and, therefore, decreases as the gas density increases.

From an experimental viewpoint, this analysis indicated that the search for RRLs would be more effective at low centimeter wavelengths where Stark broadening would be weakest and the line intensities would be the strongest. Specifically, it suggested that the search should take place at $\lambda = (2 - 5) \, \text{cm}$ in the brightest, extended H II regions, the Omega and Orion nebulae. Furthermore, at these wavelengths, the angular sizes of these bright H II regions would be well matched to the beam of typical radio telescopes available at that time, thereby ensuring maximum sensitivity for the search.

The stage had now been prepared for the main act: the actual detection of RRLs. In actuality, of course, the stories of theoretical refinements and the searches were complex and intertwined.

1.4.2 Detection of Radio Recombination Lines

The first attempt to detect radio lines emitted by highly excited atoms was undertaken at the end of 1958 in Pulkovo by Egorova and Ryzkov (1960) just after they learned about Kardashev's calculations. Utilizing the receiver developed to search for the deuterium lines ($\lambda = 91.6$ cm), and the unmovable parabolic antenna 20×15 m, they searched for hydrogen radio line corresponding to the $n_{272} \rightarrow n_{271}$, or H271$\alpha$, transition in the Galactic plane over the longitude range $l = 60°$–$115°$, but without success.

Five years later, Pulkovo radio astronomers repeated their attempt. At this time, the search was done in 1963 by Z.V. Dravskikh and A.F. Dravskikh (1964) during the testing of the new 32-m paraboloid antenna of the Space Research Center in the Crimea. A simple $\lambda = 5$ cm mixer receiver with filter width of 2 MHz and a tuning accuracy of about 1–3 MHz scanned over a 20-MHz band to search for the $n_{105} \rightarrow n_{104}$ hydrogen line at 5.76 GHz in the Omega and Orion nebulae.

According to Dravskikh (1994; 1996), a strong wind arose during their scheduled time, making it difficult to point the telescope and resulting in only eight spectrograms for the Omega nebula and five for Orion. The quality of these spectra were accordingly poor, and the Dravskikhs were reluctant to consider them further. However, a young colleague, Yuri Parijskij, insisted that spectra should be processed further, believing that the wind effects could be removed. After this processing, the lines appeared to be present in the spectra of each nebula – although too weak to convince everyone of the reality of their detection, shown in Fig. 1.3. The authors themselves estimated the detection probability to be 0.9, corresponding to a signal-to-noise (S/N) ratio of 2.

At that time, the situation was very competitive. Two Soviet groups had been preparing to search for RRLs. Besides A.F. Dravskikh and Z.V. Dravskikh in Pulkovo, the other group for detecting lines was located at the Lebedev Physical Institute in Moscow, where they had been preparing since 1963. The competition involved the quality as well as the timing of the searches. The Pulkovo group had been able to begin their observations earlier but the detections were marginal, having been achieved by necessarily salvaging the unfortunate wind-damaged spectra. On the other hand, the Lebedev group wanted to make detections that would be convincing to everyone and were willing to delay their observing until their specially designed equipment was ready.

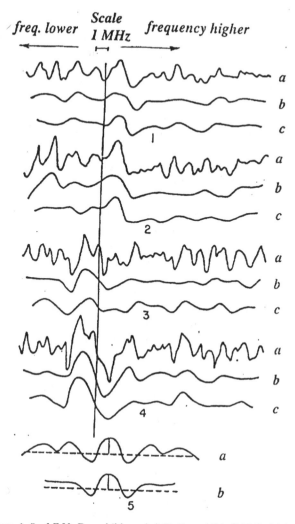

Fig. 1.3 Figures 1–5 of Z.V. Dravskikh and A.F. Dravskikh (1964): (*1a*) average of the eight spectra of the Omega nebula at 5.7 GHz, (*1b*) previous spectrum convolved with a 1-MHz filter, (*1c*) a different smoothing scheme; (*2*) same as (*1*), but only four spectra of the Omega nebula; (*3*) average of all five spectra of Orion; (*4*) average of only three spectrograms from Orion; and (*5*) the average of the left and right sides of the spectra (centered about the line proposed position) in the Omega (*a*) and Orion (*b*) nebulae. The *vertical line* marks the radial velocity position of $0 \, \mathrm{km \, s^{-1}}$ with respect to the Local Standard of Rest (LSR) for the $n_{105} \rightarrow n_{104}$ line

Based upon a closer analysis of the lines' expected properties and intensities with regard to their 22-m radio telescope (Sorochenko, 1965), the Lebedev group came to the conclusion that a new receiver would be needed – one with a sensitivity at least an order of magnitude greater than the existing spectrometers being used for observations of the $\lambda = 21$ cm line. The recombination lines were not only expected to be very weak in themselves but also expected to be weak with respect to the stronger background continuum emission emitted by the H II regions (see Fig. 1.1). In other words, the very objects in which the weak lines should appear would also be emitting strong background emission that would make detection more difficult.

To overcome these difficulties, the Lebedev staff developed a nulling-type spectral radiometer at a wavelength of $\lambda = 3.4$ cm using low noise parametric amplifiers. With great accuracy, this radiometer ensured that the noise in the 20-MHz band was the same for the source (antenna) and the reference load. In this way, it was insensitive to fluctuations of the background continuum such as pointing errors, changes in atmospheric emission, etc. At the same time, it was capable of detecting weak, narrow spectral lines superimposed upon the strong, background continuum emission.

On 27 April 1964, using this radiometer and the 22-m radio telescope of the Physical Institute in Pushchino shown in Fig. 1.4, Sorochenko and Borodzich (1965) detected the hydrogen radio line $n_{91} \rightarrow n_{90}$ (H90α) at 8,872.5 MHz in the spectrum of the Omega nebula on their first attempt. Figure 1.5 shows these spectra. Unlike the earlier observations of the Pulkovo group 4 months earlier in December 1963, this line was clearly present even in the individual spectrograms. The specially designed receiver and the better observing conditions had made a definitive difference. Observations carried out over the next 3 months showed shifts in the line frequency corresponding to the Doppler shifts expected from the Earth's orbital rotation. These frequency shifts dispelled any doubts about the cosmic origin of the line.

Nearly simultaneously with the Lebedev group, the group (Dravskikh, Dravskikh, Kolbasov, Misezhnikov, Nikulin and Shteinshleiger, 1965) at Pulkovo observatory also convincingly detected an excited hydrogen line. Only a month separated these two detections. After improving their radiometer by installing a maser amplifier for the receiver, they were able to detect the hydrogen radio line $n_{105} \rightarrow n_{104}$ at 5,762.9 MHz with the 32-m radio telescope. This time, in May and July of 1964, there was no doubt. The H104α line had definitely been detected in the Omega nebula. Figure 1.5 shows their spectra as well. The Doppler shift of line frequency due to orbital motion of Earth was also found, confirming this detection as well.

On 31 August 1964, the results of both groups were communicated to astronomers attending the XII General Assembly of the International Astronomical Union in Hamburg, Germany. In a joint session of Commissions 33, 34, and 40 organized by Westerhout, Yuri Parijskij presented a paper on behalf of Dravskikh et al. (1966), and Vitkevitch did the same for Sorochenko and Borodzich (1966). Figure 1.5 shows the first spectrograms of the excited hydrogen lines $n_{91} \rightarrow n_{90}$ and $n_{105} \rightarrow n_{104}$ with good S/R ratios that were presented to the IAU General Assembly.

At that presentation, there were a number of questions from the audience (Dravskikh, 1996). Accustomed to the much higher S/N ratios of

Fig. 1.4 The 22-m radio telescope of the P.N. Lebedev Physical Institute in Pushchino, 100 km south of Moscow. With this instrument, the excited hydrogen line $n_{91} \rightarrow n_{90}$ was clearly detected on 27 April 1964

the $\lambda = 21$ cm radio spectra of atomic hydrogen and having experienced the technical difficulties of observing weak spectral lines at that time, many astronomers were skeptical of these clearly noisy results (Price, 2002). In addition, the visual material at that time was less than ideal, and the presentations came near the end of the day, being the 23rd and 24th papers of the 26 presented. Alan H. Barrett, codiscoverer of the second known radio astronomical line (OH) during the previous October (Weinreb, Barrett, Meeks and Henry, 1963), asked Parijskij, "Are you saying that you detected the excited hydrogen line $n = 105 \rightarrow 104$?" Parijskij replied, "Yes." Barrett repeated his question, "Are you saying that these lines can exist?" Parijskij again replied, "Yes." Despite their reservations, this exchange (Parijskij, 2002) also shows

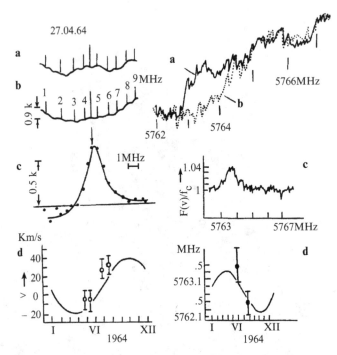

Fig. 1.5 The first spectrograms of excited hydrogen lines with good signal-to-noise ratios that were presented to the IAU General Assembly in Hamburg, Germany, in 1964. *On the left*, the $n_{91} \rightarrow n_{90}$ (H90α) hydrogen line observations from Pushchino. **a** The spectrogram toward the Omega nebula. **b** The test spectrogram with the antenna off the source. **c** The average of seven spectra toward Omega and the five test spectra made in April 1964. The abscissa is frequency; the ordinate, the antenna temperature. The *large mark* indicates the calculated line frequency; the *vertical dashes*, 1-MHz intervals. **d** The measured Doppler shift during the year 1964. This *curve* indicates the nebula's calculated radial velocity relative to Earth (Sorochenko and Borodzich, 1965). *On the right*, the Pulkovo observations of the $n_{105} \rightarrow n_{104}$ (H104α) hydrogen line. **a** The spectrogram toward Omega. **b** The test spectrogram. **c** The average of the 12 spectrograms obtained in May 1964. The abscissa is frequency; the ordinate, the ratio of nebula's line-to-continuum flux densities. **d** The measured Doppler shift during the year 1964, showing the calculated frequency shift (Dravskikh et al., 1965)

that the radio astronomers were very interested in these new results. The participants brought this news back to colleagues in their own radio astronomy groups. Yet, few of them followed up the detections immediately, possibly because of what they wrongly believed to be uncertain results.

The detections obtained by the P.N. Lebedev Physical Institute and the Pulkovo observatory groups confirmed each other. The initial detections differed by only one month or so, and the confirming Doppler observations overlapped each other. For this reason, both Soviet groups agreed to consider 31 August 1964 – the date the detections were presented to the IAU General Assembly – to be the official discovery date of radio lines emitted by excited atoms. As described earlier in Sect. 1.3, these lines result from the process of recombination of ions and electrons and, when they occur in the radio wavelength regime, they are called "radio recombination lines."

In the end, the competition between the two observational groups had produced a great success: the unambiguous detection of not just one recombination line but two. RRLs had arrived as a powerful new tool for astronomers everywhere.

1.4.3 Other Searches and Detections

Radio astronomers outside the Soviet Union had also read the Kardashev paper with interest. In particular, the German radio astronomer P.G. Mezger tried unsuccessfully to detect an excited hydrogen line at 2.8 GHz in 1960 with the 25-m Stockert radio telescope of the University of Bonn (Mezger, 1960), after having received a translation of the 1959 Kardashev paper. Mezger (1992) gives an account[7] of this search in his autobiographical book, "Blick in das kalte Weltall" ("A Look into the Cold Universe").

Some of the initial searchers for cosmic RRLs simply had bad luck. The Australian group of John Bolton, Frank Gardner, and Brian Robinson used the CSIRO 64-m telescope at Parkes to make a quick search for the lines near 5 GHz, probably in 1963 prior to the Soviet detections although the exact date has been lost (Robinson, 2001). Unfortunately, they used the n^{-3} approximation given by (1.21) to calculate the rest line frequency. The approximation error placed the actual line frequency outside of their narrowband spectrometer and, consequently, the line was not detected.

Mezger's interest persisted and, in 1963 after moving to the US National Radio Astronomy observatory (NRAO) in Green Bank, WV, he tried again to detect these lines together with B. Höglund (Höglund and Mezger, 1965), a visiting Swedish astronomer. The first series of observations were made with the NRAO 85-ft. (26-m) telescope beginning in the Fall of 1964. He used a new receiver especially designed for the detection of RRLs. The spectrometer part of the instrument had been designed and built by B. Höglund principally for observations of the then well-known $\lambda = 21$ cm hydrogen line but had been borrowed for a search for RRLs. According to Mezger (1994), these attempts gave "ambiguous results, probably because of local oscillator instabilities."

The second attempt brought success. Using the newly completed 140-ft. (43-m) telescope shown in Fig. 1.6, Höglund and Mezger (1965) were able to detect the H109α[8] lines (5 GHz) with unexpectedly high signal-to-noise

---

[7] See also an after-dinner talk given by Moran (1994) in celebration of Mezger's 65th birthday.

[8] From this point on, we shall exclusively use the standard convention (Lilley, Menzel, Penfield and Zuckerman, 1966) for naming RRLs: the elemental symbol from the Mendeleev

Fig. 1.6 The 140-ft. (42.6-m) diameter, equatorial radio telescope at the Green Bank, WV, facility of the US National Radio Astronomy observatory. The telescope saw first light in 1964 and was taken out of service in 1999

ratios on 9 July 1965, the first day of the observations. Figure 1.7 shows the astronomers and Fig. 1.8 shows their results.

Two days after their successful observations, they presented their observations to the Scientific Visiting Committee, appointed annually by NRAO's management organization, Associated Universities, Inc., to assess NRAO operations. On this committee was Ed Purcell, a Nobel laureate in physics from

Periodic Table, followed by the principal quantum number of the lower level, and followed again by a Greek letter identifying the order of the transition. For example, H109α corresponds to the hydrogen line from the transition $n_{110} \rightarrow n_{109}$. Similarly, H92$\beta$ corresponds to the hydrogen transition $n_{94} \rightarrow n_{92}$, etc.

Fig. 1.7 Control room of the NRAO 140-ft. telescope in July 1965, near the time of the detection of the H109α RRL. *Left to right*: P.G. Mezger, H. Brown (telescope operator), B. Höglund, and N. Albaugh (electronic technician). NRAO photograph

Harvard who was codiscoverer of the $\lambda = 21$ cm hydrogen line in 1951 – the first spectral line available to radio astronomers. According to Mezger (1992), Purcell was fascinated that atoms with size scales of μm could exist in interstellar space.

When, also according to Mezger (1992), Purcell returned to Harvard from the meeting, he shared the news with physics colleagues about the existence of such large atoms – to be called "Rydberg atoms" – in cosmic gases. He learned that Harvard's radio astronomy group, which had the technical capacity for the RRL detections, had not searched for the lines. Later, Mezger heard a rumor that Donald Menzel, a brilliant theoretician within the astronomy department, earlier had convinced their radio astronomy group that pressure broadening would make RRLs impossible to detect, possibly the result of reading van de Hulst's 1945 paper. Within a few days of hearing the news, the Harvard radio astronomy group returned their maser receiver to the nearest RRL frequency and detected the H156α and H158α lines near $\lambda = 18$ cm in the Omega and W51 nebulae also without difficulty (Lilley, Menzel, Penfield and Zuckerman, 1966).

However, the comparatively late participation of the Harvard astronomy group may have been due to other factors (Palmer, 2001). In 1964, OH maser lines had been discovered, and the 60-ft. (18-m) radio telescope at Harvard's

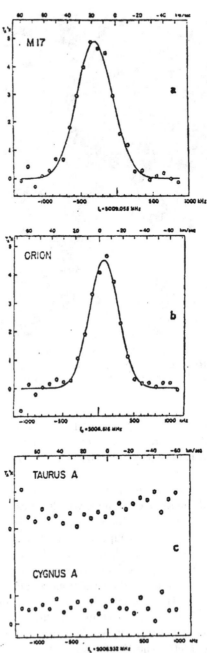

Fig. 1.8 Figure 1 of Höglund and Mezger (1965) showing the detections of the H109α
RRLs with the newly completed 140 ft. (43 m) telescope of the National Radio Astronomy
observatory in Green Bank, WV. It is significant that, whereas the RRLs appeared in the
gaseous nebulae Omega (M17) and Orion, they did not appear in the nonthermal sources
Taurus A and Cygnus A

Agassiz observing station was fully engaged exploring their characteristics. In addition, Harvard's professor of radio astronomy, Ed Lilley, was devoting most of his energy promoting the design and construction of a huge radome-enclosed radio/radar astronomy telescope (Northeast Radio Observatory Corporation – NEROC) to be used by US northeastern universities, which, unfortunately, was never funded. Finally, according to Palmer, at that time all the Harvard astronomy graduate students were involved with other research projects; none were available to pursue the new discovery. Nonetheless, Purcell did persuade the group to redirect some resources toward searching for RRLs after returning from the Green Bank meeting, which led directly to the detection of the $\lambda = 18$ cm RRLs.

With these detections, the dam had broken, and soon other radio astronomical groups were detecting RRLs with excellent signal-to-noise ratios. Lilley et al. (1966) reported the detection of RRLs from helium at $\lambda = 18$ cm. Gardner and McGee (1967) and McGee and Gardner (1967) reported the detection of a number of hydrogen α and β RRLs at frequencies of 1.4 and 3.3 GHz with the 210-ft. (64-m) telescope in Parkes, Australia. These observations gave line ratios consistent with emission in thermodynamic equilibrium. Gordon and Meeks (1967) observed 94α lines at 7.8 GHz from both hydrogen and helium in Orion, deducing a kinetic temperature of 6,600 K from the line widths and, also, detecting the H148δ line. Palmer et al. (1967) detected a recombination line with a somewhat higher frequency than the H109α line in the NGC2025 and IC1795 nebulae; it was later identified as the C109α line of atomic carbon (Goldberg and Dupree, 1967; Zuckerman and Palmer, 1968).

Observations of RRLs were quickly extended to frequencies as low as 404 MHz with the detection of the H253α line (Penfield, Palmer and Zuckerman, 1967), while Dieter (1967) detected the H158α line in 39 H II regions – thereby establishing the new lines as suitable tools for astronomical surveys.

It became evident that RRLs would be a source of vast information about the microworld of astrophysics such as the features of highly excited atoms, as well as about macroworld of the astronomical bodies such as the structure of surrounding cosmic space. Accordingly, astronomers at many observatories and institutes around the world began to observe RRLs.

Chapter 2
RRLs and Atomic Physics

Abstract This chapter derives the physics ab initio that underlie the formation of radio recombination lines in astronomical objects. It includes natural, thermal, and pressure broadening of the spectral lines; the radiation transfer of the spectra and their underlying free–free emission (Bremsstrahlung) through the ionized media; the excitation of the atomic levels; the frequency range over which lines may be detected from astronomical objects; and the sizes of excited atoms that can exist within interstellar environments.

2.1 The First Surprising Results: The Absence of Stark Broadening

The first observations of radio recombination lines (RRLs) gave surprising results. The newly detected line profiles (Sorochenko and Borodzich, 1965; Höglund and Mezger, 1965; Lilley, Menzel, Penfield and Zuckerman, 1966) showed neither the Stark broadening individually nor the variation in Stark broadening with increasing quantum number n as predicted by theory. All the RRLs observed in the Omega nebula (M17) with principal quantum numbers up to $n = 166$, within the accuracy of the measurements, had constant ratios of line width to frequency. This indicated pure Doppler broadening.

In a sense, the astronomers were pleased. RRLs would be observable. Evidently, van de Hulst's (1945) pessimistic prediction that pressure broadening of the excited hydrogen lines would render them unobservable in the radio range was incorrect. The more recent calculations of the line width based on the theory of spectral line broadening by plasmas (Sorochenko, 1965) seemed to fit the observations better, in the sense that the new theory allows their detectability.

Let us look at the data available in 1967. Figure 2.1 shows a plot of actual line widths observed for hydrogen RRLs over the years 1964–1967 together with the results of the calculations of Stark broadening. The revised theory

M.A. Gordon, R.L. Sorochenko, *Radio Recombination Lines*, Astrophysics and Space Science Library 282, DOI: 10.1007/978-0-387-09691-9_2,

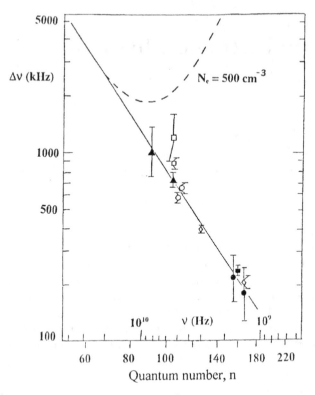

Fig. 2.1 The first observations of the RRL width $\Delta\nu$ as a function of quantum number n in Omega nebula. The *straight line* is the Doppler broadening with $\Delta\nu/\nu = 1.2 \times 10^{-4}$. The *dotted curve* shows what was expected based on the revised estimates of Stark broadening by Sorochenko (1965). The RRL observations shown are: *filled triangles* – Lebedev Physical Institute – H90α (Sorochenko and Borodzich, 1965) and H104α (Gudnov and Sorochenko, 1967); *open squares* – Pulkovo – H104α (Dravskikh et al., 1965; Dravskikh and Dravskikh, 1967); *open circles* – Green Bank – H109α (Höglund and Mezger, 1965; Mezger and Höglund, 1967); *open diamonds* – Parks – H126α and H166α (McGee and Gardner, 1967); *filled circles* – Harvard – average value of H156α and H158α (Lilley, Menzel, Penfield and Zuckerman, 1966) and H166α (Palmer and Zuckerman, 1966); and *filled squares* – University of California – H158α (Dieter, 1967)

predicts that, for $N_e = 500\,\text{cm}^{-3}$, the minimum possible value of electron density in gaseous nebulae (H II regions), Stark broadening should begin to manifest itself for $n\alpha$-type RRLs with $n > 100$. However, the measured line widths showed no Stark broadening at all. At $n = 166$, the line width should be ten times the value actually observed.

The new results were exciting. RRLs would be observable at substantially higher transitions than previously assumed. This suggested that RRLs could be a new tool for astronomers.

However, the physics of the line broadening evidently had not yet been understood. With the increasing level (n) of excitation, the outer electron becomes less connected to the atomic nucleus as shown by (1.6), (C.7), and (C.8). Correspondingly, the sensitivity of that electron to external electric fields of the ambient plasma must inevitably increase, manifesting this increasing influence by Stark broadening of the line widths. However, this was not observed. Consequently, one could suppose that the highly excited atoms have some mechanism of resisting the influence of the external fields, one that does not occur at the lower excitation levels that fit Doppler broadening very well. The solution of this problem required revision of the existing Stark broadening theory for RRLs and, probably, some additional careful experiments in atomic physics. Let us discuss this very interesting question further after a general discussion of broadening mechanisms.

2.2 The Broadening of Radio Recombination Lines

2.2.1 Natural Broadening

Several mechanisms determine the width of spectral emission lines. One is called "natural" broadening. It results from the finite length of the emitted wave train and the variation of its amplitude over the emission time. It is an intrinsic property of the atom.

2.2.1.1 Lorentz Profile

To predict the shape of a line due to natural broadening, we need to consider the broadening mechanism in more detail. Prior to quantum mechanics, physicists considered atoms as oscillating electric dipoles. From the moment that $t = 0$, the energy of an oscillator decreases as $E(t) = E_0 \exp(-\Gamma t/2)$, where the damping constant

$$\Gamma = \frac{8\pi^2 e^2}{3mc^3}\nu_0^2, \tag{2.1}$$

and ν_0 is the central frequency of oscillation and m is the mass of the oscillator.

This damping characteristic determines the frequency spectrum of the oscillator (Lorentz, 1906). If the intensity of the oscillator $f(t) = \exp(i2\pi\nu_0 t)\exp(-\Gamma t/2)$, then its complex spectrum results from the Fourier transform (Bracewell, 1965)

$$F(\nu) = \int_\infty^\infty f(t)e^{-i2\pi\nu t}\,dt \tag{2.2}$$

and the real spectrum

$$\phi(\nu) \propto \int_{-\infty}^{\infty} F^*(\nu)F(\nu)\,dt \tag{2.3}$$

$$= \frac{\Gamma/\pi}{\pi\left[4(\nu - \nu_0)^2 + (\frac{\Gamma}{2\pi})^2\right]} \tag{2.4}$$

after normalization such that

$$\int_{-\infty}^{\infty} \phi(\nu)\,d\nu = 1. \tag{2.5}$$

Equation (2.4) is then the spectrum of the oscillator known as a *Lorentz profile*. Figure 2.2 shows its shape. The full width of the Lorentz profile at half-intensity, $\Delta\nu_L$, is $\Gamma/(2\pi)$.

The Lorentz profile can also be derived from quantum mechanics. The uncertainty principle of Heisenberg states that the energy E of a system can be known only to an accuracy ΔE within a time interval Δt, where

$$\Delta E\,\Delta t \approx \frac{h}{2\pi}. \tag{2.6}$$

Accordingly, in the absence of collisions, the energy uncertainty of a quantum level n is given by

$$\Delta E_n \approx \frac{h}{2\pi}\frac{1}{\Delta t}$$

$$\approx \frac{h}{2\pi}\,\Sigma R_{nm}, \tag{2.7}$$

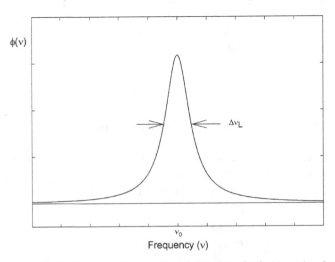

Fig. 2.2 A hypothetical Lorentz line profile described by (2.4). Note the characteristic narrow core, the steep shoulders near line center, and the broad sloping wings

where the sum is taken over all of the transition coefficients R_{nm} depopulating the level, from $n \to m$. This relationship shows that the width ($\Delta\nu \equiv \Delta E_n/h$) of a spectral line must be related to the inverse lifetime of its quantum levels, i.e., to the net transition rate out of its levels.

2.2.1.2 Natural Width of RRLs

In 1930, Weisskopf and Wigner derived the Lorentz line profile from quantum mechanics. They found the line width $\Delta\nu_L$ of (2.4) to equal the sum of the damping constants for the upper and lower quantum levels n_2 and n_1, respectively:

$$\Gamma \equiv \Gamma_{n_2} + \Gamma_{n_1} \tag{2.8}$$

$$= \frac{1}{\tau_{n_2}} + \frac{1}{\tau_{n_1}}. \tag{2.9}$$

The total spontaneous rate out of level n_2, or Γ_{n_2}, is

$$\Gamma_{n_2} = \sum_{n_1=1}^{n_2-1} A_{n_2,n_1}. \tag{2.10}$$

One writes Γ_{n_1} in a similar way.

The full width of the line at half-maximum then is

$$\frac{\Delta\nu}{\nu_0} = \frac{1}{2\pi\nu_0}(\Gamma_{n_2} + \Gamma_{n_1}) \tag{2.11}$$

$$= \frac{1}{2\pi\nu_0}\left(\sum_{n_1=1}^{n_2-1} A_{n_2,n_1} + \sum_{n_0=1}^{n_1-1} A_{n_1,n_0}\right) \tag{2.12}$$

$$\approx \frac{1}{\pi\nu_0}\sum_{n_1=1}^{n_2-1} A_{n_2,n_1}, \tag{2.13}$$

where (2.12) results from the quantum mechanical form of the Lorentz profile found by Weisskopf and Wigner (1930). Equation (2.13) obtains because the damping constants Γ_n are about the same for both upper and lower levels of normal RRLs involving large values of n and when $n_2 - n_1 \ll n_2, n_1$.

Another simplification is possible for most RRLs. Section 5.4.1 of Sobelman et al. (1995) gives a useful approximation to estimate the sum A_n of the spontaneous transition probabilities out of level n:

$$A_n \approx 2.4 \times 10^{10}\frac{\ln n}{n^5}, \quad n > 20. \tag{2.14}$$

Substituting the approximation for the RRL frequencies given by (1.22) and the summation given by (2.14) into (2.13) gives a useful expression for the natural width of RRLs:

$$\frac{\Delta \nu}{\nu_0} \approx \frac{1.2 \times 10^{-6} \ln n}{n^2}, \quad n > 20, \tag{2.15}$$

which gives a fractional natural width of 4.5×10^{-10} or $1.3 \times 10^{-4}\,\mathrm{km\,s^{-1}}$ for the H109α line, for example. This natural width is negligibly small compared to other types of broadening, as will be seen below.

2.2.2 Doppler Broadening

Observations of thermal gases in the cosmos such as H II regions involve measurements of Maxwell–Boltzmann velocity distributions if we exclude large-scale turbulence. In the absence of magnetic fields, an H II region with a kinetic temperature of $10^4\,\mathrm{K}$ and an electron density of $10^2\,\mathrm{cm^{-3}}$ will thermalize in minutes following any perturbation in the velocity distribution because of the high collision rate between the electrons and the ions. For higher densities, the thermalization proceeds even faster.

For a gas with a Maxwell–Boltzmann velocity distribution, the probability of an atom having a velocity component[1] between v_x and $v_x + dv_x$ along a line of sight through the nebula is then

$$N(v_x)\, dv_x = N \left(\frac{M}{2\pi kT} \right)^{1/2} \exp \left(-\frac{M v_x{}^2}{2kT} \right) dv_x, \tag{2.16}$$

where N is the total number of atoms contributing photons to the line and M is the mass of the atoms of that species. Using the classical Doppler formula[2] to relate the frequency observed to the line-of-sight velocity

$$\nu = \nu_0 \left(1 - \frac{v_x}{c} \right), \tag{2.17}$$

and differentiating to relate the intervals of the two domains

$$dv_x = -\frac{c}{\nu_0}\, d\nu, \tag{2.18}$$

---

[1] N.B. One axis only. See Sect. 4.11 of Chapman and Cowling (1960) for a full discussion of Maxwell–Boltzmann statistics.

[2] This formula works well when $v_x \ll c$. Gordon et al. (1992) discuss the relationship between spectral red shift and velocity for Euclidean (Special Relativity) and cosmological (General Relativity) models.

we convert (2.16) into the intensity of the line $I(\nu)$ with the assumption that the total intensity I in the line is proportional to the number of emitters N in the antenna beam and that the intensity interval $dI(\nu)$ is proportional to $dN(v_x)$, i.e., that the gas is optically thin to obtain

$$I(\nu) = I\phi_G(\nu), \qquad (2.19)$$

where the line profile $\phi_G(\nu)$ of the Doppler-broadened line is

$$\phi_G(\nu) = \left(\frac{4\ln 2}{\pi}\right)^{1/2} \frac{1}{\Delta\nu_G} \exp\left[-4\ln 2\left(\frac{\nu_0 - \nu}{\Delta\nu_G}\right)^2\right], \qquad (2.20)$$

where $\Delta\nu_G$ is the full width of the thermally broadened, Gaussian line at half-intensity and is obtained by equating the exponential arguments of (2.16) and (2.20). Using (2.17) to relate the frequency term to velocity, we obtain (Fig. 2.3)

$$\Delta\nu_G = \left(4\ln 2\frac{2kT}{Mc^2}\right)^{1/2} \nu_0, \qquad \text{or,} \qquad (2.21)$$

$$\approx 7.16233 \times 10^{-7}\left(\frac{T}{M}\right)^{1/2} \nu_0, \qquad (2.22)$$

where the mass M is in units of amu and T is in K.[3]

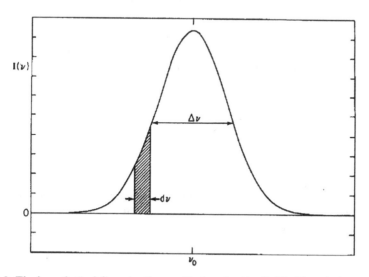

Fig. 2.3 The hypothetical Gaussian line profile described by (2.20). Note the broader core and the narrower extent of the wings with respect to the Lorentz profile

---

[3] A useful expression is that the area I of a Gaussian line is

Substituting a typical H II region temperature of $8,000\,K$ and the mass of hydrogen into (2.22) gives a value of $\Delta\nu/\nu_0 = 6.4 \times 10^{-5}$ for the H109α RRL – 10^5 times greater than the natural width estimated above. Based upon (2.18), this ratio is equivalent to a full width at half-intensity of about $19\,km\,s^{-1}$.

Microturbulence also contributes to the Doppler profile of the RRL. In many astronomical objects, emitting RRLs, such as H II regions, are cells of gas moving with respect to each other. Often, these cells are unresolved by the radio telescope; they fall within its comparatively large beam and, hence, merit the name "microturbulence." A characteristic of Doppler broadening is that the line width in units of velocity is a constant because $\Delta v = (c/\nu)\Delta\nu$. Because the velocity distribution of these cells is usually Gaussian, the observed width of the RRL results from the convolution of the thermal and turbulence Gaussians, a process (see Bracewell (1965)) that results in a new Gaussian profile with a width Δv_{Σ} given by

$$\left(\Delta v_{\Sigma G}\right)^2 = \left(\Delta v_{G-thermal}\right)^2 + \left(\Delta v_{G-turbulence}\right)^2, \qquad (2.25)$$

such that the Doppler line width can be written

$$\Delta V_G \equiv \Delta v_{\Sigma_G} = (4\ln 2)^{1/2}\sqrt{\frac{2kT}{M} + V_T^2} \qquad (2.26)$$

in terms of gas temperature and the turbulence velocity V_T.

Figure 2.4 shows how the turbulence widths $\Delta v_{G-turbulence}$ – labeled here as V_t – of RRLs increase with the angular size of the telescope beam, as seen in the Orion nebula. The width reaches a limit when the H II region becomes unresolved. The illustration shows that even at very high resolution, the turbulence component can be comparable to the thermal broadening.

2.2.3 Stark Broadening of RRLs

2.2.3.1 Early Theory

The third and, at times the most important mechanism for broadening RRLs, is a form of "pressure broadening" called the *Stark effect*. It consists of the splitting and displacement of atomic energy levels by the superposition of

$$I = \left(\frac{\pi}{4\ln 2}\right)^{1/2} I(\nu_0)\,\Delta\nu_G \qquad (2.23)$$

$$\approx 1.064\,I(\nu_0)\,\Delta\nu_G, \qquad (2.24)$$

in terms of the intensity measured at line center $I(\nu_0)$ and the full width of the line at half-intensity, the expression resulting from the integration of (2.19).

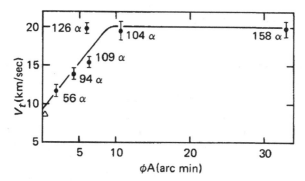

Fig. 2.4 The turbulence width of RRLs from the Orion nebula plotted against the beam size of the radio telescope used for observation (Sorochenko and Berulis, 1969). The *triangle* marks the width of a line seen at high angular resolution at optical wavelengths (Smith and Weedman, 1970)

an external electric field. When the electron is in a noncircular orbit, there is an electric dipole moment that responds to the application of an electric field. Early laboratory measurements referred to two kinds:[4] "linear" in which the displacement of the emission line is linearly proportional to the strength of the electric field and "quadratic" which has a square-law dependency on the electric field strength. The linear type occurs for weak electric fields, the quadratic type for strong fields. The effect was discovered by Stark (1913) not long before the quantum theory of atom was created.

Bohr (1923) explained the effect in the following way. If (1.14) gives the energy of an atom with a principal quantum number n, then the application of an external electric field $\mathcal{E}$ to that atom should cause a change in its energy of

$$\Delta E = \frac{3h^2 \mathcal{E}}{8\pi^2 em} n n_f, \tag{2.27}$$

where the n_f-quantum number can take the values of $0, \pm 1, \pm 2, \ldots, \pm n$. Therefore, the applied field splits the principal quantum number into $2n - 1$ sublevels because there is only one 0 in the series. From (2.27), the maximum width of each broadened energy level E_n is then $\propto n^2$.

Stark-shifted line components arise from transitions from the split upper energy levels $E_{n_2} \pm \Delta E$ to the split lower levels $E_{n_1} \pm \Delta E$. The distribution of intensities of the Stark components is complex – even for the simple hydrogen atom, as described in Sect. 65 of Bethe and Salpeter (1957). In general, the outermost line components have unobservable intensities.

---

[4] A tertiary Stark broadening can also occur.

2.2.3.2 Stark Broadening in Astronomical Plasmas

In astronomy, Stark broadening is more complicated. Astronomical plasmas include a wide range of temperatures and densities difficult to reproduce in a terrestrial laboratory. The principal characteristics of plasmas associated with the interstellar medium are low densities and low temperatures compared with, say, stellar cores and envelopes. The electric field experienced by an emitting atom might consist of a series of brief, weak, time-spaced transient fields induced by a series of successive colliding charged particles rather than by the steady-state field of a laboratory experiment.

Stark broadening in astronomical plasmas can be understood by examining the wave train emitted by the atom. Figure 2.5 illustrates the effect of sudden small phase shifts upon a sinusoidal wave train. The Fourier transform relationship between the phase shift time-line and frequency dispersion produces the profile of the spectral line. Consequently, theorists considered how charged particles could induce phase shifts in the emitted wave train. When the such encounters are transitory, they are called "collisions" or "impacts."

For example, we can calculate a Lorentz line profile by considering the effect of collisions on an oscillator with a constant amplitude. If the oscillation has the form $f(t) = \exp(i2\pi\nu_0 t)$ over the duration T between successive collisions that stop the oscillation abruptly, the frequency spectrum of the oscillation will be the Fourier transform of the wave train:

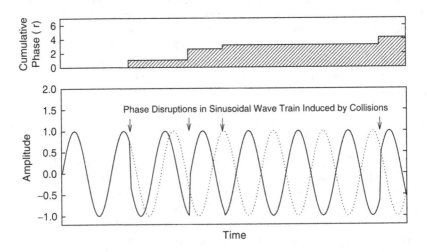

Fig. 2.5 *Top*: cumulative small phase shifts as a function of time in radians. *Bottom*: the *solid line* shows the resulting distortion of a sinusoidal wave train, the *dotted line* the undistorted wave train. The Fourier transform of the distorted wave train is a Lorentz profile

$$F(\nu) = \int_{-\infty}^{\infty} f(t)e^{-i2\pi\nu t}\, dt$$

$$= \int_{0}^{T} f(t)e^{-i2\pi\nu t}\, dt$$

$$= \frac{\exp[i2\pi(\nu_0 - \nu)T] - 1}{i2\pi(\nu_0 - \nu)}, \qquad (2.28)$$

where F is the complex spectrum of the truncated oscillator. If the collisions are distributed in time T with a mean time between collisions of τ, then the probability of a collision is $P(T) = (\exp -T/\tau)/\tau$. The spectral profile $\phi(\nu)$ results from the average of the power spectrum $|F^2(\nu, T)P(T)|$ over the distribution of T:

$$\phi(\nu) \propto \int_{0}^{\infty} F^*(\nu, T)F(\nu, T)\frac{e^{-T/\tau}}{\tau}\, dT \qquad (2.29)$$

$$= \frac{2\Delta\nu_L}{\pi\left[4(\nu - \nu_0)^2 + (\Delta\nu_L)^2\right]} \qquad (2.30)$$

after the normalization described by (2.5).

Equation (2.30) is a Lorentz profile like that shown in Fig. 2.2. As with the damped oscillator, the full width of the line profile at half-intensity is $\Gamma/(2\pi)$. It demonstrates that disruption of the wave train of an emitting atom also produces a Lorentz profile.

2.2.3.3 Weisskopf Radius

One way (Weisskopf, 1932) to characterize the size of the phase disruptions induced by individual perturbers is in terms of an "impact parameter," the closest distance between a perturber and an emitting atom.

Figure 2.6 shows a charged particle moving in a straight line past an emitting atom at a constant speed v. The impact parameter is ρ. Assume

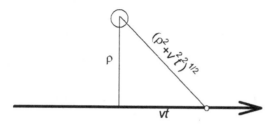

Fig. 2.6 The geometry of a perturber of speed v passing within a distance ρ (impact parameter) of an emitting atom

that, at any instant of time, the resulting shift in the angular frequency of emission $\Delta\omega$ can be modeled by a simple power law of the form

$$\Delta\omega = \frac{C_n}{R^n}, \tag{2.31}$$

where R is the instantaneous distance between the perturber and the atom, n expresses the power of the relationship between distance and frequency shift, and C_n is a constant of proportionality appropriate for the exponent n. In our rectilinear system, $R^2 = (\rho^2 + v^2t^2)$, where ρ is the closest distance (the impact parameter) between the straight-line trajectory and the atom, and vt is the distance traveled by the perturber along the trajectory. The total shift in the phase of the wave train caused by the perturbation is the integral over the entire trajectory:

$$\Delta\phi = \int_{-\infty}^{+\infty} \Delta\omega \, dt \tag{2.32}$$

$$= \int_{-\infty}^{+\infty} \frac{C_n}{(\rho^2 + v^2t^2)^{n/2}} \, dt \tag{2.33}$$

$$= \alpha_n \frac{C_n}{v\rho^{n-1}}, \tag{2.34}$$

where the coefficient

$$\alpha_n = \frac{\Gamma\left(\frac{1}{2}\right)\Gamma\left(\frac{n-1}{2}\right)}{\Gamma\left(\frac{n}{2}\right)} \tag{2.35}$$

$$= \pi, \, 2, \, \pi/2, \, 4/3, \, 3\pi/8, \, \ldots \quad \text{for} \quad n = 2, \, 3, \, 4, \, 5, \, 6, \, \ldots$$

If we ask what value of the impact parameter will result in a phase shift of 1 rad, (2.34) gives

$$\rho_0 = \left(\frac{\alpha_n C_n}{v}\right)^{1/(n-1)}, \tag{2.36}$$

a parameter known as the "Weisskopf radius." The parameters $n = 2, 3, 4$, and 6 correspond to the interaction laws of the linear Stark effect, resonance broadening, quadratic Stark effect, and van der Waals broadening, respectively.

While the Weisskopf radius holds no particular physical meaning other than to characterize an arbitrarily chosen phase shift of 1 rad in the emitted wave train, it is useful for distinguishing quantitatively between "strong" and "weak" interactions of the perturbers and the wave train as we will see below.

2.2.3.4 The Impact and Quasistatic Models of Stark Broadening

Because of widely varying conditions within interstellar plasmas and stellar atmospheres, no unified theory of Stark broadening exists for astronomical applications. Instead, astronomers select an approximation appropriate to the particular environment they are considering.

Mathematical models for astronomical Stark broadening developed up to now fall into either of two extreme approximations: "impact" or "quasistatic." We describe these briefly below, generally following the discussion in the monograph by Mihalas (1978) but see the monographs by Jefferies (1968), Griem (1974), and Sobelman (1992) for more details.

2.2.3.5 The Impact Approximation

If the total phase shift in the wave train is the result of an accumulation of discrete phase shifts within the time the atom is radiating, then the perturbation model is called "impact" because each perturbation represents a collision with the emitting atom. The simple illustration calculated early in Sect. 2.2.3 is an impact model. Its effect upon the line profile resulted from summing the results of all impacts, distributed as a probability function in that particular model. Figure 2.5 illustrates an impact situation where each discrete impact induces a rapid phase shift of less than 1 rad but the cumulative phase change in the wave train is about 4 rad.

Although the details of the impact approximation have evolved a great deal since its introduction in the early part of the twentieth century by Lorentz and others with regard to the spectrum of a damped, interrupted oscillator, its requirements are generally considered to be:

1. Each impact involves only one perturber and one atom at a time – a binary interaction. Simultaneous triple and multiparticle interactions are not considered.
2. The approximation requires a series of discrete impacts while the atom is radiating, where the effective duration τ_D of each impact must be much less than the interval τ_I between them.

   To illustrate, Mihalas (1978) defines the effective duration τ_D of an impact such that the product $2\pi\Delta\nu \times \tau_D$ is the phase shift in the wave train induced by each impact, where $\Delta\nu$ is the frequency shift in the line produced by the collision. The separation interval $\tau_I \equiv (N\pi r_0^2 v)^{-1}$; i.e., it is the reciprocal of the collision rate. The variable N is the volume density of the perturbers and r_0 is the effective interparticle distance (mean free path). The "discreteness" criterion can then be written as $\tau_D/\tau_I \ll 1$. Because $\tau_D/\tau_I \propto (\rho_0/r_0)^3$, these conditions will generally occur when the interparticle distance r_0 is much larger than Weisskopf radius, i.e., when the distance parameters can be described by the inequality $r_0 \gg \rho_0$.

3. Each weak collision causes a nearly instantaneous phase shift in the wave train. Between these phase interruptions, the oscillation of the wave train continues unperturbed, as illustrated in Fig. 2.5.

   In the rare case of strong collisions, i.e., with impact parameters $\rho < \rho_0$, a huge disruption of the wave train occurs. The impact model assumes that the oscillation stops and then restarts, with a complete loss of coherence. However, the contribution of strong collisions is small.[5] The principal contribution to the broadening comes from collisions in which the impact parameters are greater than the Weisskopf radius.

4. The time distribution of these collisions may be described by a probability function.

2.2.3.6 The Quasistatic Approximation

This model represents the other extreme situation from the impact model:

1. It assumes each atom to lie in an electric field produced by a chaotic, statistical distribution of the perturbing (charged) particles surrounding it. Appropriate corrections due to electrical shielding (Debye shielding) of the negative (electron) charges by the positive (ion) charges may be required to calculate the field accurately.

2. The charged perturbers are considered to be at rest with respect to the atom over the duration of the emission, and the perturbation model is accordingly called "quasistatic."

3. The shift in the oscillator frequency for each transition corresponds to the magnitude of the effective electrical field experienced by the radiating atom.

4. The line profile results from the probability distribution of the individual line frequency shifts, i.e., to the probability distribution of the electric fields for the radiating atoms. The amplitude at each frequency shift with respect to the center of the line is proportional to the number of atoms experiencing that particular field strength.

2.2.3.7 Application of the Stark Models to RRLs

Impact and quasistatic theories of Stark broadening give significantly different relations for the line profiles.

For a given physical situation, only one approximation can be used for a species of perturbers. *For a given kind of perturbers, the approximation must be either impact or quasistatic – not both.* Therefore, to calculate the linear Stark broadening of an RRL, one first needs to determine which of the two mathematical approximations is appropriate for a particular situation.

---

[5] Griem (1967) shows that they can be as large as 20% in some circumstances.

This test depends upon both the temperature (relative particle speeds) and density (relative interparticle distances) that determine the mean free path of the ionized particles within the plasma.

The principle of energy equipartition complicates the situation further. If the constituents of the plasma are well thermalized, the ions and electrons will have the same kinetic energy. The higher mass and, therefore, slower moving ions may require a different approximation for Stark broadening than the lighter and faster moving electrons, so that in principle *both quasistatic and impact broadening approximations may be appropriate for the same plasma*: the former describing collisions with the atoms by the ions, the latter by the electrons.

2.2.3.8 Derivation of the Weisskopf Radius for RRLs from H II Regions

To use the Weisskopf radius to distinguish between the mathematical quasistatic and impact regimes of Stark broadening, we first need to derive the coefficient C_2 in its definition given by (2.36).

As described earlier, in a weak electric field, each energy level n is linearly split into $2n - 1$ separate components. Equation (7.36) of Sobelman (1992) shows that the energy splitting ΔE of a quantum level associated with a principal quantum number n due to the linear Stark effect of an electric field $\mathcal{E}$ is

$$\Delta E = \frac{3}{2} n (p_1 - p_2) e a_0 \mathcal{E}, \tag{2.37}$$

where e is the electronic charge and a_0 is the Bohr radius given by (1.7). The term $(p_1 - p_2)$ contains the parabolic quantum numbers[6] p_1 and p_2 and is equivalent to n_f in (2.27). At a given n, $(p_1 - p_2)$ may be $n, n - 1, \ldots, -n$ to create $2n - 1$ levels because there is only one 0 in the series.

For elastic perturbations in which no energy is exchanged between perturber and the emitting atom, we calculate the difference between the frequency shift of components of the upper (n') and lower (n) principal quantum levels of an RRL transition. Equation (2.37) gives these as

$$\Delta \nu = \frac{3 h \mathcal{E}}{8 \pi^2 m e} \left[n'(p_1' - p_2') - n(p_1 - p_2) \right], \tag{2.38}$$

where we have substituted the definition of the Bohr radius a_0 and where $p_{1,2}'$ and $p_{1,2}$ are the parabolic quantum numbers of the levels n' and n, respectively.

---

[6] For calculation of perturbations induced by an external field, parabolic coordinates can be more useful than spherical coordinates because of the asymmetrical nature of the charge distribution within the atom. See Sect. 6 of Bethe and Salpeter (1957) or similar texts.

For most of the stronger Stark components of RRLs where $n \gg \Delta n \geq 1$, the differences in parabolic terms are small and $(p'_1 - p'_2) \approx n'$ and $(p_1 - p_2) \approx n$. Substitution then gives

$$\Delta \nu \approx \frac{3h\mathcal{E}}{8\pi^2 me} \left[(n')^2 - n^2 \right]$$

$$\approx \frac{3h\mathcal{E}}{4\pi^2 me} n \, \Delta n, \quad n \gg \Delta n \geq 1. \tag{2.39}$$

For convenience, we now parameterize this frequency shift into a Weisskopf radius for linear Stark broadening. The frequency form of (2.31) is

$$2\pi \Delta \nu \equiv \Delta \omega = \frac{C_2}{r^2}. \tag{2.40}$$

Substituting the definition of the electric field

$$\mathcal{E} \equiv \frac{Ze}{r^2} \tag{2.41}$$

into (2.39), substituting the resulting expression for $\Delta \nu$ into (2.40), and re-arranging gives

$$C_2 = \frac{3Zh}{2\pi m} n \, \Delta n, \quad n \gg \Delta n \geq 1. \tag{2.42}$$

In turn, substituting C_2 into the definition of the Weisskopf radius for linear Stark broadening derived from (2.36) gives

$$\rho_0 = \frac{3Zh}{2mv} n \, \Delta n, \quad n \gg \Delta n \geq 1. \tag{2.43}$$

2.2.3.9 Which Approximation for the Stark Broadening of RRLs?

Applying (2.43) to RRLs from astronomical H II regions guides us to the appropriate mathematic model for calculating Stark broadening. We compare the interparticle distance ($r_0 \approx N^{-1/3}$) of the ionized gas with the Weisskopf radius derived above. If

$$N^{-1/3} \gg \rho_0, \tag{2.44}$$

the collisions will likely be discrete and well separated, i.e., the impact approximation would be the appropriate mathematical model for calculating the line broadening.

To apply this test quantitatively, it is first necessary to adapt the Weisskopf radius of (2.43) to the specific environment of the H II region. This is done by substituting an appropriate value for the ion velocity v_i. In a Maxwell–Boltzmann gas,

$$\left\langle \frac{1}{v_i} \right\rangle = \frac{4}{\pi} \frac{1}{\langle v_i \rangle} = \left(\frac{2M}{\pi kT} \right)^{1/2} , \tag{2.45}$$

where M is the mass of the ion and $\langle v_i \rangle$ is the mean value of the ion speed.[7] The relative velocity between the hydrogen atoms and these ionized perturbers (protons) will be a factor of $\sqrt{2}$ larger because of the equal masses (via the principal of energy equipartition), and the appropriate substitution should then be

$$\left\langle \frac{1}{v_i} \right\rangle = \left(\frac{M}{\pi kT} \right)^{1/2} . \tag{2.46}$$

Combining (2.43), (2.44), and (2.46) and solving for the principal quantum number n gives the regime where the impact approximation for ion broadening will be valid:

$$n \, \Delta n \ll \frac{2m}{3ZhN_i^{1/3}} \left(\frac{\pi kT}{M} \right)^{1/2} , \quad n \gg \Delta n \geq 1, \tag{2.47}$$

$$\ll 6{,}850 \left(\frac{T}{10^4 \, \text{K}} \right)^{1/2} \left(\frac{10^4 \, \text{cm}^{-3}}{N_i} \right)^{1/3} , \quad n \gg \Delta n \geq 1. \tag{2.48}$$

For electrons, the expression will be similar except that the relative speed between electrons and atoms is

$$\left\langle \frac{1}{v_e} \right\rangle = \left(\frac{2m}{\pi kT} \right)^{1/2} , \tag{2.49}$$

and, hence, the validity region for impact broadening by electrons is

$$n \ll 208{,}000 \left(\frac{T}{10^4 \, \text{K}} \right)^{1/2} \left(\frac{10^4 \, \text{cm}^{-3}}{N_e} \right)^{1/3} , \quad n \gg \Delta n \geq 1, \tag{2.50}$$

because the ratio of the relative speeds is 30.3, as can be verified by dividing (2.46) by (2.49).

Substituting appropriate values of T and N_e, we conclude that the impact approximation is the appropriate one for calculating Stark broadening for RRLs from most H II regions for either ions or electrons.

There is an alternative way of determining which approximation to use. When the frequency shifts of the line become large, the impact approximation fails and the QS approximation must be used. In fact, one can calculate a frequency on the line profile $\Delta \nu_W$, corresponding to a collision with an impact

---

[7] See, e.g., Sect. 4.11 of Chapman and Cowling (1960) or any other text on statistical mechanics for the general technique for deriving weighted speeds of a Maxwell–Boltzmann gas.

parameter equal to the Weisskopf radius, which roughly marks the transition from the validity region of the impact approximation (discrete, separated impacts) to that of the quasistatic approximation (statistically continuous electric field).[8]

2.2.3.10 The Effect of RRL Observations on Stark Broadening Theory

The theory of impact broadening applicable to spectral lines in plasmas had been developed since the late 1950s, principally for transitions in the optical part of the spectrum (Griem, Kolb and Shen, 1959; Griem, 1960). This research showed for the first time that, at high levels of excitation, collisions with electrons might be the main cause of line broadening, even for low electron densities (Griem, 1960). These calculations served as the basis for estimating the Stark broadening of RRLs just before they were detected. They showed that Stark broadening would be important for RRLs with transitions with $n > 100$ (Sorochenko, 1965).

As we have described in Sect. 2.1, the early observations of RRLs disagreed with the broadening theory. Stark broadening theory available at that time predicted line widths many times what was actually observed. For example, Lilley et al. (1966) in their announcement of their detections of the H156α and H158α lines at 1.7 GHz stated, "If the Stark equations as given by Kardashev were actually valid, ... for the source M17... we obtain [electron densities] of $10 \, \text{cm}^{-3}$" which, of course, would be ridiculously low for that H II region by about two or three orders of magnitude. They suggested that the Stark broadening theory developed for high densities may be responsible for the overestimate of the Stark effect. Mezger (1965) wrote to Sorochenko: "the line widths of your, our, and the Harvard observations plotted against line frequency can be fitted with a straight line... This means that the line broadening is entirely due to Doppler broadening and that no Stark broadening is effective... This contradicts all theories of Stark broadening."

These observations indicated the necessity of revising Stark broadening theory for hydrogenic atoms in an astrophysical plasma. Examination of the available impact theory showed that, while it had been tested for low-n transitions, it did not consider an important characteristic of interaction with atoms at high levels of excitation. During elastic interactions with highly excited atoms, a compensation mechanism occurs. The distortions of the

---

[8] Griem (1967) used this method by comparing the threshold QS frequency shift from the center of an RRL with the Doppler width. If this ratio $\gg 1$, then the validity region of the QS approximation falls far from the Doppler core, and the QS approximation cannot be used. Substitution of $T = 10^4 \, \text{K}$ – typical for an astronomical H II region – into his expression gave a ratio of several thousand for both ions and electrons. Griem thereby concluded that the QS approximation was an inappropriate model for the Stark broadening of RRLs by either ions or electrons. This is the same conclusion we reach above with (2.48) and (2.50).

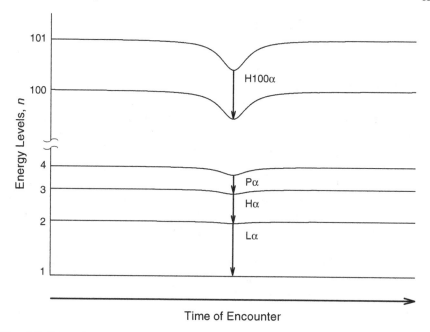

Fig. 2.7 Distortions in the energy of Stark components of quantum levels n caused by elastic collisions. This sketch – not to scale – illustrates the situation for the H100α RRL and the optical lines: Paschen α, Hα, and Lyman α lines. Unlike the optical lines, similar distortions of high-n levels compensate for each other to reduce the Stark broadening of RRLs enormously

energy levels produced by the Stark effect occurred in the same sense for closely neighboring levels as shown in Fig. 2.7. The *difference* between the upper and lower energy levels thereby changes much less than the energy levels themselves. As a result, the frequencies and the widths of the RRL line profiles, determined just by this difference, are not changed very much (Griem, 1967; Minaeva, Sobelman and Sorochenko, 1967).

For Stark broadening of the Balmer and Paschen series in the optical range, the compensation mechanism does not play a significant role. The Stark perturbation of the upper energy level substantially exceeds the perturbation of the lower one, the compensation mechanism is insignificant, and the theoretical calculations agree with experimental data within the measurement errors. Through experiments with the Balmer and Paschen series, some authors noticed that the theory needed to be improved to apply to the higher-order lines of hydrogen (Ferguson and Shlüter, 1963). Vidal (1964; 1965) noticed the slight difference between the Stark theory and the measured broadening of the optical lines. In spite of this, most laboratory researchers at that time were happy that the agreement of theory and data was so good. In fact, the puzzling RRL observations provided the impetus for a re-examination of the impact theory in astronomical plasmas.

2.2.3.11 Elastic and Inelastic Impact Broadening: A Closer Look

The collision of a charged particle with an atom can be either elastic or inelastic. "Elastic" means the interaction between perturber and the emitting atom does involve an exchange of energy; i.e., the interaction is adiabatic. The interaction splits and shifts the energy levels but does not change the n values. The collision efficiency is inversely proportional to the relative velocity of the atom and the charged particle. Because of their higher masses and correspondingly lower velocities, ions interact mainly through elastic collisions.

Inelastic collisions have a very different physical nature. They can induce nonradiative atomic transitions and thereby change the principal quantum number n. Therefore, they can be nonadiabatic. RRL broadening for these kinds of collisions involves not just through the splitting and shifting of the n levels but by decreasing the lifetime of the atomic quantum levels.

For inelastic collisions, the minimum impact radius ρ_{min} (or effective distance ρ_{eff}) becomes the critical parameter rather than the Weisskopf radius, which is used in elastic collisions. According to Griem (1967), this minimum impact radius is

$$\rho_{min} \approx \sqrt{\frac{5}{6} \frac{hn^2}{2\pi m v_e}} \approx \frac{n^2}{v_e}. \tag{2.51}$$

Substituting ρ_{min} into the inverse of (2.44), we obtain a criterion for the impact approximation in case of inelastic collisions:

$$n \ll \sqrt{\frac{1}{N_e^{1/3} < v_e^{-1} >}}. \tag{2.52}$$

For the same conditions ($Te = 10^4\,\mathrm{K}$, $Ne \le 10^4\,\mathrm{cm}^{-3}$), the impact approximation is valid for inelastic collisions with electrons if $n \ll 1{,}500$ and for inelastic collisions with ions if $n \ll 300$.

Therefore, both for elastic and inelastic collisions of highly excited atoms with electrons in the conditions of H II regions, the impact approximation is appropriate through the entire range of RRLs. The question is the relative contribution of a given type of collisions to the RRL broadening. For collisions with ions, the impact approximation is valid only to moderate values of n. But, as will be shown below for interactions with ions, more significant is the question of whether inelastic collisions are possible at all.

The criterion that determines the division between elastic and inelastic collisions depends on the relationship of the collision time τ_c with the angular frequency $\omega_{n'n}$ of the transition $n' \to n$, where n' is the upper principal quantum number. Inelastic collisions are possible if (Griem, 1967; Kogan, Lisitsa and Sholin, 1987)

$$\omega_{n'n}\tau_c = \frac{\omega_{n'n}}{v/\rho_{min}} \ll 1, \tag{2.53}$$

or, in other words, when the transition frequency is significantly lower than the perturbation frequency (the reciprocal of the time of flight of the particle at the distance ρ_{min}).

Substitution into (2.53) of ρ_{min} from (2.51) and the frequency of the hydrogen α line from (1.22) gives the principal quantum number n of hydrogen atom where inelastic transitions between neighboring levels can occur:

$$n \gg \frac{1.8\, R_H ch}{mv^2} \approx \frac{4.5 \times 10^{16}}{v^2}, \tag{2.54}$$

where R_H is the Rydberg constant for hydrogen. At the average electron velocity of $v_e \approx 6 \times 10^7\,\mathrm{cm\,s^{-1}}$ (for $T = 10^4\,\mathrm{K}$), inelastic transitions are possible for $n \gg 12$. A parallel situation occurs for interactions with ions. Because the relative atom–ion velocity is approximately 30 times lower than the electron velocity, the inequality of (2.54) changes to $n \gg 10^4$ and, accordingly, the threshold value of n for inelastic collisions increases by a factor of about 900.

Therefore, for Stark broadening of RRLs in H II regions, the impact approximation is the appropriate one. In collisions of excited atoms with electrons with transitions $n' \rightarrow n$, both elastic and inelastic types can occur. In contrast, only elastic collisions can occur for ion–atom collisions. Inelastic collisions with ions are unlikely and are negligible for Stark broadening of RRLs.

Note that the inelastic collisions with electrons cannot occur for atoms with low-n values and, therefore, cannot broaden recombination lines in the optical range.

2.2.4 Elastic and Inelastic Impact Broadening: Calculated Line Widths

Griem (1967) produced a revised theory of impact broadening specifically directed toward RRLs. For impact broadening by ions, Griem (1967) neglects the inelastic collisions to obtain an expression for the "ion" width of an RRL:

$$\Delta \nu_L^i = \frac{3}{\sqrt{\pi}} \left(\frac{h}{2\pi m} \right)^2 \sqrt{\frac{M}{kT_e}} N_i n^2 \left[\frac{3}{2} + \frac{2}{e^2} \ln\left(\frac{2n}{3} \right) \right]$$
$$\times \left[\frac{1}{2} + \ln\left(\frac{\lambda\, m}{n\, h} \sqrt{\frac{kT_e}{M}} \right) \right], \tag{2.55}$$

where $\Delta \nu_L^i$ is the full line width at half-intensity, λ is the spectral line wavelength, N_i is the volume density of the ions, e is the base of the natural logarithm, and M is the mass of the ion. After substitution of the numerical values of the constants into (2.55), the expression simplifies, indicating dependence of Stark broadening on level number, ion density and temperature only:

$$\Delta\nu_L^i = \frac{6.7 \times 10^{-5}}{\sqrt{T_e}} N_i n^2 \ln(172\,n) [\ln(9.4 \times 10^{-3} n^2 \sqrt{T_e})] \tag{2.56}$$

in units of Hz when N_i is in units of cm^{-3} and T_e is in K.

For the interaction of the highly excited atoms with electrons, the inelastic collisions are the most important. The cancellation of the perturbations of the upper and lower states by elastic collisions greatly reduces their contributions to line broadening – as mentioned earlier. The inelastic contribution (Griem, 1967) is

$$\Delta\nu_L^e \approx \frac{10}{3} (2\pi)^{-5/2} \left(\frac{h^4}{m^3 k T_e}\right)^{1/2} N_e \, n^4 \left[\frac{1}{2} + \ln\left(\frac{2\pi}{3} \frac{k T_e \lambda}{h c n^2}\right)\right], \tag{2.57}$$

which Griem (1974) evaluates as

$$\Delta\nu_L^e \approx \frac{5.16^{-6} N_e n^4}{\sqrt{T_e}} \ln\left(8.25 \times 10^{-6} T_e n\right) \tag{2.58}$$

by substitution of numerical constants. Figure 2.8 shows these predictions for the inelastic broadening by ions and electrons. The ratio of (2.56) to (2.58) is $\ll 1$ for $20 < n < 200$ for $N_i = N_e$ and $T_e = 10^4$ K, indicating that only inelastic electron collisions are important for Stark broadening of RRLs from H II regions – as can be seen in the figure.

Unlike Griem (1967; 1974), who utilized a modified classical approach of perturbers interacting inelastically with the emitting atoms, Brocklehurst and Leeman (1971) calculated RRL broadening with quantum theory applicable to the inelastic collisions of excited atoms with electrons. In this case, the line width for the transition $n_2 \to n_1$ is determined by the total cross sections of inelastic collisions:

$$\Delta\nu_L^e = \frac{1}{2\pi} N_e \left[< \sigma(n_1) v_e > + < \sigma(n_2) v_e >\right], \tag{2.59}$$

where $< \ldots >$ denotes the average over a Maxwellian distribution of the electron velocities and cross sections determined by summing over all possible transitions $\sigma_n = \sum_{\Delta n > 0} \sigma_{n, n \pm \Delta n}$.

Later, various authors defined the cross sections for inelastic collisions more exactly. The most exact are probably the cross sections calculated through semiempirical formulae by Gee et al. (1976). The rates of collision transitions from level n to level $n \pm \Delta n$ are $\alpha(n, n+\Delta n) = < \sigma(n, n+\Delta n) v_e >$, obtained by averaging over the Maxwellian distribution of electron velocities, have an error less than 20% in the region of temperatures $100\,(100/n)^2 < T_e \ll 3 \times 10^9$ and quantum numbers $n, n + \Delta n \geq 5$.

Using the transition rates of Gee et al. (1976), Smirnov (1985) obtained the rather simple approximate expression for the RRL broadening by electron collisions in Hz when N_e is in units of cm^{-3}:

$$\Delta \nu_L^e = 8.2 N_e \left(\frac{n}{100} \right)^{\gamma} \left(1 + \frac{\gamma}{2} \frac{\Delta n}{n} \right). \tag{2.60}$$

The factor γ is the growth rate of RRL collisional broadening as a function of n. In the range $n = 100 - 200$, γ increases while temperature decreases, but the value of its variation $\Delta \gamma$ is small. For example, $\Delta \gamma \le 0.16$ when temperature decreases from 10^4 to 5×10^3 K and the average value of γ is 4.5.

Because the probability of a collisional transition is a sharply decreasing function of Δn, it is adequate to sum only over $\Delta n \le 10$ to calculate the rates $\alpha(n, n + \Delta n)$.

Figure 2.8 shows the broadening by electrons calculated from (2.60). For completeness, the figure also shows broadening by inelastic ion collisions. These ion calculations utilized a classical approach for the collision of ions

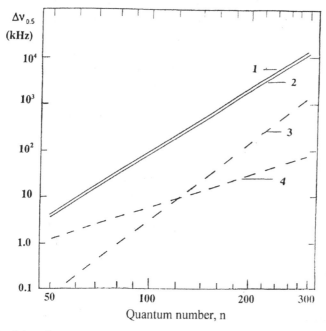

Fig. 2.8 Stark broadening of hydrogen RRL in plasmas as a function of quantum number n for $N_e = N_i = 10^4 \, cm^{-3}$ and $T_e = 10^4$ K. *Solid lines* show the broadening by electron collisions, *dashed ones* by ion collisions. *Curve 1* is from inelastic electron collisions calculated by a classical approach from (2.58). *Curve 2* is the same, but calculated with an aid of inelastic collisions with cross sections from (2.60). *Curve 3* is from inelastic ion collisions from (2.61). *Curve 4* is from elastic ion collisions from (2.56)

with excited atoms and appropriate cross sections (Beigman, 1977). The complete expression for the ion broadening is complex but can be simplified to a more compact form obtained by Smirnov (1985):

$$\Delta\nu_L^i = (0.06 + 0.25 \times 10^{-4}T_e)\left(\frac{n}{100}\right)^{\gamma_i}\left(1 + \frac{2.8\,\Delta n}{n}\right)N_i, \qquad (2.61)$$

where $\gamma_i = 6 - 2.7 \times 10^{-5}T_e - 0.13(n/100)$. The value of $\Delta\nu_L^i$ obtained from (2.61) differs no more than 5% from the exact values (Beigman, 1977) in the range of temperatures $5,000\,\mathrm{K}< T_e <15,000\,\mathrm{K}$ and quantum numbers $50 < n < 300$.

Inspection of Fig. 2.8 shows that the dominant mechanism for the Stark broadening of RRLs is inelastic electron collisions by a large margin. The similarity of curves 1 and 2 also demonstrates that the broadening calculations made with classical physics agree well with those made with quantum theory.

The Stark broadening by ions is significantly less important for RRLs than broadening by electrons for two reasons. For elastic collisions, the energy levels shift in the same direction, thereby compensating each other and reducing the contribution to line broadening from this mechanism. The result is that the dependence of the broadening on principal quantum changes from $\Delta\nu \propto n^4$ to $\Delta\nu \propto n^2$, the lower exponent becoming significant at $n \sim 100$ and more so for $n > 100$. For inelastic collisions, the larger ion mass means smaller velocities and a correspondingly smaller probability of collision with the atoms than the electrons. The overall result is that inelastic electron collisions dominate the Stark broadening of RRLs and, in subsequent theoretical calculations and comparison with observations, we shall use (2.60) to estimate Stark broadening of RRLs.

In (2.30), $\Delta\nu_L$ is the full width at half-maximum of the line profile $I(\nu)$ and ν_0 is the center frequency of the line. Strictly speaking, since line profile is a summation over individual Stark components, the sum of Lorentzian profiles does not have a Lorentzian line shape. But, according to Smirnov (1985), the Stark-broadened profile of the RRLs may be considered to be Lorentzian with an accuracy sufficient for practical use. Compared to the "pure" Lorentzian, the summed profile is about 2–3% higher at maximum and about 1–2% lower at half-maximum.

2.2.5 Combining Profiles: The Voigt Profile

In general, RRLs will have line profiles influenced by both thermal effects (Gaussian profiles) and linear Stark effects (Lorentz profiles). If the two mechanisms are independent, the two-component profiles can be convolved to produce a composite line profile. The thermal broadening involves only the velocities of the atoms. The impact Stark broadening involves only the velocities of the colliding electrons because, in energy equipartition, their masses are much smaller and, hence, their velocities are much greater than the emitting atoms. Therefore, the two velocity distributions are independent, and we can convolve the two-line profiles to obtain the composite line profile:

$$\Psi(\nu, \eta) \propto \int_{-\infty}^{\infty} \phi_G(\nu')\phi_L(\nu - \nu')\, d\nu', \tag{2.62}$$

where η indicates the relative weights of the line profiles G and L in the mix. The proportionality sign signifies that the resulting function $\Psi(\nu)$ will be normalized such that

$$\int_{-\infty}^{\infty} \Psi(\nu)\, d\nu = 1, \tag{2.63}$$

so that the resulting composite profile can be written

$$I_V(\nu) = I_0 \Psi(\nu). \tag{2.64}$$

Combining by substitution and convolution equations (2.19), (2.20), (2.30), and (2.60) and normalizing produces the composite line profile:

$$
\begin{aligned}
I_V(\nu) &= \left(\frac{4\ln 2}{\pi}\right)^{1/2} \frac{1}{\Delta\nu_G} H(a, v) \\
&\approx \frac{1.665}{\Delta\nu_G \sqrt{\pi}} H(a, v),
\end{aligned}
\tag{2.65}
$$

where the a parameterizes[9] the mix of Gaussian and Lorentzian profiles:

$$
\begin{aligned}
a &= \frac{(\ln 2)^{1/2} \Delta\nu_L}{\Delta\nu_G} \\
&\approx \frac{0.833\, \Delta\nu_L}{\Delta\nu_G},
\end{aligned}
\tag{2.66}
$$

and the parameter v parameterizes[10] the distance from line center:

$$
\begin{aligned}
v &= \frac{(4\ln 2)^{1/2}(\nu - \nu_0)}{\Delta\nu_G} \\
&\approx \frac{1.665\,(\nu - \nu_0)}{\Delta\nu_G},
\end{aligned}
\tag{2.67}
$$

and the function H is

$$H(a, v) \equiv \frac{a}{\pi} \int_{-\infty}^{\infty} \frac{e^{-t^2}\, dt}{a^2 + (v - t)^2}\, dt. \tag{2.68}$$

---

[9] Some authors define a differently.

[10] The parameter v is sometimes listed as b.

In spectroscopy, this composite profile $I_V(v)$ is called a *Voigt profile* because it uses the function $H(a, v)$ named after the German spectroscopist Voigt (1913). Hjerting (1938) describes this function in detail, and Davis and Vaughan (1963) and Finn and Mugglestone (1965) tabulate it in convenient forms.

Examination of (2.65) shows the characteristics of the Voigt profile. Near the Doppler core, i.e., where $v \approx t$, the function H behaves like an exponential as one would expect from a Gaussian profile. In fact, when the collisional broadening becomes very small such that $a \Rightarrow 0$, the function $H(a, v) \Rightarrow 1$, and the intensity at the center of the line ($\nu_0 = 0$) is

$$I_V(\nu_0) = \frac{1.665}{\Delta\nu_G\sqrt{\pi}}, \tag{2.69}$$

which is the peak intensity of a purely Doppler-broadened line as can be seen from (2.20).

On the other hand, well away from the line core where a^2 is small with respect to $(v-t)^2$, the function $H \propto v^{-2}$ and falls rapidly with increasing $|v|$ and, hence, with increasing distance ν from line center as is characteristic of the wings of a Lorentz profile. In fact, when $\Delta\nu_L \gg \Delta\nu_G$, the broadening is a pure Lorentzian one, with the intensity at line center:

$$I_V(\nu_0) = \frac{2}{\pi\Delta\nu_L}, \tag{2.70}$$

as shown by (2.30). When the two-line widths are the same, i.e., $\Delta\nu_L = \Delta\nu_G$, the ratio of the intensities at line center show the Gaussian to be the stronger

$$\frac{I(\nu_0)_G}{I(\nu_0)_L} = 1.476, \tag{2.71}$$

because the normalized Lorentzian line profile has more emission in the line wings than the normalized Doppler profile.

Smirnov (1985) gives a useful simple approximation for the width of the Voigt profile described by (2.65):

$$\Delta\nu_V = 0.5343\,\Delta\nu_L + \sqrt{\Delta\nu_G^2 + (0.4657\,\Delta\nu_L)^2}, \tag{2.72}$$

where $\Delta\nu_V$, $\Delta\nu_G$, and $\Delta\nu_L$ are the full widths at half-intensity of the Voigt, Gaussian (Doppler), and Lorentz (impact) profiles, respectively. This expression gives an error of less than 0.08% for $0.1 < a < 10$.

Figure 2.9 compares three profiles of equal area: Gaussian, Lorentz, and Voigt profiles. Both Gaussian and Lorentz profiles have full widths at half-intensity of 20. The Voigt profile results from the convolution of these two profiles, as given by (2.65). Its corresponding width is 35, as given by (2.72). The convolution process of the Gaussian and Lorentz profiles shifts radiant

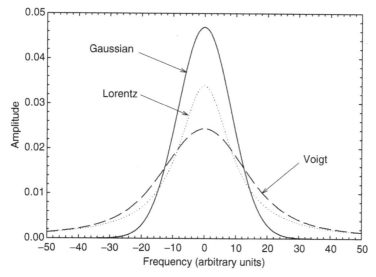

Fig. 2.9 Comparison of Gaussian, Lorentz, and their convolution (Voigt) line profiles. Each profile has the same area. The full widths at half-intensity are 20 for the Gaussian and Lorentz profiles, and 35 for the resulting Voigt profile

energy from their line cores to the near wings of the composite profile, thereby decreasing the intensity of the composite core near the line center and widening the core. The far wings of the Voigt profile are the same as those of the Lorentz profile.

2.2.6 Observational Test of the Revised Theory

The revised theory of RRL broadening in plasma decreased the calculated values of the line widths, thereby explaining the first observations of RRLs with n up to 166 from H II regions.

However, there were still problems. The revised theory conflicted with observations. Transitions up to $n = 220$ were observed in Orion nebula – the most studied H II region – but no Stark broadening was detected (Pedlar and Davies, 1972). Figure 2.10 shows the measured line widths in the Orion nebula plotted from data in the RRL catalogue of Gulyaev and Sorochenko (1983). The observations agreed with theory only for $N_e = 100\,\mathrm{cm}^{-3}$, which is much smaller than the electron density of $N_e = 10^4\,\mathrm{cm}^{-3}$ determined from the optical forbidden lines (Osterbrock and Flather, 1959).

Something fundamental was being overlooked. As discussed earlier, theoretical calculations of Stark broadening by various methods produced similar results. Differences in the estimates of cross sections for electron impact broadening agreed within 30% (Griem, 1974) even though the calculations

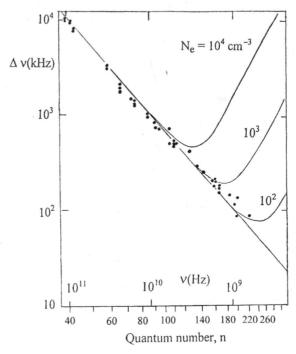

Fig. 2.10 The full widths of H$n\alpha$ RRLs from the Orion nebula plotted as a function of principal quantum number n. The straight line shows pure Doppler broadening with $\Delta\nu_G/\nu = 10^{-4}$. The *curved lines* show the expected line widths $\Delta\nu$ calculated for different values of N_e from (2.26), (2.60), and (2.72)

were made by different researchers. Yet, the observations of Stark broadening of the RRLs disagreed with theory by an order of magnitude.

The explanation of the discrepancy of theory and observation was found by closely examining the characteristics of the astronomical targets. The theory of spectral line broadening in plasmas was based upon the *homogeneous* densities, corresponding to laboratory conditions. However, in the interstellar medium, such as H II regions, this density is almost always inhomogeneous.

Some hints had already appeared in the literature but many in the radio astronomical community seemed not to be paying attention. Hoang-Binh (1972) specifically suggested that the low-density gas within the H II regions was determining the line shapes such that Stark broadening would not be seen. Brocklehurst and Seaton (1972), using a spherically symmetrical model of the Orion nebula with the density decreasing from the center outward toward the borders, were able to fit the widths and intensities of all RRL observations up to $n = 220$. Simpson (1973a; 1973b) suggested a mix of low and high electron densities from her radio and optical studies of the Orion nebula. Gulyaev and Sorochenko (1974) came to similar conclusions as Brocklehurst and Seaton after modeling the Orion nebula as a dense central core ($N_e = 10^4 \, \text{cm}^{-3}$)

surrounded by a rarefied ($N_e = 200 \rightarrow 700\,\mathrm{cm}^{-3}$) extended envelope, based upon earlier continuum observations at a wavelength of $\lambda = 8\,\mathrm{mm}$ (Berulis and Sorochenko, 1973).

Still, at that time, not all astronomers were convinced that the explanation for the enigmatic Stark broadening – or rather, the absence of – lay solely in the density structure. Lockman and Brown (1975a) suggested that the densest parts of the Orion nebula must be somehow cooler than the surrounding lower density gas. They considered the nebula essentially to consist of three regions, each with a unique temperature, density, and size. And, Shaver (1975) also worked out a model involved varying densities but with spatially varying temperatures as well.

The physical reasons for the absence of Stark broadening in the observed profiles in the case of an inhomogeneous distribution of electron density in the nebula are rather evident. They are a consequence of the fact that absorption in a plasma increases as frequency decreases, gradually leading to complete opaqueness. Figure 2.10 suggests this effect. The decrease in the observed line widths with increasing n is correspondingly a decrease with frequency. Here, the increase in gas opacity means that the lines increasingly reflect conditions in the more transparent, lower density outer regions of the H II region. For the high-n RRLs that occur at lower frequencies, the core of the H II region becomes opaque. The more rarified envelope gas would contribute much less if any Stark broadening even at the same electron temperature.

To complicate analysis further, the observations shown in Fig. 2.10 were made with radio telescopes of differing physical diameters and, accordingly, differing beamwidths – even for the same lines. Moreover, the beamwidths of these telescopes increased with decreasing frequency and, hence, with increasing n. The result was that the data involved different regions of the H II region except for the small telescopes whose large beamwidths included the entire nebula.

The changing opacity as a function of frequency and location in most H II regions and the variations in the beamwidths of the radio telescopes prevented the detection of Stark broadening as a function of principal quantum number n.

The new models of H II regions incorporating density inhomogeneities explained the absence of Stark broadening in astronomical RRLs. But, these "negative" results were insufficient to verify the refined theories of Stark broadening. Such verification had to await specially designed observations.

Minaeva et al. (1967) suggested observing RRLs of increasing order. Here, as Δn increases from 1 to 2, 3, and 4, the corresponding Hα, Hβ, Hγ, and Hδ RRLs would occur near the same frequency with suitable choices of n and, therefore, with the same beamwidth. If the observations are made at sufficiently high frequency where the gas is optically thin, the lines will be emitted by the same volume of gas. The idea was that the higher-order lines in the series would exhibit increasing amounts of Stark broadening compared with the pure Doppler broadening of the $n\alpha$ line.

The basic Bohr theory predicts the series. Rewriting (1.17) for $\Delta n \equiv n_2 - n_1 \ll n$ gives

$$\nu \cong \frac{2cR_H\,\Delta n}{n^3}, \quad n_1, n_2 \gg 1. \tag{2.73}$$

If $\Delta n = 1, 2, 3, 4, \ldots$ and values of n are selected to make $\Delta n/n^3 \approx$ constant, then (2.73) will give a sequence of RRLs of increasing order but with similar rest frequencies. For example, the frequencies of the H110α, H138β, H158γ, and H173δ RRLs all lie between 4.87 and 4.91 GHz.[11]

The problem with this technique is that the intensities of the RRLs weaken with increasing order, varying approximately as $(\Delta n)^{-2}$. The decreasing intensities make it impossible to observe a series with a single telescope to detect with adequate precision a long sequence of higher-order lines with increasing amounts of Stark broadening.

A special observing technique overcame this difficulty by using two large radio telescopes of different diameters (Smirnov, Sorochenko and Pankonin, 1984). Rather than one telescope observing a complete series of higher-order lines, the different diameters of two telescopes observed at different frequencies such that a large range of high-order lines were observed at the same beamwidths. Combining the line observations from both telescopes as a function of beamwidth allowed a wide range of quantum numbers to be sampled at a single beamwidth.

In this experiment (Smirnov et al., 1984), the smaller telescope, the 22-m telescope of the Lebedev Physical Institute in Pushchino, Russia, shown in Fig. 1.4, observed RRLs at 22 and 36.5 GHz. The larger telescope, the 100-m telescope of the Max-Planck-Institut für Radioastronomie at Effelsberg, Germany, shown in Fig. 2.11, observed RRLs at 5 and 9 GHz. The RRLs observed at 5 GHz with the 100-m telescope used the same beamwidth (2.′6) as those observed at 22 GHz with the 22-m telescope. Similarly, those observed at 9 GHz with the 100-m telescope used the same beamwidth ($\approx$1.′7) as those observed at 36.5 GHz with the 22-m telescope. It was then possible to assemble two series of line profiles for the same volume of gas over a wide range of principal quantum numbers – with signal-to-noise ratios adequate to search for Stark broadening.

In each beamwidth series, the smallest-n Hα line was assumed to be dominated by Doppler broadening and, hence, Gaussian except for small deviations due to possible large-scale gas flows in the nebular gas subtended by the telescope beams. These transitions were the H56α and H66α lines for the series associated with the small and large beamwidths, respectively. Fitting each line with a Gaussian established the Doppler component assumed to be part of the Voigt profile for each of the higher-n RRLs in the series for that particular beamwidth and, hence, for that particular volume of nebular gas. Fitting Voigt profiles to each spectrum composed of the now known Doppler

---

[11] See also very recent observations described at the end of this section.

Fig. 2.11 The 100-m radio telescope at Effelsberg, Germany

component would determine the residual Lorentz profile created by the Stark broadening.

The expected small size of the Lorentz residual required that the RRL line observations be processed carefully. Averaging individual spectrograms increased the signal-to-noise ratios of the composite line profile. Particular attention was paid to removing the inevitable instrumental baselines from the composite spectra, including the weak sinusoidal ripples generated by multiple reflections (standing waves) between parts of the telescope surfaces.

The results clearly showed Stark broadening in RRLs. Figure 2.12 shows the variation in line widths as a function of frequency for the smaller beamwidth observations. The line widths increase systematically with principal quantum number n as expected. Figure 2.13 shows the widths of the Lorentz component of the fitted Voigt profiles, after extraction of the Gaussian component contributed by Doppler effects.

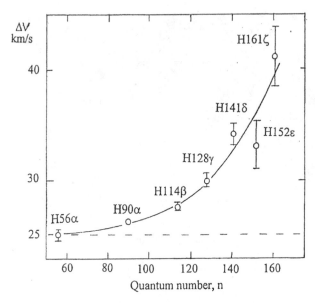

Fig. 2.12 The full width at half-intensity of the Voigt profiles plotted against principal quantum number. The width of the H56α line is presumed to be entirely Doppler. The *solid line* is an approximation fitted to the observations obtained by Smirnov et al. (1984)

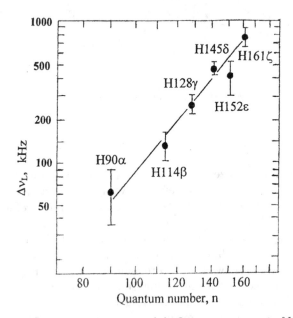

Fig. 2.13 The full widths at half-intensity of the Lorentz component of Voigt profiles of RRLs observed at 9 GHz from the Orion nebula (Smirnov et al., 1984). The *slope of the regression line* corresponds well to theoretical predictions of Stark broadening theory

The power law dependence of principal quantum number n derived from the data agreed well with theory. Fitting the observed Lorentz widths to the simple expression

$$\Delta \nu_L = A\, N_e \left(\frac{n}{100}\right)^{\gamma} \left(1 + \frac{\gamma}{2}\frac{\Delta n}{n}\right) \tag{2.74}$$

over the observed range of n and Δn gives a γ of 4.4 ± 0.6. The constant $A = 8.2\,\mathrm{Hz\,cm^3}$ if the electron density is approximately $10^4\,\mathrm{cm^{-3}}$ as suggested from observations of the O III optical lines from the core of the nebula (Osterbrock and Flather, 1959). Similar electron densities have been derived from more recent observations of the O II, O III, and Cl III emission lines from the Orion nebula (Peimbert and Torres-Peimbert, 1977). This value of γ agrees well with the value of 4.4 derived theoretically by Brocklehurst and Leeman (1971).

The second series of RRLs corresponding to the larger $2\rlap{.}'6$ beamwidths obtained the smaller value of $\gamma = 3.8$. This value may result from the inclusion of more rarified gas in the telescope beam, thereby contributing less Stark broadening to the line profiles. This de-emphasis of Stark broadening in the large-beam observations could explain why Churchwell (1971), Davies (1971), Simpson (1973a), and Lang and Lord (1976) were unable to detect the broadening.

An example of new information is a report of observations of line broadening inconsistent with current theory. Bell et al. (2000) used an innovative observing technique (Bell, 1997) to observe many orders of hydrogen recombination lines within a single spectral window. Observations near 6 GHz showed RRLs over the range $1 \le \Delta n \le 21$ from the Orion and W51 H II regions the measured line width to increase with the principal quantum number n, reaching a maximum near $n = 200$ before decreasing. This result conflicts with theoretical predictions of Stark broadening.

Two explanations were considered and rejected. Watson (Watson, 2007) examined electron impact broadening for $\Delta n \le 70$ and found that the widths of lines with different n and Δn must increase up to $n \ge 300$. Oks (Oks, 2004) suggested that increasing collisions with ions at higher values of n might explain the Bell's observations. However, Griem (2005) found the contributions of ion collisions too small to create a maximum broadening at $n \approx 200$.

What could be wrong? Central to these results is Bell's technique of producing radio spectra by rapidly switching the frequency window by frequency intervals large compared with the expected line widths but small with respect to sinusoidal "ripples" in the baseline produced by variations in impedance match as a function of frequency. Essentially, this process is a Fourier convolution of the spectrum with the function $F(\nu) = \delta(\nu) - \pi(\nu)$, using nomenclature developed by Bracewell (1965). Recovering the astronomical spectrum requires a Fourier deconvolution. However, a rigorous deconvolution is not possible because the process would involve division by 0 at points within the transform domain. Therefore, Bell (1997) developed an iterative, guess-and-subtract technique he calls LINECLEAN. As with similar deconvolution algorithms used in radio astronomy, the accuracy of the deconvolved spectra is sensitive to the distribution of noise (signal-to-noise ratio) in the observations, which is unknown a priori.

The Bell et al. (2000) results are so different from what had been expected, and the observing technique is so new, that we suggest waiting for an independent confirmation of the observations before accepting a fault in the present theory of RRL Stark broadening.

We conclude this section by noting that observations of the line widths of RRLs not only provided new information regarding the density structure of H II regions, but also offer an opportunity for testing and refining theories of elastic and inelastic Stark broadening in low-density plasmas.

2.3 Intensity of Radio Recombination Lines

2.3.1 Radiation Transfer

Radiation from astronomical bodies is the primary source of information for astronomers. Not only does it carry spatial information that locates the position of object in the cosmos, but also the radiation itself carries characteristics that identify its origin and describe its environment. In addition, the interstellar medium (ISM) between its source and the telescope also impresses information on the radiation. Disentangling the characteristics of these natural messengers from the cosmos, identifying the nature of the radiation source, and sharing their findings through publication in journals is what astronomers do for a living.

The special physics used to analyze these characteristics is called "radiation transfer." In its simplest form, radiation transfer involves two basic assumptions. It assumes symbolic loss and gain mechanisms for the radiation along the path of propagation without regard to the detailed processes involved – initially, at least. Usually, it assumes all processes associated with the radiation to be "stationary," i.e., that the parameters do not change over the timescale of the observations.

Neither of these assumptions is realistic. Eventually, the astronomer needs to understand the detailed physics of the loss and gain mechanisms, which can change within the timescales of the observations. But, all analyses must have a beginning. It is better to see how much understanding can be achieved with a simple model before adding complexities.

Consider the situation sketched in Fig. 2.14. As the radiation intensity $I(0)$ moves through the cloud from back to front, it loses intensity to the cloud through absorption and by scattering photons into directions other than the observer. On the other hand, the radiation intensity increases from additional radiating elements dx within the cloud. The net change in the intensity dI is

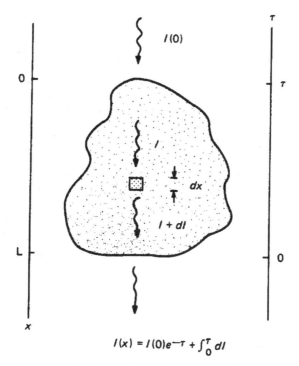

$$I(x) = I(0)e^{-\tau} + \int_0^\tau dI$$

Fig. 2.14 The background radiation intensity $I(0)$ travels from the back of a cloud toward the observer where it is strengthened or weakened by the incremental amount dI in each distance interval dx. The equation shows the intensity seen by the observer. From Gordon (1988). Reproduced with permission of Springer-Verlag

$$dI = -I\kappa\, dx + j\, dx, \qquad (2.75)$$

where x is the distance the radiation travels *toward* the observer, κ is called the *linear absorption coefficient* and accounts for *all* depletions from the radiation in the direction of the observer, and j is called the *linear emission coefficient* and accounts for *all* gains in intensity in the direction of the observer. The first term on the right accounts for the weakening in the intensity by extinction, i.e., by scattering or absorption within the differential distance element, the second for the strengthening by emission.

Integrating (2.75) from the far side of the nebula to the observer[12] gives the intensity seen by the observer as

$$I(\tau = 0) = \int_{\text{cloudback}}^{\text{observer}} dI$$

$$= I(\tau)e^{-\tau} + \int_0^\tau \frac{j}{\kappa}e^t\, dt, \qquad (2.76)$$

---

[12] The integration uses the integrating factor $e^{-\tau}$ and integration by parts.

where the parameter τ is called the "optical depth":

$$\tau \equiv -\int_{x1}^{x2} \kappa(x)\, dx. \tag{2.77}$$

The first term of (2.76) is the background radiation attenuated as it passes through the medium. The second term describes the contribution to I by the medium itself, including emission and extinction in the direction of the observer. Under some conditions, a third term might be appropriate to include a radiation source between the emitting "cloud" and the observer.

The parameter τ reduces the number of variables by concealing the spatial variation of the extinction coefficient $\kappa(x)$ along the sight line x that, in most circumstances, cannot be known. Because factor $e^{-\tau}$ is a direct measure of the fractional loss of photons propagating through a medium, the factor $(1 - e^{-\tau})$ is the fraction of photons surviving that passage.

For those unusual circumstances where the sight line conditions are known, (2.76) becomes

$$I(x = L) = I(0)e^{-\hat{\kappa}L} + \int_{0}^{L} \frac{j}{\kappa} e^{\kappa(x-L)}\, dx, \tag{2.78}$$

where $\hat{\kappa}$ is the mean value of the extinction along the sight line calculated as $[\int_{L} \kappa(x)\, dx]/L$.

The symbol I specifies a radiant quantity known as the "specific intensity" (Chandrasekhar, 1950; Jefferies, 1968). Its general definition is radiant energy per unit time per unit collecting area per unit bandwidth interval per unit solid angle. Sometimes also called "brightness," here I means I_ν in units of ergs per second per Hertz per square centimeter per steradian.

Because of their different bandwidth dependencies, the relationship between the frequency (I_ν) and wavelength (I_λ) forms of I is

$$I_\nu\, d\nu = I_\lambda\, d\lambda, \tag{2.79}$$

and, hence,

$$I_\nu = I_\lambda \frac{c}{\nu^2}. \tag{2.80}$$

What does I mean physically? The specific intensity is a simple proportionality constant between the radiant energy E and all of the factors that affect its detected strength, such as the time length of the observations dt, the collecting area of the surface or telescope $d\sigma$, the bandwidth of the detector $d\nu$ or $d\lambda$, and the solid angle into which the energy is collected $d\Omega$. In this sense, I is the essence of radiation that is independent of the observing process. Figure 2.15 shows the geometry.

By design, the specific intensity has some peculiar properties. Through its definition, I is constant along any ray in free space. Even if the radiation cone diverges from source to observer, I will not weaken between source

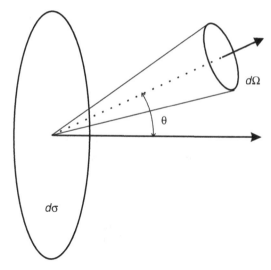

Fig. 2.15 The geometry of specific intensity I passing through a surface element $d\sigma$ and an angle θ to the surface normal n

and observer in the absence of absorption or scattering. Also, I does not change by *perfect* reflections at any mirror or combination of mirrors. Using a mirror to focus the Sun's rays to heat an object does not change I but, rather, increases the solid angle of the cone containing I and, correspondingly, the angular energy density falling on the object. In fact, the spectral flux density of radiation S_ν is "energy/unit collecting area/unit frequency interval/unit time," defined as

$$S_\nu \equiv \int_{\text{solidangle}} I_\nu \, d\Omega, \tag{2.81}$$

and showing that S_ν can be varied by changing the solid angle range of the integral without changing I. Appendix F discusses the relationship of S_ν to parameters directly observed by radio telescopes.

2.3.1.1 Source Function

Both (2.76) and (2.78) contain the ratio j/κ. These linear absorption and emission coefficients, κ and j, are physically related. Their ratio j/κ defines the ratio of photons being emitted to those being lost from I to each point along the radiation path. Consequently, this ratio is called the "source function," usually symbolized by S. The source function is an intrinsic property of the emitting medium. Its specification as a function of location and frequency (or wavelength) allows the solution of the equation of radiation transfer.

In some circumstances, the source function is easy to specify without detailed calculations of the linear coefficients j and κ. In an enclosure from which no photons can escape, the radiation field will reach a state called "thermodynamic equilibrium" that is fully specified by the *single parameter* T called "temperature." Here,

$$j = \kappa B(T) \tag{2.82}$$

from Kirchhoff's law of thermodynamics. The function $B(T)$ is known as the *Planck function* and is written in the frequency form

$$B_\nu(T) \equiv \frac{2h\nu^3}{c^2} \frac{1}{e^{h\nu/kT} - 1} \tag{2.83}$$

$$= \frac{1.4743 \times 10^{-47} \nu^3}{e^{4.7993 \times 10^{-11}\nu/T} - 1} \tag{2.84}$$

or the wavelength form (in CGS)

$$B_\lambda(T) \equiv \frac{2hc^2}{\lambda^5} \frac{1}{e^{hc/\lambda kT} - 1} \tag{2.85}$$

$$= \frac{1.1909 \times 10^{-5}}{\lambda^5} \frac{1}{e^{1.4388/\lambda T} - 1} \tag{2.86}$$

where h and k are Planck's and Boltzmann's constants, respectively. Equations (2.83) and (2.85) are not the same but relate to each other by

$$B_\nu \, d\nu = B_\lambda \, d\lambda \tag{2.87}$$

and completely specify radiation fields known as "black-body radiation." Here, the word "black" means that all emitted photons are subsequently absorbed by the medium, i.e., none escape from the enclosure. True black-body radiation does not exist in nature although some environments closely approximate it in certain frequency or wavelength regions.

If $h\nu \ll kT$, which in astronomy usually occurs in the radio range,

$$B_\nu(T) \approx \frac{2\nu^2 kT}{c^2} = \frac{2kT}{\lambda^2}, \tag{2.88}$$

and is known as the "Rayleigh–Jeans approximation" to the Planck function. If $h\nu \gg kT$, which in astronomy usually occurs in the optical range,

$$B_\nu(T) \approx \frac{2h\nu^3}{c^2} \exp(-h\nu/kT), \tag{2.89}$$

and is known as the "Wien approximation" to the Planck function.

Note that neither approximation may be appropriate for the millimeter and submillimeter wavelength ranges in astronomical conditions. Usually, a test must be made for each situation in these ranges.

2.3.2 Continuum Emission

Before applying the transfer equation to predict the intensities of RRLs, we will examine the characteristics of the continuum emission from H II regions underlying these lines.

Continuum radiation from H II regions is a mixture of thermal emission from heated dust particles and "free–free" emission or Bremsstrahlung (braking radiation) from unbounded charged particles. Figure 2.16 shows the emission from the two components in the typical H II region W3 as a function of frequency. Ultraviolet radiation from embedded stars heats the dust and ionizes the gas. Because the ionized particles are free, their energy states are not quantized, and the free–free radiation is continuous over the spectrum.

In gaseous nebulae, the spectrum of the dust emission is similar to that of a black body. However, it differs somewhat. The short-wave side (Wien) of the peak emission is optically thick but the long-wave side (Rayleigh–Jeans) can be optically thin in the millimeter wavelength range. At very high frequencies, the dust emission can be more intense than the free–free emission and, because its effective temperature is about $100\,\mathrm{K}$, peaks in the infrared part of the spectrum.

With regard to RRLs, the free–free emission is the more important of the two continuum components. This emission measures the amount of ionization in the H II region. As we shall see, the ratio of the line emission in the RRLs to the continuous free–free emission can be a good measure of the thermodynamic state of the gas. It indicates the ratio of bound to unbound electrons in the H II region. Because most H II regions include heated dust, Fig. 2.16 shows that observations of free–free emission are best made at frequencies $\leq 10^{11}\,\mathrm{Hz}$ where the dust emission is small compared with the free–free emission.

2.3.2.1 Continuum Absorption Coefficient

Calculating the continuum absorption coefficient κ_C is challenging.[13] The calculations require modeling not only the electrical interaction between two charged particles, but also the velocity distribution of the particles. Classically, the encounter of two moving charged particles involves changes in their directions – either toward each other for unlike charges or away from each

---

[13] Oster (1961) discusses the history and problems of calculating the free–free emission coefficient in detail.

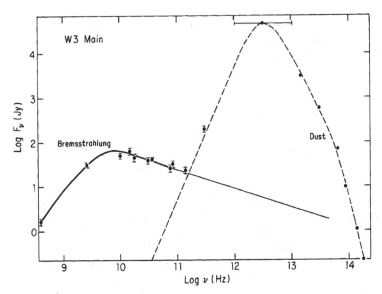

Fig. 2.16 The flux density in Janskys plotted against frequency from approximately 400 MHz ($\lambda = 75$ cm) through 178 THz ($\lambda = 1.7$ μm) for the H II region W3. The *solid curve* marks the free–free or Bremsstrahlung emission and the *broken curve* marks the thermal emission from warm dust embedded in the H II region. The *filled circles* show the observations. From Gordon (1988). Reproduced with permission of Springer-Verlag

other for like charges. The early work of Hertz showed that the acceleration from these direction changes causes radiation. Close encounters involve substantial Coulomb forces and large accelerations, leading to radiation in the X-ray range. Correspondingly, emission in the radio range involves more distant encounters where the Coulomb forces are weaker, the accelerations are much smaller, and the particles can be considered to continue moving in almost a straight line. In any case, the free–free absorption coefficient is determined by integrating the emission produced during each encounter over the velocity distribution of the particles, which is usually taken to be Maxwellian.

Some approximations are required to perform the integration. The free electrons tend to shield the electric field of the ions, and thus the force field is effective only over some finite distance that depends upon the density. Usually, the calculations assume (1) that the radiated energy is small compared with the kinetic energy of the electrons moving past the ions and (2) that the reciprocal of the radiated frequency is small compared with the time for the electron to undergo a 90° deflection. Physically, these assumptions imply that the electron–ion encounter is nearly adiabatic and that the period of the emitted wave train is short compared with the duration of the encounter.

Equation (162) of Oster (1961) gives an expression for the free–free absorption coefficient valid for the Rayleigh–Jeans – usually the radio – domain ($h\nu \ll kT$):

$$\kappa_C = \left(\frac{N_e N_i}{\nu^2}\right)\left(\frac{8Z^2 e^6}{3\sqrt{3}\, m^3 c}\right)\left(\frac{\pi}{2}\right)^{1/2}\left(\frac{m}{kT}\right)^{3/2}\langle g\rangle, \qquad (2.90)$$

where N_e and N_i are the volume densities of electrons and ions, respectively. The factor $\langle g\rangle$ is the Gaunt factor averaged over a Maxwellian velocity distribution characterized by the kinetic temperature[14] T. Equation (2.90) results from the fundamental relationship between the absorption and emission coefficients:

$$\kappa_C = \frac{j_C}{B_\nu(T)} = j_C \frac{\lambda^2}{2kT}, \qquad (2.91)$$

allowing recovery of j_C for the general frequency domain. All units are CGS, giving units of κ_C in cm^{-1}.

The choice of the Gaunt factor depends upon the environment. For temperatures less than 550,000 K where classical physics approximations obtain,

$$\langle g\rangle \approx \frac{\sqrt{3}}{\pi}\ln\left[\left(\frac{2kT}{\gamma m}\right)^{3/2}\frac{m}{\pi\gamma Z e^2 \nu}\right] \qquad (2.92)$$

and for temperatures greater than 550,000 K where quantum effects obtain,

$$\langle g\rangle \approx \frac{\sqrt{3}}{\pi}\ln\left(\frac{4kT}{\nu h\gamma}\right), \qquad (2.93)$$

where γ is an Euler's constant in the form $\exp(0.557) = 1.781$. Both approximations for the Gaunt factor apply only when the wave frequency greatly exceeds the plasma frequency, or $\nu \gg 10^4 \sqrt{N_e}$ Hz, which is the usual situation for radio waves from H II regions.

Combining (2.90) and the Gaunt factor appropriate for H II regions (2.92) gives an expression for the free–free absorption coefficient that can be evaluated numerically as

$$\kappa_C = 9.770 \times 10^{-3}\frac{N_e N_i}{\nu^2 T^{3/2}}\left[17.72 + \ln\frac{T_e^{3/2}}{\nu}\right]. \qquad (2.94)$$

in CGS units, which gives κ_c in units of cm^{-1} when the densities are in units of cm^{-3}, T in K, and ν in Hz.

Altenhoff et al. (1960) suggested a simple approximation for (2.94) in units appropriate for many radio astronomical observations:

$$\kappa_C \approx \frac{0.08235\, N_e N_i}{\nu^{2.1} T^{1.35}}, \qquad (2.95)$$

---

[14] In thermodynamic equilibrium, all temperatures are the same. In reality, various temperatures can differ from each other.

where ν is in units of GHz, N_e in units of cm^{-3}, T_e in units of K, and κ_C in units of pc^{-1}. Its simplicity makes this formula often used in the analysis of observational data from H II regions. The numerical version of (2.94) in CGS units is

$$\kappa_C \approx \frac{0.2120\, N_e N_i}{\nu^{2.1} T^{1.35}} \tag{2.96}$$

to give κ_C in units of cm^{-1} when ν is in units of Hz.

Any approximation involves some loss of accuracy. Altenhoff et al. claimed an accuracy within 5% for the traditional radio range, which is better than the usual accuracy of observations. Other approximations exist with higher accuracies but often are restricted to specific circumstances. For example, Hjellming et al. (1979) give an alternative approximation but this is restricted to $8 < \nu < 11$ GHz, a narrow range of frequency.

Under conditions of low radio frequencies and low temperatures, (2.94) and (2.95) no longer hold because the basic assumptions (1) and (2) are violated. Here, the Gaunt factors must be evaluated for each particular case. Oster (1970) gives expressions appropriate for a range of low temperatures and wave frequencies.

2.3.2.2 Continuum Emission Coefficient

As was the case for the line emission coefficient, Kirchhoff's radiation law gives the emission coefficient for the continuum emission per unit volume as

$$j_C = \kappa_C\, B(T), \tag{2.97}$$

where $B(T)$ is the Planck radiation function that obtains in thermodynamic equilibrium as discussed in Sect. 2.3.1.

2.3.3 Transfer Equation for Continuum Radiation

From the basic relationship between κ_C and j_C given by the Rayleigh–Jeans form of Kirchhoff's law of (2.91) and from the general form of the transfer equation given by (2.78), we can model the intensity of the free–free radiation from an H II region at a frequency ν as

$$I_C(x) = I(0)e^{-\tau_C(0)} + \int_0^{\tau_C(0)} \frac{2kT\nu^2}{c^2} e^{-t}\, dt + I(x > L), \tag{2.98}$$

with reference to Fig. 2.14. The continuum optical depth $\tau_c \equiv \int \kappa_C\, dx$. The terms $I(0)$ and $I(x > L)$ refer to emission on the far side of the H II region and between the H II region and the observer, respectively.

If the H II region is homogeneous in density and temperature, and if the foreground emission is zero, (2.98) becomes

$$I_C(x) = I(0)e^{-\tau_C(0)} + \frac{2kT\nu^2}{c^2}\left(1 - e^{-\tau_C(0)}\right) \qquad (2.99)$$

to describe the free–free continuum emission observed at the radio telescope.

In practice, astronomers use units of temperature to measure the intensity of radiation. At radio wavelengths where the Rayleigh–Jeans approximation usually holds, the brightness temperature corresponding to the observed specific intensity at the telescope is $T = I\,c^2/2k\nu^2$, where T is the temperature of the emitting gas such that the transfer equation for the continuum emission from the nebula becomes

$$T(x) = T(0)e^{-\tau_C(0)} + T\left(1 - e^{-\tau_C}\right), \qquad (2.100)$$

where $T(0)$ is background emission attenuated by passage through the nebula.

The actual antenna temperature[15] T_A is less than the brightness temperature by an efficiency factor η and a beam dilution factor W, defined as the ratio of the solid angle subtended by the radio source to that of the radio beam. If we think of the telescope as a thermometer, this temperature is the effective rise in temperature of the antenna as it points to the H II region. Specifically,

$$T_A = T\,\eta W. \qquad (2.101)$$

2.3.4 Comparison with Continuum Observations

Because RRLs are measured with respect to the free–free emission underlying their spectra, it is necessary to understand the characteristics of the continuum emission to interpret the observations of the spectral lines.

Over a wide range of frequencies, the continuum emission from an H II region has a unique spectrum. Using the spectral flux density defined by (2.81) because it is integrated over a solid angle and ignoring the background term, we rewrite the transfer equation for continuum emission as

$$S_\nu \equiv S\left(1 - e^{-\tau_C}\right), \qquad (2.102)$$

where S_ν is the total flux density emitted by the nebula and S is the source function of the radiation. At high frequencies, the gas is optically thin, and

---

[15] At millimeter and submillimeter wavelengths where calibration procedures involve measurements of atmospheric emission, the units of antenna temperature can be quite different and unintuitive – often symbolized by T_A^*, T_R^*, or T_{mb}^*. Appendix F describes their relationship to astrophysical units.

$$S_\nu \approx S\tau_C \propto \nu^{-0.1} \tag{2.103}$$

because $S \propto \nu^2$ in the Rayleigh–Jeans range and $\tau_c \propto \nu^{-2.1}$ as shown in (2.95).

At low frequencies where the gas is optically thick, $\tau_c \gg 1$ and

$$S_\nu \approx S \propto \nu^2. \tag{2.104}$$

The range of middle frequencies is usually called the "turnover" range. One can describe the continuum spectrum from an H II region in terms of its "turnover frequency," usually defined as the frequency where $\tau_c = 1$. A few authors use $\tau_c = 1.5$. Here,

$$S_\nu = S\left(1 - e^{-1}\right) = 0.632\,S. \tag{2.105}$$

Identifying the frequency where $\tau_C = 1$ (or 1.5) and estimating the electron temperature allows an astronomer to determine from (2.95) an important characteristic of an H II region known as the "emission measure" (EM). Defined as $\int_{\text{source}} N_e^2 \, dx$, EM is a measure of the ionized mass of the nebula emitting the free–free radiation. Its usual units are $\text{cm}^{-6}\,\text{pc}$.

Figure 2.17 illustrates how well these equations describe actual observations. The spectral flux density rises at low frequencies and falls very slightly at high frequencies. The broken vertical line marks the turnover frequency.

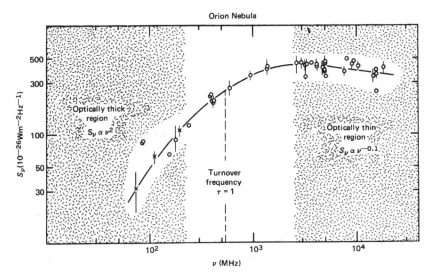

Fig. 2.17 The spectral flux density is plotted against frequency for observations of the Orion nebula (Terzian and Parrish, 1970). The *vertical broken line* marks the turnover frequency. *Points* indicate observations. The *solid line* marks the best fit of (2.102). Figure from Gordon (1988). Reproduced with permission of Springer-Verlag

The figure shows the optical thin and thick regions as shaded areas. Agreement between theory and observation is good.

Although the salient characteristics of the theory describe the continuum emission well, there are practical limitations. The analysis presumes an H II region to be homogeneous in density and isothermal. This is almost never the case. The distribution of newly formed stars that ionize the gas causes gradients in density and, probably, temperature as well. Furthermore, the nebular gas usually contains large-scale flows that appear as widenings or asymmetries in the spectral profiles of the RRLs. Optical studies of the Orion nebula made with high angular resolution show detailed structure on scales that would be within the beam of many radio telescopes (Osterbrock and Flather, 1959). As we shall see, these inhomogeneities can be important with regard to understanding RRLs.

2.3.5 Line Absorption and Emission Coefficients

Like free–free emission, calculating line emission begins with the linear emission and absorption coefficients, and the transfer equation.

2.3.5.1 Line Absorption Coefficient

The linear absorption coefficient for an RRL is

$$\kappa_L = \frac{h\nu}{4\pi}\phi_\nu(N_{n_1}B_{n_1,n_2} - N_{n_2}B_{n_2,n_1}), \qquad (2.106)$$

where n_2 and n_1 are the principal quantum numbers of the upper and lower levels, respectively, N is the number density of atoms in the subscripted levels, and B_{n_1,n_2} and B_{n_2,n_1} are the Einstein coefficients for absorption and stimulated emission, respectively, in the indicated direction in units of inverse specific intensity per unit time.[16] The units of κ_L are inverse length. As defined earlier, the factor ϕ_ν is the line profile in units of Hz^{-1}. Note that the term $h\nu/4\pi$ has units of energy per steradian – a part of the definition of I. The first term within the brackets is stimulated absorption, the second term is stimulated emission considered here as "negative absorption" because we are discussing an absorption coefficient.

---

[16] Some authors (cf. Spitzer (1978)) define the Einstein coefficient B in terms of *energy density* rather than *specific intensity*, the latter being the form commonly used by astronomers and the form we use here (cf. Sect. 90.1 of Chandrasekhar (1950), Sect. 4.1 of Mihalas (1978), or Sect. 6.5.3 of Allen's Astrophysical Quantities (Hjellming, 1999)). These two definitions are not interchangeable. In brief, for a transition of $n = 2 \to 1$, B_{21}(specific intensity) $= (4\pi/c) \times B_{21}$(energy density).

The Boltzmann formula gives the relative populations of two quantum levels in terms of volume densities:

$$\frac{N_{n_2}}{N_{n_1}} = \frac{\varpi_{n_2}}{\varpi_{n_1}} e^{-h\nu/kT}, \tag{2.107}$$

where ϖ is the statistical weight of the subscripted level. For hydrogen or hydrogenic ions, $\varpi = 2n^2$. Furthermore, $\varpi_m B_{m,n} = \varpi_n B_{n,m}$. Substituting into (2.106), we derive

$$\kappa_L = \frac{h\nu}{4\pi} \phi_\nu N_{n_1} B_{n_1,n_2} \left[1 - e^{-h\nu/kT}\right] \tag{2.108}$$

without approximations. The term in brackets of (2.108) is the correction for stimulated emission in thermodynamic equilibrium.

To obtain the Rayleigh–Jeans form that is usually appropriate for the radio range in astronomy, we expand the exponential term by a MacLaurin expansion to obtain

$$\kappa_L \approx \frac{h^2\nu^2}{4\pi kT} \phi_\nu N_{n_1} B_{n_1,n_2}, \qquad h\nu \ll kT. \tag{2.109}$$

Keeping the first two terms of the expansion series gives this form of κ_L, which underestimates the exponential term by an error of about $(h\nu/kT)^2/2$.

It is convenient to rewrite the stimulated transition coefficients in terms of the oscillator strength f_{n_1,n_2}.[17] The absorption oscillator strength relates to the emission oscillator strength and to the specific intensity form of the B coefficient as

$$f_{n_1,n_2} = -\frac{\varpi_{n_2}}{\varpi_{n_1}} f_{n_2,n_1} \tag{2.110}$$

$$= \frac{mch\nu}{4\pi^2 e^2} B_{n_1,n_2}. \tag{2.111}$$

Note that emission oscillator strength is negative.

Goldwire (1968) and Menzel (1969) give oscillator strengths appropriate for RRLs from hydrogenic atoms. In addition, Menzel (1968) gives a useful approximation for the absorption oscillator strength:

$$f_{n_1,n_2} \approx n_1 M_{\Delta n} \left(1 + 1.5\frac{\Delta n}{n_1}\right), \tag{2.112}$$

---

[17] This term evolved from the "Ladenburg f," a vestige of classical physics when the intensity of a spectral line was characterized in terms of the number of dispersion electrons per atom or, "oscillators." See Appendix D for detailed information. Also, see the discussion relating the quantum mechanical to the classical form of the oscillator strength (Kardashev, 1959).

where $M_{\Delta n} = 0.190775$, 0.026332, 0.0081056, and 0.0034918 for $\Delta n = 1$, 2, 3, and 4, respectively.

Using the temperature, we can relate the number density of atoms in the lower bound level N_{n_1} to the population of the electrons and ions of the unbound states of hydrogenic atoms, N_e and N_i, by means of the Saha–Boltzmann ionization equation:

$$N_{n_1} = \frac{N_e N_i}{T^{3/2}} \frac{n_1^2 h^3}{(2\pi mk)^{3/2}} \exp\left(\frac{Z^2 E_{n_1}}{kT}\right), \tag{2.113}$$

where E_{n_1} is the energy of level n_1 below the continuum. For hydrogen, dividing (1.17) by kT gives $E_n/kT = 1.579 \times 10^5/n^2/T$, where T is in K.

Substituting (2.111) and (2.113) into (2.108) gives a general expression for the line absorption coefficient for an RRL:

$$\kappa_L = \frac{\pi h^3 e^2}{(2\pi mk)^{3/2} mc} n_1^2 f_{n_1, n_2} \phi_\nu$$
$$\times \frac{N_e N_i}{T^{3/2}} \exp\left(\frac{Z^2 E_{n_1}}{kT}\right) \left(1 - e^{-h\nu/kT}\right). \tag{2.114}$$

If $h\nu \ll kT$ (the Rayleigh–Jeans regime) and if the first two terms of (1.21) are substituted for the line frequency, the absorption coefficient becomes

$$\kappa_L \approx \frac{2}{\sqrt{\pi}} \frac{e^2}{m} \left(\frac{h^2}{2mk}\right)^{3/2} \frac{h}{k} R \phi_\nu Z^2 \Delta n \frac{f_{n_1, n_2}}{n_1}$$
$$\times \left(1 - \frac{3\Delta n}{2n_1}\right) \frac{N_e N_i}{T^{5/2}} \exp\left(\frac{Z^2 E_{n_1}}{kT}\right) \tag{2.115}$$

$$\approx 3.469 \times 10^{-12} \phi_\nu Z^2 \Delta n \frac{f_{n_1, n_2}}{n_1}$$
$$\times \left(1 - \frac{3\Delta n}{2n_1}\right) \frac{N_e N_i}{T^{5/2}} \exp\left(\frac{Z^2 E_{n_1}}{kT}\right), \tag{2.116}$$

where κ_L is in cm^{-1} if N_e and N_i are in cm^{-3}, T is in K, and E_{n_1} is in ergs.

Equation (2.116) – and (2.114) for the more general case – can be very useful. Because of the definition of the line profile given by (2.5), choose $\phi = 1\,\mathrm{Hz}^{-1}$ for the total absorption over the line. To obtain the absorption coefficient for just the center ($\nu = \nu_0$) of a purely thermally broadened line, substitute the expression given by (2.20), $\phi_\nu = 1/(1.064\,\Delta\nu_G)$ where the full width of the Gaussian line at half-intensity is in Hz.

Figure 2.18 compares the variation of the free–free absorption coefficient, given by (2.94), with the Hnα line absorption coefficient at line center given by (2.116) with a Gaussian substitution for ϕ. For these particular conditions of $T = 10^4$ K and $N_i = N_e = 10^4$ cm^{-3}, the peak line absorption exceeds the continuum absorption at $n > 33$, or $\nu > 192\,\mathrm{GHz}$.

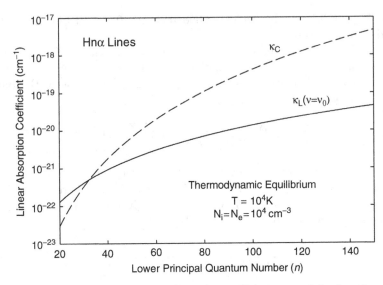

Fig. 2.18 Plots of the linear free–free absorption coefficient κ_C and the line absorption coefficient κ_L (at line center) as a function of the lower principal quantum number for H$n\alpha$ lines. Here, the line broadening is entirely thermal; it contains neither turbulence nor Stark broadening

The line absorption coefficient increases approximately as n^3 but decreases inversely as $T^{5/2}$. This n-dependence results from $\kappa_L \propto \nu^{-1}$ at a given temperature, a variation principally due to the increase of the thermally broadened line width with the increasing line frequency and the corresponding decreasing of $\phi \propto \nu^{-1}$ (see (2.20) and (2.22)).

The absorption coefficient varies directly with $N_i N_e$ because, at any given temperature and principal quantum number n, the number of bound atoms available for absorption must be proportional to the ionization products $N_i N_e$ along any sight line through the plasma, i.e., matter is conserved in the plasma when summed over both bound and unbound constituents.

2.3.5.2 Line Emission Coefficient

In thermodynamic equilibrium, the emission coefficient per unit volume is defined as

$$j_L = \kappa_L B_\nu(T) \tag{2.117}$$

$$\approx \kappa_L \frac{2kT\nu^2}{c^2}, \qquad h\nu \ll kT, \tag{2.118}$$

just as it was defined by (2.91) for free–free emission. Since we have defined κ_L earlier, this equation is sufficient to calculate the emission from the RRLs.

However, the line emission coefficient could also have been derived in terms of the spontaneous Einstein coefficient $A_{n_2 n_1}$:

$$j_L = N_{n_2} \frac{h\nu}{4\pi} A_{n_2 n_1} \phi, \qquad (2.119)$$

where ϕ_ν is the normalized line profile. Applying the relationship[18] between the A and B coefficients,

$$A_{n_2 n_1} \equiv \frac{2h\nu^3}{c^2} B_{n_2 n_1} \qquad (2.120)$$

would allow us to work backward to derive the form of κ_L given by (2.118). *Note*: Whereas spontaneous emission occurs isotropically, stimulated emission occurs with the same angular distribution as I.

2.3.6 Transfer Equation for RRLs

Having the coefficients for both the line and continuum emission, we are now able to calculate the intensity of the RRL relative to the underlying free–free continuum. Because the linear opacities are additive, at any frequency within the RRL, the intensity of the emission will be

$$I = I_L + I_C = B_\nu(T) \left[1 - e^{-(\tau_C + \tau_L)} \right]. \qquad (2.121)$$

The intensity of the line itself must then be

$$I_L = I - I_C = B_\nu(T) e^{-\tau_C} (1 - e^{-\tau_L}) \approx B_\nu \tau_L, \qquad (2.122)$$

when $\tau_C, \tau_L \ll 1$ as is the usual case for H II region gas at centimeter wavelengths. Under the same conditions, $I_C \approx B_\nu \tau_C$, and the ratio of the line-to-continuum emission for the same object must be

$$\int_{\text{line}} \frac{I_L}{I_C} d\nu = \int_{\text{line}} \frac{\tau_L}{\tau_C} d\nu = \int_{\text{line}} \frac{\kappa_L}{\kappa_C} d\nu. \qquad (2.123)$$

Equation (2.123) is particularly useful because beam efficiencies, dilution factors, and calibration factors apply more or less equally to numerator and denominator and, therefore, more or less cancel.

Direct substitution of the absorption coefficients derived in (2.95) and (2.116) gives an approximate expression[19] relating the quantities observed

---

[18] Here, B is again defined in terms of specific intensity rather than energy density. See the earlier footnote 16 for details.

[19] Equation (2.124) does not include the second term of the expansion for the line frequency given by (1.21). This omission will lead to an overestimate of I_L by a few percent, depending

for *hydrogen* RRLs to the average physical conditions of the emitting gas in thermodynamic equilibrium:

$$\int_{\text{line}} \frac{I_L}{I_C} \, d\nu \approx 1.301 \times 10^5 \Delta n \frac{f_{n_1 n_2}}{n_1} \frac{\nu^{2.1}}{T^{1.15}} F \exp\left(\frac{1.579 \times 10^5}{n_1^2 T}\right). \quad (2.124)$$

The integration element $d\nu$ is measured in kHz and ν is in GHz. The ratio I_L/I_C allows intensities to be given in any convenient units, such as antenna temperature, because the units will cancel each other.

The continuum emission from gaseous nebulae will contain free–free emission from all ionized atoms, especially from hydrogen and helium because of their dominant abundance in terms of number density. Therefore, this observed emission will overestimate the contribution of hydrogen alone and underestimate the value of the integral. The factor F in (2.124) corrects the observed free–free emission for the contribution of ionized helium:

$$F \equiv \left(1 - \frac{N_{He}}{N_H}\right). \quad (2.125)$$

Observations have established that the cosmic ratio N_{He}/N_H is approximately 0.075 ± 0.006 (Gordon and Churchwell, 1970), giving a value for F of 0.925 that will compensate for the observed value of continuum.

In (2.124), note the relationship of the line-to-continuum ratio with temperature and frequency. The line-to-continuum ratio increases with frequency because of the decreasing intensity of the free–free emission. It decreases with temperature because fewer atoms occupy a particular principal quantum level – as described by the Boltzmann equation. It varies directly with line intensity as expected from quantum mechanics and inversely with n_1 as described by the Boltzmann equation.

Figure 2.19 illustrates the ratio of the line amplitude to the underlying free–free continuum as a function of principal quantum number for α-type RRLs. These calculations result from (2.124) when the line width is entirely thermal broadening from a gas with a temperature of 10^4 K, calculated from (2.22). In this example, the line amplitude exceeds the continuum emission for quantum numbers less than 43 (frequencies greater than 79 GHz).

2.3.7 The First Measurements of RRL Intensity

Early observations of RRLs led to surprising results regarding the line intensities. Figure 2.20 shows the temperatures calculated from the better detections reported through 1966 for the M17, Orion, and W51 nebulae

---

on the frequency. The reader can improve the accuracy by adding a multiplicative factor like $(1 - 3\Delta n/2n_1)$ to the right-hand side of (2.124).

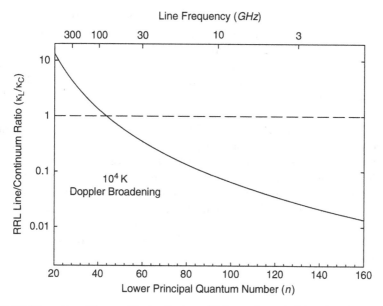

Fig. 2.19 The ratio of the amplitude of α-type RRLs to the underlying free–free emission as a function of n_1. The gas temperature is 10^4 K and the line broadening is assumed to be entirely thermal. The *upper, nonlinear, abscissa* is marked with the corresponding frequency. *Note*: The slight difference from Fig. 2.18 in the value of n where $I_L/I_C = 1$ is due to the Altenhoff approximation for κ_C used in (2.124)

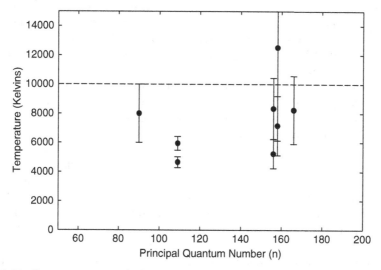

Fig. 2.20 Gas temperatures calculated from the initial detections of RRLs plotted against lower principal quantum number. The *broken line* marks the canonical 10,000-K temperature of H II regions accepted at that time on the basis of optical emission lines

(Sorochenko and Borodzich, 1965; Höglund and Mezger, 1965; Lilley, Menzel, Penfield and Zuckerman, 1966; Palmer and Zuckerman, 1966), calculated from (2.124). The temperature usually attributed to H II regions in our Galaxy on the basis of forbidden optical emission lines was 10,000 K. Yet, the temperatures derived from RRL observations with the best signal-to-noise ratios averaged to ≈5,000 K – half of the value generally accepted for nebula gas.

Additional observations supported the surprisingly low gas temperatures. Mezger and Höglund (1967) obtained $T = 5,820$ K by averaging results from observations of the H109α (5 GHz) line 16 H II regions. Dieter (1967) found a still lower value of 5,200 K from observations of the H158α (1.6 GHz) line in 39 sources. At the same time, new observations of optical emission lines reaffirmed the characteristic gas temperatures of 10,000 K (O'Dell, 1966; Peimbert, 1967). Were these RRL temperatures really the correct ones? Or, was there something wrong with the RRL theory with regard to the line intensities?

2.3.8 Departures from LTE

The explanation came from the astrophysics developed decades earlier to explain anomalous intensities of optical lines from nebulae and stellar atmospheres. In a series of papers, Menzel and coworkers (cf. Baker and Menzel (1938)) had explained these intensities as consequences of departures from thermodynamic equilibrium. Using his experience interpreting similar line spectra from the Sun, Goldberg (1966) showed that the intensities of the newly detected RRLs were also a consequence of departures from thermodynamic equilibrium.

Physically, the term "thermodynamic equilibrium" (TE) describes a situation in which the energy exchange between the radiative and kinetic energy domains of a gas is so efficient that a single parameter, temperature, describes exactly the characteristics of both domains. While this situation cannot occur in the open systems found in astronomy, there are localized situations that are so close that the TE equations may be used. The term "local thermodynamic equilibrium" (LTE) describes these situations. Spectroscopically, LTE refers to circumstances in which one kind of rate into a level exactly balances a similar rate out of a level, i.e., the radiative rates into a level exactly balance the radiative rates out of that level, and, similarly, with the collisional rates.

Accordingly, the equations developed above, based upon thermodynamic equilibrium, require a correction factor. Goldberg noted that the "excitation temperature" T_{ex} that describes the relative population of the bound quantum levels was not the same as the temperature T_e of the ionized gas in the nebula. Consequently, the factor $h\nu/kT_e$ in (2.108) is not the proper correction for stimulated emission in this environment.

To demonstrate this, Goldberg equated the exponential of the Boltzmann distribution of (2.107) to a corrected form involving T_e:

$$e^{-h\nu/kT_{ex}} = \frac{b_n}{b_{n-1}} e^{-h\nu/kT_e}, \qquad (2.126)$$

where the factor b_n is the ratio of the actual number of atoms in a level n to the number which would be there if the population were in thermodynamic equilibrium at the temperature of the ionized gas. Choosing $T_e = 10^4\,\mathrm{K}$ and estimating the ratio $b_{110}/b_{109} = 1.00072$ for the H109α line at $5\,\mathrm{GHz}$ (Seaton, 1964) gives $T_{ex} = -360\,\mathrm{K}$. He concluded that the very small relative difference $(b_{110} - b_{109})/b_{109} = 7 \times 10^{-4}$ "takes on great importance at radio frequencies because it is large compared with $h\nu/kT_e = 2.40 \times 10^{-5}$." In this way, ignoring the small departures from LTE in (2.108) significantly underestimates the amount of the stimulated emission, overestimates the absorption coefficient κ_L, overestimates the line intensity, and underestimates the equivalent T required to account for the intensity of the RRL relative to the underlying continuum. This effect would not occur for lines in the optical range because the energy difference between upper and lower quantum levels is much greater, i.e., the quantity $h\nu/kT_e$ is much larger than in the radio range.

Earlier, to develop the general theory appropriate to describe RRL line intensities in an LTE (or TE) environment, we used a single temperature, specifically T, to characterize the statistical state of bound *and* unbound levels of the atoms.

Now, to transform these LTE equations into non-LTE forms, we insert correction factors to account for the differences between the thermodynamic characteristics best described by an excitation temperature and those best described by the electron temperature. To emphasize this difference, we select T_e – the electron temperature that characterizes the free electrons – as *the* reference temperature for our calculations.

Because the non-LTE form of Boltzmann equation (see (2.107)) gives the relative level populations as

$$\frac{N_{n_2}}{N_{n_1}} = \frac{b_{n_2}}{b_{n_1}} \frac{\varpi_{n_2}}{\varpi_{n_1}} e^{-h\nu/kT_e} \left[= \frac{\varpi_{n_2}}{\varpi_{n_1}} e^{-h\nu/kT_{ex}} \right]. \qquad (2.127)$$

The rightmost term in brackets shows how the population ratio can also be expressed in terms of an excitation temperature T_{ex}. Following Goldberg (1968), we multiply the line absorption coefficient given by (2.108) by b_{n_1} to correct for the population available to absorb photons and insert a second factor, involving the ratio b_{n_2}/b_{n_1}, to correct for the non-LTE amount of stimulated emission:[20,21]

$$\kappa_L = \kappa_L^* b_{n_1} \left[\frac{1 - (b_{n_2}/b_{n_1})e^{-h\nu/kT_e}}{1 - e^{-h\nu/kT_e}} \right] \qquad (2.130)$$

$$= \kappa_L^* b_{n_1} \beta \qquad (2.131)$$

$$\approx \kappa_L^* b_{n_1} \left(1 - \frac{kT_e}{h\nu} \frac{d\ln b_{n_2}}{dn} \Delta n \right), \qquad h\nu \ll kT_e, \qquad (2.132)$$

where κ_L^* refers to the LTE form developed earlier, given by (2.116).

The non-LTE emission coefficient is

$$j_L = \kappa_L^* b_{n_2} B_\nu(T_e) \qquad (2.133)$$

by definition.

2.3.9 Non-LTE Line Intensities

As before, the intensity within the line is

$$I = I_L + I_C \qquad (2.134)$$

$$= S \left[1 - e^{-(\tau_c + \tau_L)} \right], \qquad (2.135)$$

where the non-LTE source function S for the RRL emission can be written

---

[20] The original mathematical expansion of β given by Goldberg (1968) as his equation (22) has

$$\beta \approx \frac{b_{n_2}}{b_{n_1}} \left(1 - \frac{kT_e}{h\nu} \frac{d\ln b_{n_2}}{dn} \Delta n \right), \qquad (2.128)$$

which would create a cofactor of b_{n_2} in (2.132). Note the difference in subscript. Because that would be unphysical for an absorption coefficient, because $b_{n_1} \approx b_{n_2}$, and because it is an approximation in any case, we write (2.132) as it stands.

Also, the definition of the correction factor β has evidently evolved slightly. Brocklehurst and Seaton (1972) (BS) use

$$\beta \equiv \beta_{n_1,n_2} = \left(1 - \frac{kT_e}{h\nu} \frac{d\ln b_{n_1}}{dE_{n_1}} \right), \qquad (2.129)$$

which differs in the argument of the logarithm. Because $n_2 - n_1 \equiv \Delta n \approx 1$ in many RRL observations of interest, it is a common practice to ignore the difference between b_{n_1} and b_{n_2} in applications of the corrective factors. Using the generic symbol n as a subscript usually involves no significant loss of accuracy and, in fact, BS adopt this simplification. Here, we include all subscripts for completeness, however.

[21] Goldberg (1968) introduced the symbol $\gamma \equiv d\ln b_{n_2} \Delta n/dn$, which has now passed from common usage.

$$S = \frac{j_C + j_L}{\kappa_C + \kappa_L} \tag{2.136}$$

$$= \frac{\kappa_C + \kappa_L^* b_{n_2}}{\kappa_C + \kappa_L^* b_{n_1} \beta} B_\nu(T_e) \tag{2.137}$$

$$= \eta \, B_\nu(T_e), \tag{2.138}$$

where the factor η corrects the Planck function for departures from LTE and is

$$\eta = \frac{1 + b_{n_2}(\kappa_L^*/\kappa_C)}{1 + b_{n_1}(\kappa_L^*/\kappa_C)\beta}. \tag{2.139}$$

The non-LTE correction factors apply only to the line coefficients because, by definition, T_e describes the thermodynamic state of the ionized gas and, therefore, the continuum coefficient needs no correction.

The intensity of the line relative to the continuum can be written

$$\frac{I_L}{I_C} = \frac{I - I_C}{I_C} = \frac{I}{I_C} - 1 = \frac{\eta(1 - e^{-\tau_\nu})}{(1 - e^{-\tau_C})} - 1, \tag{2.140}$$

where τ_ν is the sum of the actual line and continuum opacities:

$$\tau_\nu \equiv \tau_C + \tau_L = \tau_C + \tau_L^* b_{n_1} \beta. \tag{2.141}$$

If $|\tau_L|$ and τ_C are much less than one[22] as is the case for H II regions in the centimeter wave radio range:

$$I_L \approx I_L^* \, b_{n_2} \left(1 - \frac{\tau_C}{2}\beta\right), \qquad h\nu \ll kT_e, \tag{2.142}$$

where we have expanded the exponentials to three terms. This approximation shows that line amplification involves competition between a weakening due to the depopulation of the lower level expressed by b_{n_1} and a strengthening due to the joint function of τ_C and β (usually negative) representing the enhancement of stimulated emission. In this approximation, the factor τ_C parameterizes the column density of material along the sight line.

Substitution of the approximation equation (2.142) into (2.123)) gives

$$\int_{\text{line}} \frac{I_L}{I_C} \, d\nu \approx \int_{\text{line}} \frac{I_L^* b_{n_2} \left(1 - \frac{\tau_C}{2}\beta\right) d\nu}{I_C}, \qquad h\nu \ll kT_e, \tag{2.143}$$

and, in turn, this term into (2.124) and rearranging gives

$$T_e^{1.15} \exp\left(\frac{-1.579 \times 10^5}{n_1^2 T_e}\right) \approx$$

$$1.299 \times 10^5 \frac{\tau_C}{P} \, \Delta n \frac{f_{n_1 n_2}}{n_1} \, \nu^{2.1} \, F \times b_{n_2} \left(1 - \frac{\tau_C}{2}\beta\right), \tag{2.144}$$

---

[22] Equation (2.132) shows that τ_L will be negative in non-LTE situations where $\beta < 0$. Hence, we use the absolute value of τ_L in the criteria for this approximation.

in "observational units" for the Rayleigh–Jeans regime where the integrated power in the line $P \equiv \int T_L \, d\nu$ is in units of K kHz and ν is in GHz.[23] For observations of H II regions in the centimeter wave range, the exponential term is effectively unity.

Inserting the departure coefficients estimated by Goldberg (1966) into (2.144) raises the derived LTE electron temperatures by approximately 32 and 122% for the observed H109α and H165α RRLs to approximately $6{,}500 \pm 550$ and $11{,}000 \pm 3{,}000$ K, respectively. These corrections moved the derived electron temperatures for H II regions closer – but not to – the canonical 10,000 K determined from optical emission lines. Rather than using the RRLs to determine accurate electron temperatures for H II regions, Goldberg suggested that astronomers accept the 10,000 K and use the RRLs to derive departure coefficients that, in turn, would determine the thermodynamic state of the nebular gas.

At this time, more needed to be done to interpret the intensities of the RRLs quantitatively. Astronomers had established that the line intensities were probably enhanced in the centimeter wavelength range because of the non-LTE environment of gaseous nebulae. The principal contribution to the line enhancement appeared to be stimulated emission, creating a "partial maser effect." Further progress would require a wide range of improved observations and more accurate calculations of departure coefficients.

2.3.10 Calculating Departure Coefficients

2.3.10.1 Statistical Equilibrium

The departure coefficients result from solving a system of equations describing the equilibrium of the gas. All of the ways out of a quantum level n are equated to all of the ways into that level:

$$N_n \sum_{n \neq m} P_{nm} = \sum_{n \neq m} N_m P_{mn}, \tag{2.146}$$

where the rates P are in the directions indicated by the subscripts and include both radiative and collision processes. Such an environment is often called "statistical equilibrium."

It is important to understand that this situation differs from thermodynamic equilibrium or even the more usual astrophysical situation known as

---

[23] A more accurate correction factor (the term following the $\times$ symbol in (2.144)) is

$$\frac{b_{n_2} \left(1 - \frac{\tau_C}{2} \beta\right)}{1 - \tau_C}, \qquad \tau_c, \ |\tau_L|, \ (h\nu / kT_e) \ll 1. \tag{2.145}$$

local thermodynamic equilibrium that applies to a restricted locale or a restricted quantum mechanical environment. In TE or LTE, each rate into a quantum level must balance exactly with the *same kind* of rate out of that level, hence the term "detailed balance." For example, TE would require

$$N_m N_p C_{mn} = N_n N_p C_{nm} \tag{2.147}$$

and

$$N_m J B_{mn} = N_n \left[A_{nm} + J B_{nm} \right], \tag{2.148}$$

where the collision (C) and radiative (A and B) rates between levels n and m exactly balance the inverse rates of their own kind. The parameter J is the mean intensity, i.e., the time-averaged integral of specific intensity I over all directions with units of $\mathrm{ergs\,s^{-1}\,Hz^{-1}\,cm^{-2}}$. The volume density N_p refers to the number of particles available for collisions. Here, arbitrarily, n refers to the upper quantum level. These equations describe a thermodynamic environment where one temperature T is sufficient to characterize everything: the relative populations of bound levels of the atomic constituents of the gas through the Boltzmann equation, the intensity and spectral distribution of the ambient radiation field through the Planck equation, and the nature of the associated ionized gas through the Saha–Boltzmann equation. True TE does not occur in nature, which is why astronomers think in terms of LTE.

For most astronomical problems, both thermodynamic models assume time-invariant (often called "stationary") models, i.e., they assume that the "equilibrium" achieved by the gas does not change over timescales long compared with the inverse of each rate. This assumption does not hold for all situations, however.

Specifically, the equations for statistical equilibrium are

$$N_n \left(\sum_{\substack{m=n_0 \\ m \neq n}}^{\infty} (C_{nm} + B_{nm}) + \sum_{m=n_0}^{n-1} A_{nm} + C_{ni} + B_{ni} \right)$$

$$= \sum_{\substack{m=n_0 \\ m \neq n}}^{\infty} N_m (C_{mn} + B_{mn}) + \sum_{m=n+1}^{\infty} N_m A_{mn} + N_e N_+ (\alpha_n^r + \alpha_n^3). \tag{2.149}$$

The left-hand side includes all processes that depopulate level n. The first summation on the left accounts for collisional transitions up and down, and stimulated radiative decay. The second summation includes spontaneous transitions down, collisional ionization by electrons, and radiative ionization.

The right-hand side of (2.149) includes all processes that populate level n. The first term accounts for collisional and stimulated radiative transitions from other bound levels into n. The second term includes the spontaneous radiative terms and the three-body terms associated with radiative recombination into the bound level and collisional recombination. The level n_0 is the lowest level considered in the calculation.

The system of equations described by (2.149) is not closed. There is one more unknown variable than equations. The necessary extra equation for closure results from normalizing the level populations N_i to the populations N_i^* expected in thermodynamic equilibrium by using the Boltzmann equation. This process creates the dimensionless variables $b_i \equiv N_i/N_i^*$, referenced above as departure coefficients. Each resulting set of departure coefficients is a function of a specific electron temperature and volume density through the collision coefficients and the Boltzmann equation. For convenience, the reference temperature is taken to be the electron temperature T_e, which describes the characteristics of the ionized gas in the Saha–Boltzmann equation.

2.3.10.2 Evolution of the b_n Calculations

The first calculation of the population of hydrogen atomic levels was that of Baker and Menzel (1938) for the Balmer series. The solution was carried out for an infinite number of levels by the so-called n-method, which assumes that the population $N(\ell n)$ of the azimuthal sublevels ℓ is in equilibrium, i.e., proportional to their statistical weights:

$$N_{\ell n} = \frac{2\ell + 1}{n^2} N_n. \tag{2.150}$$

Only radiative transitions were considered, i.e., the quantum levels were assumed to be populated by recombination and cascade transitions and depopulated by cascade transitions to the lower levels. The object of these calculations was to determine the b_n coefficients. These calculations showed that the values of b_n for the first 30 levels differed significantly from unity. For $T_e = 10^4$ K and depending on the Case (see below), they found values ranging from $b_3 = 0.03 \rightarrow 0.1$ and $b_{30} = 0.45 \rightarrow 0.62$.

In general, calculations of departure coefficients are made for two situations: "Case B" in which the Lyman lines of the nebula are assumed to be optically thick (in LTE) but all other lines are optically thin and "Case A" in which all lines are presumed to be optically thin[24] (Baker and Menzel, 1938). Case B is the appropriate choice for H II regions from which RRLs have been detected.

In the years that followed, the accuracy of the calculations improved. For example, Seaton (1959) introduced a cascade matrix into the calculation technique, which enabled consideration of transitions into a given level by *all* possible routes. Such calculations showed that radiative processes were important only for small quantum levels (Seaton, 1964). For larger values of n, collisions dominate, causing the level population at, say, $n \geq 40$ (at a density of $Ne = 10^4$ cm^{-3}) to move toward the Boltzmann distribution referenced

---

[24] Actually, Baker and Menzel (1938) subdivided Case A into a Case A_1 and a Case A_2, distinguished by the Gaunt factors that are used.

to the electron temperature, i.e., toward $b_n = 1$. These results showed that, in H II regions, collisions of atoms with electrons play a major role in the level populations, as well as radiation-less collisions, both of which excite and de-excite quantum levels.

These calculations were made before the detection of RRLs. Their purpose was to facilitate the understanding of the intensities of optical lines in stellar atmospheres and in other astronomical environments – the optical lines themselves involving only small quantum levels.

Following the detection of RRLs, the calculations resumed but now with emphasis on extending them to the much larger quantum numbers directly associated with the RRL transitions. Seaton's (1964) extant calculations had considered only transitions between adjacent levels, $n \rightarrow n\pm1$. Now, processes between additional levels were necessary to determine departure coefficients sufficiently accurate for the interpretation of the RRL observations. For example, Hoang-Binh (1968) considered not only transitions between adjacent levels but also for $\Delta n = 2$ and 3. And, a year later, Sejnowski and Hjellming (1969) considered transitions up to $\Delta n = \pm20$ using a matrix condensation technique.

The most important step in calculating the population of highly excited atoms was the one taken by Brocklehurst and Salem (1977) and Salem and Brocklehurst (1979) (SB). Unlike earlier calculations, theirs included the external radiation field, which causes stimulated transitions that profoundly influence the level populations. The influence of stimulated transitions increases with n and, hence, is very important for calculating the intensities of the high-n RRLs, especially for environments with low electron densities. The SB calculations consider collisional transitions using semiempirical values for the cross sections of excited atom–electron collisions (Gee, Percival, Lodge and Richards, 1976). The SB departure coefficients are the most generally accepted as well as the most often used for the interpretation of RRL observations from H II regions (emission nebulae) at wavelengths exceeding $\lambda = 1\,\text{cm}$ or principal quantum numbers $n > 40$ and at moderate to high electron densities.

Appendix E.1 lists FORTRAN code adapted from SB, converted to the FORTRAN 77 standard, and modified to extend to quantum numbers down to $n = 10$ (Walmsley, 1990). The results also include values of the β parameter. The departure coefficients in many of our figures (below) resulted from this code. This code executes in seconds on a modern PC when compiled under, e.g., the MS-Fortran Powerstation compiler.

Note: For transitions involving small quantum numbers $n < 40$ *and* low gas densities, Storey and Hummer (1995) have calculated even more accurate departure coefficients. Unlike the SB calculations, these calculations consider collisions with the angular momentum states of hydrogen, which can be important under conditions described below in more detail. However, under conditions common to many H II regions, these departure coefficients agree well with those of SB.

2.3.10.3 Tables of Departure Coefficients

Tables of departure coefficients appropriate for RRLs from H II regions exist
in the literature (Seaton, 1964; Sejnowski and Hjellming, 1969; Brocklehurst
and Salem, 1977; Salem and Brocklehurst, 1979; Walmsley, 1990; Storey and
Hummer, 1995).[25] Departure coefficients are also available for the cold, partially ionized interstellar gas (Gulyaev and Nefedov, 1989; Ponomarev and
Sorochenko, 1992).

Determining appropriate collision cross sections is a difficult part of the calculations. Approximations are generally required that cannot apply equally
effectively to atoms in all levels. For example, visualize an electron orbiting
the nucleus at a radius of $r = 0.529(n^2/Z)$ Å, an equation derived from Bohr
theory. An atom in a principal quantum level of 200 has a target area 10^4
times larger than one in level 20. In general, the most recent calculations of
departure coefficients employ the most accurate cross sections.

2.3.10.4 Characteristics of Departure Coefficients

Figure 2.21 shows departure coefficients calculated from the modified SB code
for $T_e = 10^4$ K and a range of electron densities. The figure plots the two new
factors in (2.132) that represent a weakening factor (b_n) due to the depletion
of the level population available for absorption and a factor (β) describing an
enhancement of the stimulated emission because of the enhanced population
gradient across the principal quantum levels.

The upper panel of the figure shows asymptotes to illustrate the physics
involved with the radiation processes. In a dense medium where the collision rates dominate the level populations, the departure coefficients will be
unity because the electron temperature T_e accurately characterizes the relative populations of the bound levels. In a tenuous medium, on the other hand,
the radiative rates dominate, the temperature T_e no longer characterizes the
populations well, and the departure coefficients correspondingly lie below 1.
As the principal quantum number becomes smaller, the effective electronic
radius decreases, the target area of the atom decreases, the influence of collisions wanes, and the departure coefficient decreases from unity toward the
radiative asymptote. With respect to changes in density, the transition region between collision domination and radiative domination moves to lower
principal quantum numbers as density increases and collision rates become
increasingly effective.

The lower panel of the figure is also instructive. For conditions typical of
H II regions, the largest correction for stimulated emission – and, correspond-

---

[25] The Storey and Hummer calculations of departure coefficients for a wide range of T_e and
N_e are available from the Centre de Données Astronomique de Strasbourg via anonymous
FTP to the directory /pub/cats/VI/64 of cdsarc.u-strasbg.fr. These files do not contain
β_n values, however.

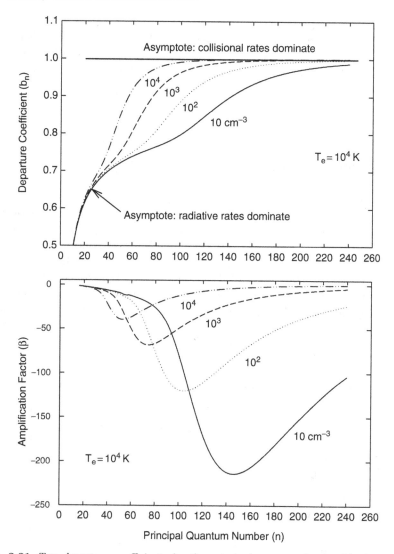

Fig. 2.21 *Top*: departure coefficients for the principal quantum levels of hydrogen for a range of electron densities, Case B. *Bottom*: the corresponding correction factor for stimulated emission for α-type RRLs plotted in the form β defined by (2.132)

ingly, the greatest line amplification – occurs for the lower densities. This results from the slope of the departure coefficient $d\ln(n)/dn$ being inversely weighted by the photon energy $h\nu$ in the correction term β. Perhaps, a better way of understanding the physics is to think of the derivative term as the relative (and normalized) population gradient with respect to the energy between levels, i.e., to think of the principal factor of β written as $d\ln(n)/dE_n$ because $h\nu\, dn \equiv dE_n$. This is why $|\beta|$ is largest at principal quantum numbers larger than those where db_n/dn maximizes for a given density, as seen in the top panel. Very specifically, the denominator dE_n in the derivative becomes smaller with increasing n. In other words, the correction for stimulated emission involves the gradient of the normalized population across the quantum levels *and* the energy of the photons associated with those levels, peaking at principal quantum numbers larger than the region of the maximum population gradient $d\ln(n)/dn$.

2.3.10.5 Differences Between Calculations of b_n

Compared to the earlier calculations, the modern collision cross sections and the inclusion of many orders of transitions increased the effect of collisions with respect to radiative processes. Figure 2.22 compares departure coefficients (Case B) calculated for the canonical H II region conditions of $T_e = 10^4\,\mathrm{K}$ and $N_e = 10^4\,\mathrm{cm}^{-3}$ by Seaton (1964), by Sejnowski and Hjellming (1969), by the code contained in Appendix E.1 modified from Salem and Brocklehurst (1979), and by Storey and Hummer (1995). Not only do the most recent calculations extend the influence of collisions to lower quantum numbers, but also the slope of the transition region differs greatly from Seaton's 1964 calculation used by Goldberg (1966).

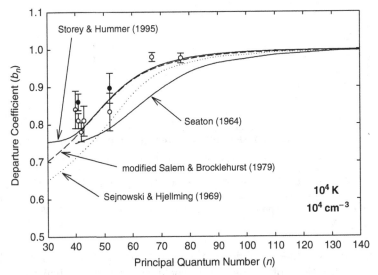

Fig. 2.22 Evolution of calculations of departure coefficients for hydrogen for $T_e = 10^4\,\mathrm{K}$ and $N_e = 10^4\,\mathrm{cm}^{-3}$ for Case B. Of the four, the Storey and Hummer calculation is the most sophisticated and, at $n \geq 40$ and this electron density, is nearly identical to the modified Salem and Brocklehurst calculation performed from the code listed in Appendix E.1. Juxtaposed are several departure coefficients determined from observations by Sorochenko et al. (1988) (*open circles*) and Gordon and Walmsley (1990) (*filled circles*)

Also shown in Fig. 2.22 are departure coefficients determined from observations of several RRLs made in the Orion nebula by Sorochenko et al. (1988) and Gordon and Walmsley (1990). Initially, to derive the departure coefficients observationally, the gas was assumed to have the 8,000-K temperature and 10^4-cm^{-3} density of Orion. Here, the values have been scaled to 10^4 K to allow comparison with the theoretical values plotted in Fig. 2.22. Agreement is good within the experimental errors but, unfortunately, not good enough to select the best theoretical calculations of b_n at $n < 40$. Nonetheless, the observationally determined points confirm the general correctness of the non-LTE transfer theory.

The difference between the Case B departure coefficients calculated by Storey and Hummer (1995) (SH) and the modified Salem and Brocklehurst (1979) code in Appendix E.1 is significant at small quantum numbers, as can be seen in Fig. 2.22. Although Salem and Brocklehurst limited their published tables of departure coefficients to $n \geq 50$, Walmsley (1990) extended their code to $n \geq 20$ by including collisions between low-n levels and collisional ionizations from low-n levels.

This difference results from the way the angular momentum sublevels are handled in the Storey and Hummer code, as described in detail by Strelnitski, Ponomarev and Smith (1996). At densities appropriate to H II regions, the angular momentum sublevels ℓ are completely degenerate at moderate to high principal quantum numbers n because of proton collisions, and they can be neglected in the calculations of b_n. However, the collision cross section of the (Bohr) atom varies approximately as n^4 and, at small principal quantum numbers, the influence of the collisions becomes much less. The ℓ quantum levels emerge from degeneracy and become locations in quantum-mechanical space that the hydrogen atoms can occupy. Strelnitski et al. refer to this transition as an "unblurring" of those angular momentum quantum levels as $n \rightarrow 1$. At a given density, the unblurring makes an increasing number of ℓ levels available as n decreases. Because the decay rates increase with decreasing ℓ, the low-ℓ sublevels of a given n are unblurred first (Strelnitski, Ponomarev and Smith, 1996). Because the effective departure coefficient is calculated as

$$b_n = \sum_\ell \frac{2\ell + 1}{n^2} b_{n\ell}, \tag{2.151}$$

the result of unblurring is an increase in the b_n value of the entire n level at low quantum numbers, as is seen at $n \leq 40$ in the Storey and Hummer calculation plotted in Fig. 2.22. Of course, eventually, the departure coefficients return to the radiative asymptote as n continues to decrease, because (1) the proton collisions lose their influence as the diameter of the (Bohr) atom decreases and (2) the number of ℓ levels at any given value of n diminishes as $2n + 1$, i.e., with the statistical weight of the principal quantum level.

The n-location and b_n-range of the unblurring are also a function of density at a given electron temperature. Figure 2.23 plots hydrogen departure

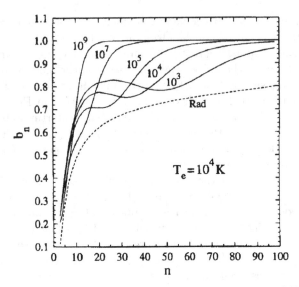

Fig. 2.23 Departure coefficients for hydrogen calculated by Storey and Hummer (1995) for $T_e = 10^4$ K and a range of electron densities N_e. The *broken line* marks the radiative asymptote calculated by V. Ponomarev. Figure from Strelnitski, Ponomarev and Smith (1996)

coefficients for $n \leq 100$ for a range of N_e and a temperature of 10^4 K, calculated by Storey and Hummer (1995). Note that, at high densities, proton collisions blur the angular momentum states so there is no visible inflection in the b_n curve. As the density decreases, the blurring becomes increasingly important, moving toward higher-n as collisions become less effective.

We conclude by noting that the b_n code contained in Appendix E.1 is often adequate for centimeter and meter wave RRLs from H II regions; in fact, the resulting values of b_n agree well with the Storey and Hummer results. For the submillimeter and millimeter regimes where RRLs involve small principal quantum numbers, it is more accurate to use the departure coefficients calculated by Storey and Hummer.

2.3.11 Line Intensities in Terms of Transfer Theory

Figure 2.24 shows the change in line intensity described by the approximation given by (2.142). For comparison, it also shows the exact calculation to illustrate where the approximation fails. The calculations are for α-type RRLs from a fictitious H II region with a diameter of 0.025 pc, an electron temperature of 10^4 K, and an electron density of 10^4 cm^{-3}– a hypothetical small H II region not unlike the Orion nebula.

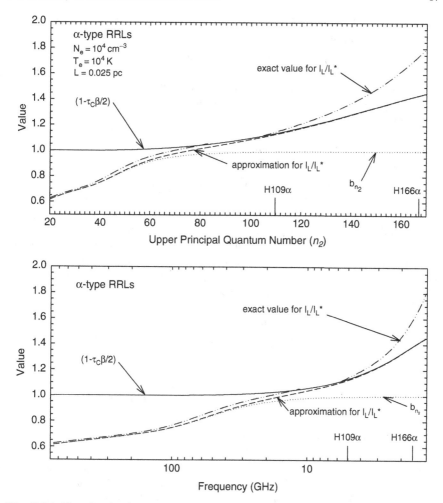

Fig. 2.24 *Top*: the simple approximation for line enhancement (2.142) for α-type RRLs plotted against the upper principal quantum number for conditions of a hypothetical H II region. For comparison, the exact enhancement is also shown, except for the range $n = 92 \rightarrow 100$ where the correction for stimulated emission goes through a discontinuity. The departure coefficient b_{n_2} is important at small quantum numbers and the factor $(1 - \tau_C \beta/2)$ at large quantum numbers. *Bottom*: the same quantities plotted against frequency. The positions of the H109α and H166α RRLs are indicated. The frequency range corresponding to the discontinuity is not shown in curve for the exact calculation. Departure coefficients were calculated from the code of Appendix E.1

At small quantum numbers (high frequencies), the departure coefficient dominates the line intensity though the underpopulation of the upper level relative to that expected from a Boltzmann population with $T_e = 10^4$ K. Although Fig. 2.21 shows β to be significant in this range of quantum numbers, the amplification term $(1 - \tau_C \beta/2)$ is unity because of the small value of the free–free optical depth.[26] LTE temperatures calculated from RRLs in this region will be too high.

At large quantum numbers, collisions ensure that the departure coefficient b_{n_2} is near unity. However, the larger value of τ_C enhances the gradient term β, resulting in a large amplification of the RRL intensity. In this illustration, the line enhancement becomes quite significant at large quantum numbers even though there is no actual inversion of the level population. LTE temperatures calculated from RRLs in this region will be too low.

Although the H109α and H166α lines considered by Goldberg (1966) both fall in the enhancement region of Fig. 2.24, the sizes of the line enhancement are less than his estimates. He calculated amplification factors of 1.4 and 2.8 for the H109α and H166α lines, respectively, on the basis of departure coefficients calculated by Seaton (1964). The newer calculations (Salem and Brocklehurst, 1979) shown in Fig. 2.24 give amplification factors of only 1.15 and 1.73. Figure 2.22 shows why. The slope of the modern b_n-curve at $n = 109$ is significantly lower than the slope calculated by Seaton. Yet, the departure coefficients b_{109} are not very different for the two curves and the values of τ_C should be the same. Consequently, the line enhancement is less because the amplification factor $b_{n_2}(1 - \tau_C\beta/2)$ in (2.142) is less, because $|\beta|$ is significantly lower in the modern calculations. The same is true at $n = 166$ although we have not plotted the calculations to this value of n.

2.3.12 Line Enhancement: A More General View

2.3.12.1 Non-LTE Line Intensities in Terms of Opacities

To understand the enhancement of RRL intensities more generally, it is useful to examine the detailed influence of the line and continuum opacities on the radiation transfer. The correction factor η to the Planck function must be considered at the same time. Equation (2.140) illustrates the relationship of these factors to the line intensity without simplifying assumptions, and we therefore repeat it below:

$$\frac{I_L}{I_C} = \frac{\eta(1 - e^{-\tau_\nu})}{(1 - e^{-\tau_C})} - 1. \tag{2.152}$$

Figure 2.25 plots the component optical depths τ_C and τ_L, the net optical depth $\tau_\nu = \tau_C + \tau_L$, and the correction factor η as functions of principal quantum number for the same fictitious Orion-like H II region described earlier. The departure coefficients for these parameters result from the code listed

---

[26] In the equations above, the free–free optical depth τ_C is also a measure of the free–free *emission* because of Kirchhoff's law of thermodynamics.

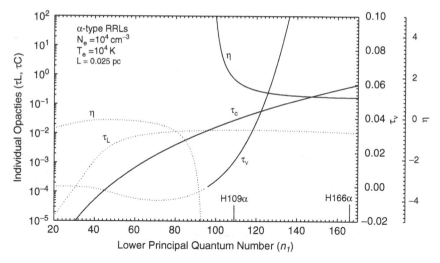

Fig. 2.25 Referenced to the *left ordinate* is the variation of the free–free (τ_C) and actual line (τ_L) opacities as a function of lower principal quantum number for the fictitious H II region. Referenced to the *inner right ordinate* is the sum of these opacities (τ_ν) and, to the *outer right ordinate*, the correction to the Planck function for stimulated emission (η). *Dotted segments of the lines* indicate where the quantities are negative. The positions of the H109α and H166α RRLs are indicated

in Appendix E.1. In this figure, the solid curves mark the regions where the variables are positive and the dotted curves where they are negative.

At principal quantum numbers $n > 100$, the opacities and η are positive for our model H II region. The net optical depth in the line τ_ν increases more slowly with n than the free–free optical depth τ_C while η approaches unity. The result is that the ratio I_L/I_C decreases *slowly* with increasing n.

At principal quantum numbers $n < 90$, the net line opacity and η are both negative, with both factors staying at small but – very, very roughly – constant values, thereby confining the $\eta\left(1 - e^{-\tau_\nu}\right)$ term of (2.152) to a small positive value. However, in this regime of n, combining (1.22) and (2.95) shows the free–free absorption τ_C to vary roughly as n^6, starting from a small value, thereby *sharply* decreasing the ratio I_L/I_C as n increases.

Figure 2.26 shows these results graphically for the hypothetical H II region. Note the logarithmic scale of the ordinate. The ratio I_L/I_C decreases[27] with n but at different rates in the low and high value regimes of n for the reasons we have just discussed above.

---

[27] In this particular model, because η and τ_ν change sign at $n \approx 95$, our simple calculations of I_L/I_C exhibit a computational discontinuity due to insufficient significant figures, and we therefore show this region as a gap in the plot.

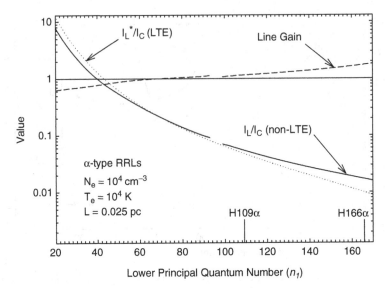

Fig. 2.26 The variation of the ratios I_L/I_C (non-LTE) and I_L^*/I_C (LTE) as a function of principal quantum number for a model H II region. A *broken line* indicates the ratio of I_L/I_L^* – the line gain – as a function of principal quantum number. The *horizontal line* is a reference for a line gain of 1. The positions of the H109α and H166α RRLs considered by Goldberg (1966) are marked. Stark broadening has not been included. *Gaps* indicate a region of insufficient computational precision

For comparison with the non-LTE value, Fig. 2.26 also shows the behavior of the LTE ratio I_L^*/I_C. For this model of an H II region, the line gain (defined here as I_L/I_L^*) is less than one at small values of n, increases to unity near $n = 66$, and then exceeds 1 as n increases further.

The definition of κ_L, given by (2.132), and the values of β in Fig. 2.21 show why. At small quantum numbers, the line intensity I_L lies below the LTE value because of the level depletion b_n, i.e., the correction factor $b_n|\beta| < 1$. The level depletion offsets the increased population gradient N_{n_2}/N_{n_1} parameterized by β, and the line intensity I_L falls below the LTE value of I_L^*. At larger quantum numbers, $b_n \to 1$ as $n \to \infty$ but $|\beta| > 1$ such that the line intensity I_L exceeds its LTE value. In other words, the population gradient increasingly dominates the effect of the level depopulation.

Figure 2.26 also illustrates that this model is close to what is required to explain observations of the H109α and H166α RRLs from the Orion nebula. The H109α RRL would exhibit a small gain over the LTE value and the H166α line would exhibit a somewhat larger gain. Goldberg referred to this as the "partial maser" effect because it does meet the maser gain criterion described below.

2.3.12.2 RRL Masers

Actual RRL masers[28] are possible.

Conditions for an RRL maser are stringent. To amplify, the populations of the principal quantum levels must be inverted such that $\tau_L < 0$. However, with respect to RRLs, this condition is necessary but insufficient for a maser. An intrinsic characteristic of the medium in which RRLs arise is continuum emission due to free–free radiation. For this reason, the line optical depth must not only be negative but its absolute value must exceed the optical depth of the free–free emission, which is always positive. The condition for an RRL maser is that the net absorption coefficient and, hence, the net optical depth in the line be (Ponomarev, 1994; Strelnitski, Ponomarev and Smith, 1996)

$$\tau_\nu \equiv \tau_C + \tau_L < 0. \tag{2.153}$$

For example, if $\tau_L < 0$ but $\tau_\nu > 0$, the medium does not amplify – as masers are required to do by definition. This is the situation for the H109α and H166α lines considered by Goldberg (1966) as a "partial maser effect" and illustrated in Figs. 2.24–2.26. For these lines, τ_ν is positive. This situation is then not a "maser" but, instead, a line enhancement caused by a decrease in the net absorption with respect to the LTE values. For this particular physical model, Fig. 2.25 shows that the condition of $\tau_\nu < 0$ will only be fulfilled for $n < 95$. Only in this region of quantum numbers, the RRL will actually be amplified.

In special circumstances, great amplification can occur. Specifically, when $|\tau_\nu| \gg 1$, (2.152) becomes (Strelnitski, Ponomarev and Smith, 1996)

$$\frac{I_L}{I_C} \approx \frac{|\eta|}{1 - e^{-\tau_C}}\, e^{|\tau_\nu|}, \quad \eta,\, \tau_\nu \ll 0. \tag{2.154}$$

Figures 2.25 and 2.26 do not show this regime.

Because (2.154) contains $|\tau_\nu|$ as an exponent, the ratio I_L/I_C can become very large if $\exp(|\tau_\nu|) \gg 1$. Because $\tau_\nu \equiv \tau_C + b_n\beta\tau_L^*$, the maser requires a physical environment where $\beta \ll 0$. Figure 2.21 shows that these conditions can occur when the density of an H II region is not very high. In general, $|\tau_\nu|$ does not reach maximum at the inflection of β because both τ_L^* and β have different dependencies on the electron density N_e. Section 2.4.1 discusses the conditions for RRL masers in more detail.

2.3.13 Classification of a Non-LTE Transition

Strelnitski, Ponomarev and Smith (1996) give an interesting way to classify the effects of non-LTE on the population of RRL levels in terms of the exci-

---

[28] Microwave amplification by stimulated emission radiation.

tation temperature. From (2.127), we express the ratio of the weighted level populations N's in terms of an excitation temperature T_{ex}:

$$\frac{N'_2}{N'_1} \equiv \frac{N_2/\varpi_2}{N_1/\varpi_1} = \exp\left(-\frac{h\nu_0}{kT_{ex}}\right), \tag{2.155}$$

where ϖ_i is the statistical weight of the principal quantum level i and ν_0 is the line frequency. The variable N'_i is then the population of level i per degenerate sublevel. We designate the upper level of the transition to be 2 and the lower to be 1. We can then write

$$T_{ex} = \frac{h\nu_0/k}{\ln(N'1/N'2)}. \tag{2.156}$$

The ratio β_{12} of the source function S_ν to the Planck function $B_\nu(T_e)$ is defined for this transition in the usual way:

$$\beta_{12} \equiv \frac{1 - \exp(-h\nu_0/kT_{ex})}{1 - \exp(-h\nu_0/kT_e)}. \tag{2.157}$$

Figure 2.27 shows the variation of T_{ex} and β_{12} as a function of the population ratio N'_1/N'_2 when $T_e = 10^4\,\mathrm{K}$ and, for this illustration, $h\nu_0/k = 100\,\mathrm{K}$. As the population ratio N'_1/N'_2 decreases from ∞ (all the atoms in the lower

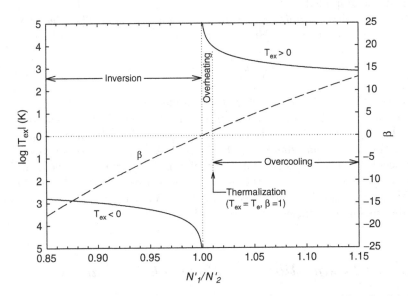

Fig. 2.27 Classification of non-LTE states of a quantum transition. After Strelnitski, Ponomarev and Smith (1996). N'_1 is the population of the lower quantum level per degenerate sublevel, T_{ex} is the excitation temperature for the two levels, and β is the correction to the Planck function at ν_0

level) to 0 (all the atoms in level 2), the excitation temperature T_{ex} increases from 0 to ∞, jumps to $-\infty$ at $N'_1 = N'_2$, and then increases to 0. Through this range of $N'_1/N'_2 = \infty \rightarrow -\infty$, β_{12} decreases monotonically from $[1 - \exp(-h\nu_0/kT_e)]^{-1}$ to $-\infty$, passing through 0 at $N'_1 = N'_2$.

As described by Strelnitski et al., these excitation regimes may be easily understood in terms of temperature. The right side regime where most of the population lies in the lower level can be called the "overcooled" regime, and $T_{ex} > 0$; the left side where most of the population lies in the upper level can be called the "inversion" regime, and $T_{ex} < 0$. At $T_{ex} = T_e$, the levels are thermalized and $\beta_{12} = 1$. In the narrow regime where $T_e < T_{ex} < \infty$, the gas can be called "overheated" because the excitation temperature exceeds the kinetic temperature of the H II region. Here, the source function S_ν is positive but less than the Planck function.

Figure 2.28 shows these excitation regions with respect to specific b_n and β curves characteristic of a typical H II region. As noted earlier, the overcooled condition occurs where the b_n curve increases toward small values of n, an inflection that appears at low densities in the departure coefficients calculated by Storey and Hummer (1995) because of their consideration of the population of the degenerate levels. The inversion region occurs toward larger values of n, where the slope $d \ln b_n/dn$ is large and, hence, β is correspondingly large and negative. Thermalization corresponds to the region where $b_n \approx 1$, usually at high electron densities where collisions dominate the population of the upper and lower quantum levels.

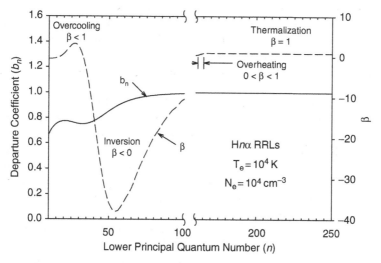

Fig. 2.28 Classification of the excitation regions of a quantum transition with respect to a b_n and β curve for hydrogen α-type RRLs. Departure coefficients from Storey and Hummer (1995). The *abscissa* is broken from $n = 100 \rightarrow 150$ to illustrate the distinct regions

The departure coefficients plotted in Fig. 2.23 illustrate these regions with respect to a larger range of electron densities. The overcooling regime disappears at high densities.

2.4 The Range of RRL Studies

2.4.1 High-Frequency RRLs

RRL observations began at centimeter and decimeter wavelengths but quickly spread into the millimeter wave range as receiver sensitivity improved. Observations in this range have several advantages:

1. The ratio of the line-to-continuum intensities grows with frequency because of the decreasing free–free emission, as shown by (2.124) and (2.143).
2. It is perhaps easier to interpret millimeter wave RRLs because one can neglect stimulated emission because of the decreasing free–free emission and Stark broadening because of the steep dependence of the line width on principal quantum number ($n^{4.4}$) shown by (2.74). Knowing the appropriate departure coefficient remains a problem, however.
3. For filled-aperture radio telescopes, the beamwidth decreases and, hence, angular resolution increases with increasing frequency. This is not a significant limitation for radio interferometers, however.

The first observations of an RRL in the millimeter wave range were carried out with the 22-m radio telescope in Pushchino, Russia, equipped with an 8-mm maser receiver. Figure 2.29 shows the H56α line from the Omega nebula (Sorochenko, Puzanov, Salomonovich and Steinschleiger, 1969) at 36.5 GHz. This was the first spectral line of any kind detected in the Galaxy at millimeter wavelengths.

With improving technology, RRL detections moved to even shorter wavelengths. Waltman et al. (1973) detected the H42α line at 85.7 GHz from the Orion nebula with the 36-ft. telescope of the National Radio Astronomy Observatory at Kitt Peak, AZ. Wilson and Pauls (1984) detected the H41α and H39α lines in Orion at 99.0 and 106.7 GHz, respectively, with the 7-m offset telescope of Bell Laboratories in New Jersey. With the resurfacing of the Kitt Peak telescope and a new 3-mm receiver, Gordon (1989) detected the H40α line in seven Galactic sources.

To a large extent, further advancements toward shorter wavelengths involved observations of a single object, the strong RRL maser emission from MWC349 (Martín-Pintado, Bachiller, Thum and Walmsley, 1989). Figure 2.30 shows spectra from MWC349 at $\lambda \approx 1$ mm and $\lambda \approx 3$ mm. While the H41α ($\lambda = 3.3$ mm) line has a Gaussian profile, the profiles of the H31α ($\lambda = 1.42$ mm), H30α ($\lambda = 1.29$ mm), and H29α ($\lambda = 1.17$ mm) lines have double peaks. The dependence of intensity of observed lines on n was also

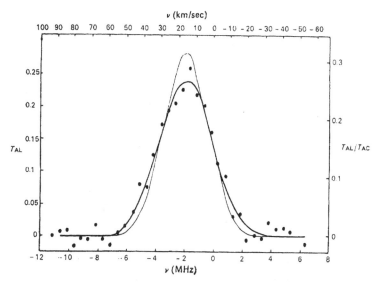

Fig. 2.29 Radio recombination line H56α from the Omega nebula – the first RRL detected in the millimeter range. The *thick line* shows the observed profile. The *thin line* shows the profile corrected for the bandwidth of the spectrometer. The *left ordinate* is antenna temperature and the *right* is the ratio of the line-to-continuum intensity

unusual. Instead of the line intensity decreasing as the quantum number decreased as expected from thermodynamic equilibrium, the opposite occurred. The line intensity increased as the quantum number decreased. The line intensities of the $\lambda = 1$ mm lines were at least 50 times greater than those of the $\lambda = 3.3$ mm line.

Observations of RRLs at still shorter wavelengths soon followed. Thum et al. (1994) detected the H21α ($\lambda = 0.45$ mm) line from MWC349 with the James Clerk Maxwell telescope on Mauna Kea, HI. This line had the same double-peaked profile as those observed in the $\lambda = 1$ mm range. The Earth's atmosphere prevents observations at shorter wavelengths from the ground, but Strelnitski, Haas, Smith, Erickson, Colgan and Hollenbach (1996) overcame this limitation by observing from the Kuiper Airborne Observatory and detecting the H15α, H12α, and H10α from MWC349 in the infrared portion of the spectrum. Technically, these were no longer "radio" recombination lines but "infrared" recombination lines (IRLs). Finally, Smith et al. (1997) made observations of the H6α ($\lambda = 12.4 \, \mu m$), H7α ($\lambda = 19.1 \, \mu m$), H7β ($\lambda = 8.2 \, \mu m$), and H8γ ($\lambda = 12.4 \, \mu m$) recombination lines from the same object in the middle infrared range (MIRLs). The spectral resolution was inadequate to determine the details of the line profiles but the line amplitudes showed MWC349 still to be masing in these lines.

The double-peaked profile of the RRLs from MWC349 seems to be consistent with a model of maser emission from the border of an edge-on, rotating

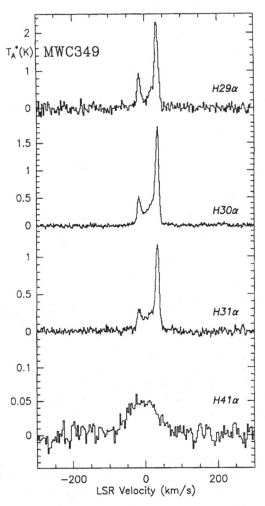

Fig. 2.30 Radio recombination lines from the binary star system MWC349 at millimeter wavelengths (Martín-Pintado et al., 1989). The $\lambda = 1\,\mathrm{mm}$ lines are the first RRL maser ever detected. Note the double-peaked profile of the high-frequency lines

circumstellar disk. For this reason, the two components are believed to be shifted symmetrically relative to the system velocity (Gordon, 1992; Thum, Martín-Pintado and Bachiller, 1992). The next part of this book will consider the MWC349 emission in more detail.

The detection of the hydrogen RRL maser in the MWC349 source stimulated a closer look at the theory of masing RRLs.

Section 2.3.12 showed the necessary conditions for a maser in an H II region to be large negative values of the net optical depth (τ_ν). Because $\tau_\nu \equiv \kappa_\nu L$, we can examine the net absorption coefficient κ_ν independently of the path length L through the nebula. The absolute value of k_ν determines the maser gain (Strelnitski, Ponomarev and Smith, 1996). From (2.141),

$$\kappa_\nu = \kappa_C + b_{n_1}\beta k_L^*, \tag{2.158}$$

where n_1 is the lower level of the transition. The coefficient β is evaluated at the lower principal quantum number n_1 as defined by (2.132). Since all quantities in (2.158) except β are positive, β has to be sufficiently large and negative to fulfill the maser condition of $\kappa_\nu \ll 0$, i.e., $\beta \ll -\kappa_C/(b_{n_1}\kappa_L^*)$.

Within (2.158), each parameter depends on the physical conditions within the H II region – temperature and density – as well as upon the principal quantum number and the frequency of corresponding transition. These dependencies are different. For example, the density corresponding to maximum maser gain differs from the one corresponding to the maximum of the energy-weighted population gradient characterized by β.

Figure 2.31 shows the net absorption coefficient κ_ν as a function of electron density for a number of hydrogen α-lines at millimeter wavelengths (Strelnitski, Ponomarev and Smith, 1996). The temperature for the calculations was taken to be 10^4 K, a canonical value for H II regions. The departure coefficients are those calculated by Storey and Hummer (Storey and Hummer, 1995).

The picture shows that there are optimal values of density for maximizing the amplification of each line, corresponding to minimum value of the net absorption coefficient. These minimums shift toward higher densities with decreasing quantum number, while absorption coefficient itself increases. Therefore, for each small group of adjacent lines, there is a relatively narrow interval of densities where the maximum maser gain can be realized. For example, the optimum density is $N_e = 5.6 \times 10^5\,\mathrm{cm}^{-3}$ for the group of lines near H55α, $N_e = 4 \times 10^7\,\mathrm{cm}^{-3}$ near H30α, and $N_e = 1.6 \times 10^9\,\mathrm{cm}^{-3}$ near H15α.

However, the density corresponding to minimum of the non-LTE *line* absorption coefficient does not coincide with the density of the maximum maser gain. Compare the dashed curve ($\kappa_L = b_{n_1}\beta\kappa_L^*$) for H36$\alpha$ line shown at the upper right diagram with the solid line (κ_ν). The minima of the two curves occur at different densities. Because the total absorption coefficient (κ_ν) seen by the photons also includes free–free absorption that is also a function of density, the extrema of the line and total absorption coefficients occur at different densities for any given line – as shown by (2.158). The effect of κ_C is to decrease the absolute value of the negative line absorption and to shift its minimum toward lower densities.

With specific regard to the RRL maser in the MWC349 star system, the calculations (Strelnitski, Ponomarev and Smith, 1996) show that the maximum gain should occur in the range H15α to H34α, which corresponds to the range of electron densities $N_e \approx 10^9 \to 10^6\,\mathrm{cm}^{-3}$, respectively.

The existence of a high-gain hydrogen radio recombination maser requires a second condition in addition to the large negative value of κ_ν. The photons

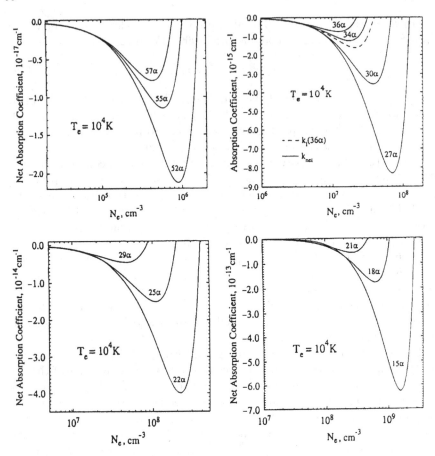

Fig. 2.31 The net absorption coefficient plotted as a function of electron density for a number of α-type RRLs (Strelnitski, Ponomarev and Smith, 1996). In this figure, k_{net} and k_l correspond to our κ_ν and κ_L, respectively

along the path must maintain excellent phase coherence with each other, which puts stringent limits for homogeneity of density and hydrogen atoms velocity along the lie of sight. This requirement usually means exceptionally narrowly focused maser beams in astronomical sources. For example, if the accumulated phase difference $\Delta\phi$ along two lines of sight differing by a small angle θ is

$$\Delta\phi = \frac{2\pi\ell}{\lambda}(1 - \cos\theta) \approx \frac{\ell}{\lambda}\pi\theta^2, \qquad (2.159)$$

the path length in wavelengths through a hypothetical astronomical masing region, $\ell/\lambda \approx 10\,\mathrm{AU}/1\,\mathrm{mm} \approx 10^{15}$, would require a beam angle of $\theta < 2 \times 10^{-8}$ rad to maintain the phase difference of $\Delta\phi = 1$ rad. Perhaps, because of this coherence requirement,[29] strong hydrogen RRL masers have only been detected in MWC349 up to present time; this discovery could have been a matter of luck for astronomers. The only other possible source of strongly masing RRLs is η Carina, where Cox et al. (1995) found peculiar millimeter recombination lines of hydrogen.

2.4.2 Low-Frequency RRLs

The solution to the Stark broadening problem explained why it was possible to observe RRLs at significantly higher atomic levels than originally presumed (see Sect. 2.2.4). In the early years, the highest atomic level associated with an observed RRL was $n = 301$. The H300α line was detected from both the Sgr A region and W43 (Casse and Shaver, 1977; Pedlar, Davies, Hart and Shaver, 1978).

Can atoms with still higher levels of excitation exist in the cosmos? To answer this question, a number of observatories searched for RRLs at $n = 350 \rightarrow 650$ lying in the meter and decameter wavelength ranges.

At the Pushchino Radioastronomical Observatory after several years of attempts, and with continuous improvements of the equipment to increase the detection sensitivity to 4×10^{-4} of the background continuum (Ariskin, Kolotovkina, Lekht, Rudnitskij and Sorochenko, 1982), success was achieved. RRLs with $n > 400$ were detected, but from carbon rather than hydrogen. It has not been possible to detect hydrogen RRLs at these quantum levels. The C427α ($\lambda = 3.56$ m), C486α ($\lambda = 5.25$ m), C538α ($\lambda = 7.12$ m), and C612β ($\lambda = 3.56$ m) lines were observed with the North–South line 864×40 m size of the DKR-1000 cross-axis radio telescope. The lines were observed in absorption toward the powerful source of low-frequency radio emission Cassiopeia A (Ershov, Iljsov, Lekht, Smirnov, Solodkov and Sorochenko, 1984).

Carbon lines with still higher transition numbers were detected toward Cassiopeia A by Konovalenko and Sodin (Konovalenko and Sodin, 1980; Konovalenko and Sodin, 1981) of the Radioastronomical Institute in Kharkov. Observations of the C630α ($\lambda = 11.4$ m), C631α ($\lambda = 11.5$ m), and C640α ($\lambda = 12$ m) were made with the decameter radio telescope UTR-2 – the world's largest with a size of $1,800 \times 900$ m, shown in Fig. 2.32. It is worth noting that the decameter carbon lines were detected earlier than the meter ones, after unsuccessful attempts to detect excited hydrogen lines at long wavelengths (Konovalenko and Sodin, 1979).

---

[29] Moran (2002) notes that this argument can also be reversed. The equation illustrates that "the phase requirements are so stringent that no cosmic maser can operate as a spatially coherent amplifier. As a result, cosmic masers have virtually no intrinsic beaming properties ... [and] are essentially temporally incoherent and spatially incoherent, unlike laboratory lasers with parallel mirrors."

Fig. 2.32 The 1,800 × 900 m UTR-2 radio telescope of the Radioastronomical Institute of the Ukraine at Kharkov. With this instrument, the longest wave carbon RRLs were detected at decameter wavelengths

The C631α line, which was first detected in the decameter range, was initially identified by authors (Konovalenko and Sodin, 1980) as the hyperfine transition $F = 5/2 \rightarrow 3/2$ of atomic nitrogen ($\nu = 26.127$ MHz) predicted by Shklovsky (1956b). This interpretation, however, encountered large difficulties. The observed optical depth of the line would require more than an order of magnitude increase of the nitrogen abundance relative to the accepted value.

Blake et al. (1980) had shown that observed line could be identified as a C631α line in absorption, whose frequency ($\nu = 26.126$ MHz) is very close to the frequency of nitrogen line. This explanation eliminated the necessity of revising the nitrogen abundance in interstellar medium.

Subsequent observations confirmed this interpretation. Konovalenko and Sodin (1981) detected two more lines toward Cassiopeia A with frequencies that corresponded exactly to the carbon RRL C630α ($\nu = 26.250$ MHz) and C640α ($\nu = 25.0396$ MHz). This frequency correspondence confirmed that all three lines were carbon RRLs.

During few next years in Kharkov (Konovalenko, 1984), in Green Bank (Anantharamaiah, Erickson and Radhakrishnan, 1985), and in Pushchino (Ershov et al., 1987), astronomers detected a number of carbon RRLs in the meter and decameter wavelength range up to C732α ($\lambda = 18$ m) toward Cas A. Figure 2.33 shows the profiles of the low-frequency carbon lines obtained in Pushchino and Kharkov. These lines were observed in absorption with low ($\approx 10^{-3}$) line-to-continuum ratios. The integration time of each spectrogram was 20–40 h.

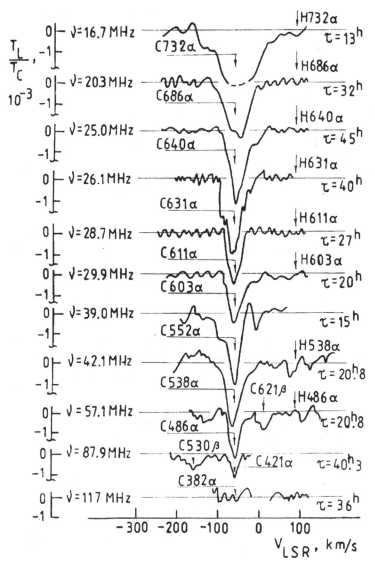

Fig. 2.33 Low-frequency carbon RRLs toward Cas A. $\nu < 30\,\mathrm{MHz}$, observations in Kharkov (Konovalenko, 1984); $\nu > 30\,\mathrm{MHz}$, observations in Pushchino (Ershov et al., 1984; Ershov et al., 1987). The *ordinate* is the amplitude of the line relative to the continuum emission. On the *left* is the frequency and on the *right* is the integration time. The *arrows* indicate the calculated positions of the carbon and hydrogen lines. From Sorochenko and Smirnov (1990)

The detection of the low-frequency lines was quite unexpected and extremely interesting both for physics and astronomy. A number of questions immediately arose:

1. Why are the carbon lines reliably observed at the highest excitation levels $(n > 400)$ while the lines of the more abundant hydrogen are not detected? Why were all attempts to detect hydrogen lines with $n > 300$ toward Cas A and other sources unsuccessful (H352α: Shaver et al. (1976); H400α: Ariskin et al. (1979); H630α–H650α: Konovalenko and Sodin (1979); H351α: Hart and Pedlar (1980); H392α–H394α: Ariskin et al. (1982))?

2. Why are carbon RRLs for the excitation levels $n = 530$–700 ($\nu = 44$–17 MHz) stronger than the lines of less excited atoms for $n = 420$–480 ($\nu = 88$–59 MHz), while the lines for $n = 380$–400 ($\nu \cong 100$ MHz) are in general impossible to detect?

3. If one can succeed in detecting lines corresponding to the 732th level and, in this case, their intensity does not fall with increasing n, then where is the limit to the formation of RRLs and what defines it?

At present, the answers to these questions appear to be quite clear. The answer to the first question – why carbon rather than hydrogen lines are observed at the highest excitation levels – is connected with the special nature of the regions where these lines originate. Analysis revealed that, to emit RRLs corresponding to transitions between extremely high excitation levels, the emitting region in the ISM must (a) have a low electron density and (b) be sufficiently cold.

The first requirement stems from the fact that the sensitivity of atoms to collisions with charged particles, mainly electrons, dramatically increases with excitation level. For H II regions with the temperature $T_e = 5 \rightarrow 10 \times 10^3$ K, the line width depends on such collisions as described by (2.74), i.e., $\Delta \nu_L \propto N_e n^{4.4}$.

For the lower temperatures found in H I regions, $T_e = 20 \rightarrow 200$ K. There, Stark broadening has a similar dependence on the principal quantum number of the level and the electron density (Ershov et al., 1984):

$$\Delta \nu_L = 1.16 N_e \left(\frac{n}{100} \right)^{5.1} \left(\frac{T_e}{100} \right)^{0.62} \text{ Hz.} \tag{2.160}$$

If we accept as a criterion that a line is observable when its width does not exceed 30% of the separation between adjacent RRLs, i.e., if

$$\Delta \nu_{\text{lim}} = \frac{6 \times 10^{15}}{n^4} \text{ Hz,} \tag{2.161}$$

then we can examine some general relationships between the density N_e and the principal quantum number associated with detectable low-frequency RRLs.

At these large quantum numbers, simplifications are possible. For example, Stark broadening dominates thermal broadening, i.e., $\Delta \nu_L \gg \Delta \nu_G$. Furthermore, at these large quantum numbers, carbon atoms can be considered to

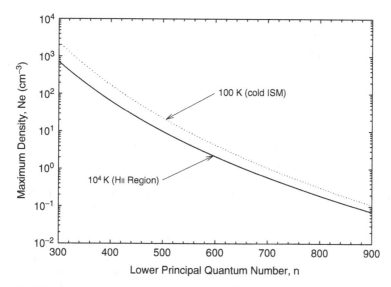

Fig. 2.34 Maximum values of the electron density N_e as a function of the lower principal quantum number n for detectable RRLs

be hydrogenic, i.e., these atoms have effectively the same electronic structure of a hydrogen atom – a single electron bound to a unitary, positively charged nucleus.

Therefore, combining (2.74) and (2.161) gives us a density limit of $N_e = 4.6 \times 10^{23} n^{-8.4}$ for detectable RRLs in H II regions with $T_e = 10^4$ K, and combining (2.160) and (2.161) gives a limit of $N_e = 8.2 \times 10^{25} n^{-9.1}$ for detectable RRLs in H I regions with $T_e = 100$ K. Figure 2.34 shows the detection limits for these two environments as a function of n, the "hot" environment of the H II region and the "cold" one of the ISM.

The figure shows the dependence of the maximum density for detectability of RRLs to be a strong function of principal quantum number. For example, the 400α line from the cold medium can be detected at densities up to $170\,\mathrm{cm^{-3}}$, whereas the 800α line cannot be detected at densities exceeding $0.3\,\mathrm{cm^{-3}}$. The limitation for the cold medium comes from the fact that the probability of recombination of ions and electrons increases as temperature decreases. Correspondingly, lower temperatures increase the likelihood of highly excited atoms existing and, also, increase the intensity of the line itself. Mathematically, (2.116) and (2.124) describe this behavior exactly, showing that T_L is approximately proportional to $N_e N_i / T_e^{3/2}$. In other words, in the ISM where the ionization products N_e and N_i are small – only partial ionization of the medium – detectable low-frequency RRLs can only occur when the temperature is very low.

In the ISM, the requirements of partial ionization and low temperatures are fulfilled in regions far from hot stars, in the so-called H I regions. Here, the hydrogen is mostly neutral because of its high ionization potential of 13.6 eV and a weak ambient radiation field. In this environment, other elements with lower ionization potentials can be more easily ionized, as described by the Saha–Boltzmann equation (2.113). Note that the ionization energy enters as an exponential. Relatively, small differences in ionization potential can make enormous changes in $N - eN_i$, especially at low temperatures where the argument of the exponential term can be large.

Among the constituents of the ISM with ionization potentials lower than hydrogen is carbon, with an ionization potential of 11.3 eV. Of these candidates with lower ionization potentials, carbon is the most abundant. Consequently, the combination of a greater degree of ionization and a significant cosmic abundance is why carbon RRLs can be observed at very high excitation (n) levels but not hydrogen (Sorochenko and Smirnov, 1987).

The answer to the second question is connected to the specific dependence of the line intensities upon the level population. Since the lines form in a cool low-density environment, the level populations are not in thermodynamic equilibrium – even for $n > 400$. In the specific case of carbon, the level populations are influenced by a low-temperature dielectronic-like recombination (DR).

Watson et al. (1980) have shown that in the cold ISM ($T_e \approx 100$ K), the recombination of ionized carbon and free electrons to highly excited levels can occur simultaneously with the $^2P_{1/2} -^2 P_{3/2}$ fine-structure excitation of the C^+ core, where the energy (ΔE_{fs}) associated with this fine structure is 92 K, 1.27×10^{-14} erg, or $\lambda = 158$ μm depending on your preference for units. In this case, the reverse process – autoionization – to a large extent is suppressed by rapid ℓ-changing collisions (ℓ is the quantum number for orbital angular momentum). For $\ell > 10$, autoionization is unlikely and the highly excited atom is stabilized.

This low-temperature dielectronic recombination[30] is an emission-free process. It occurs when the kinetic energy of the recombining electron is insufficient to excite the ion fine-structure level C^+ $^2P_{3/2}$. This energy deficit exactly compensates the energy of the bound level n, i.e., the kinetic energy of the electron divides according to

$$\frac{mV^2}{2} = \Delta E_{fs} + E_n, \qquad (2.162)$$

where $E_n = -2.18 \times 10^{-11}/n^2$ erg is the energy of the bound level n as described by (1.14) and V is the velocity of the recombining electron. Because the kinetic energy of the electron is always positive, (2.162) is executed if $n \geq 42$. For this reason, only highly excited levels of carbon experience can gain additional population through dielectronic recombination – a very difference situation than for hydrogen (Walmsley and Watson, 1982).

---

[30] Watson et al. (1980) suggested the term dielectronic "capture" for this low-temperature process.

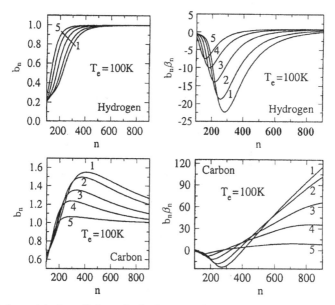

Fig. 2.35 b_n and $b_n\beta$ coefficients for hydrogen and carbon at $T_e = 100$ K as a function of n. The *numbers* correspond to following values of electron densities N_e:1–0.05, 2–0.1, 3–0.3, 4–1.0, 5–3.0 cm^{-3}. From Ponomarev and Sorochenko (1992)

Figure 2.35 shows values of b_n and $b_n\beta$ coefficients calculated for hydrogen and carbon for $T_e = 100$ K and various electron densities (Ponomarev and Sorochenko, 1992). The curves for hydrogen are similar to those shown in Fig. 2.21. The b_n values smoothly increase to 1 with increasing quantum number, and the $b_n\beta$ terms are mostly negative for the large quantum numbers.

The carbon curves have a very different character. Unlike the case of hydrogen, the departure coefficients rise to values exceeding one and then decrease toward the LTE value of one, owing to the effects of dielectronic recombination. Consequently, the values of the term $b_n\beta$ cross into the positive domain in the range $300 < n < 500$, where their values considerably exceed their absolute values in the negative domain.

These characteristics explain the intensities of carbon RRLs in the meter and decameter wavelength ranges. In the region where $b_n\beta_n = 0$, (2.132), (2.140), and (2.141) show that line optical depth (τ_L) will be zero, the total optical depth will be only that of the free–free continuum, and the carbon lines will not appear. This is why the C382α line was not detected and why an earlier attempt to detect the C400α line toward Cas A also failed (Ershov, Lekht, Rudnitskij and Sorochenko, 1982).

At $n > 420$, carbon lines are observed in absorption with the depth of the absorption line increasing with n – exactly as would be expected from the positive values of $b_n\beta_n$ shown in Fig. 2.35. In the range of negative values of $b_n\beta_n$, the medium within the source amplifies the background radiation, and the lines appear in emission. Figure 2.36 shows C RRLs over a

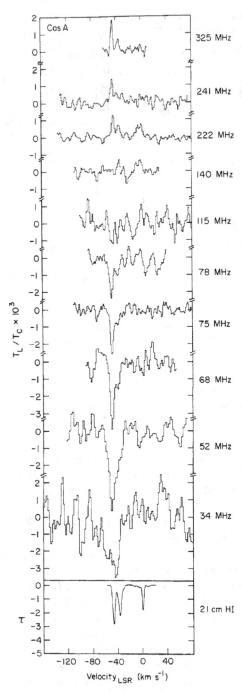

Fig. 2.36 Carbon α radio recombination lines observed toward Cas A at ten frequencies in the range 34–325 MHz. The quantum numbers corresponding to these frequencies are $n = 565, 502, 450, 446, 436, 385, 360, 310, 300,$ and 272. The bottom spectrum is the Hɪ absorption in units of τ. From Payne et al. (1989)

large frequency range observed at the NRAO observatory in Green Bank, WV (Payne, Anantharamaiah and Erickson, 1989). Note that the spectra change from absorption to emission with increasing frequency just as theory predicts. The quantum numbers $n = 272, 300$, and 310 involve carbon lines in emission; the numbers $n = 436, 446, 450, 502$, and 565, absorption; and the intermediate numbers $n = 360$ and 385, the transition region from emission to absorption where lines are not detected. The H I spectrum at the bottom shows the distribution of the line-of-sight interstellar gas as a function of radial velocity.

Therefore, the theoretical calculations of the population of the excited levels of carbon in the "cold" ISM, which take into account low-temperature dielectronic recombination, completely explain the dependence of the carbon RRL intensities in all wavelength ranges where they have been observed.

2.5 How Many Atomic Levels Can Exist?

The detection of RRLs in the meter and decameter wavelength ranges – particularly, the detection of carbon lines up to C732α – prompts the questions: what is the maximum quantum number of stable atomic levels and what are the restrictions on this level? These questions are very interesting both for atomic and for elementary-particle physics. According to Bohr theory (see (1.6)), the diameter of an atom in a distinct quantum level n is $1.06 \times 10^{-8} n^2$ cm. Therefore, the carbon atom in the ISM with $n = 732$ has a diameter of $d \approx 50\,\mu m$. Can atoms exist in the cosmos with an excitation level of 1,000, corresponding to a diameter of $100\,\mu m$? Or, with $n = 3,000$, corresponding to $d \approx 1\,mm$?

Until the detection of meter and decameter RRLs, astronomers had assumed that the limit of highly excited atomic levels was determined by collisions with electrons, i.e., only by the electron density. It became evident that in interstellar medium, large-diameter atoms can occur at very low electron densities $N_e = 10^{-1}, 10^{-2}$, and $10^{-3}\,cm^{-3}$. At these densities, electronic collisions would place a limit on highly excited atoms and, correspondingly on RRL emission, to values of n equal to many thousands.

The limiting effect of neutral particles is even weaker. Cross sections for the interactions of the excited atoms with atoms in ground state increase to a limit at $n = 20 \rightarrow 30$, after which this limit remains constant (Mazing and Wrubleskaja, 1966). As a result, for $n \approx 1,000$, these cross sections are almost ten orders of magnitude less than the ones for electronic collisions.

What determines the maximum number of distinguishable atomic levels and, hence, the limiting dimensions of cosmic atoms? It turns out to be the background Galactic, nonthermal radiation.

2.5.1 Radiation Broadening of RRLs

This background radiation stimulates transitions between highly excited levels, reducing the lifetimes of these levels, and consequently causing Lorentz line broadening described earlier by (2.9) and (2.10). In a thermalized medium, the stimulated emission is isotropic because the mean intensity J has no preferential direction; it is the time-averaged integral of I over all angles (see footnote 31). Accordingly, the full width of the Lorentz profile at half-intensity is

$$\Delta\nu_L \equiv \frac{\Gamma}{2\pi} = \frac{\Gamma_n + \Gamma_m}{2\pi}, \tag{2.163}$$

where the total rate out of level n, or Γ_n, is

$$\Gamma_n = \underbrace{\sum_{m=1}^{n-1} A_{n,m} + \sum_{m=1}^{n-1} J_\nu B_{n,m}}_{emission} + \underbrace{\sum_{k=n+1}^{\infty} J_{\nu'} B_{n,k}}_{absorption}. \tag{2.164}$$

In this environment, at large principal quantum numbers, the contribution of spontaneous emission (giving the natural width component of the line) will be small in comparison to the stimulated terms and can be neglected. The line width due to radiation broadening is then

$$\Delta\nu_L \approx \frac{1}{2\pi}\left(\underbrace{2\sum_{m=1}^{n-1} J_{n,m} B_{n,m}}_{depopulation\ of\ n} + \underbrace{2\sum_{l=1}^{m-1} J_{m,l} B_{m,l}}_{depopulation\ of\ m}\right) \tag{2.165}$$

$$\approx \frac{2}{\pi}\left(\sum_{m=1}^{n-1} J_{n,m} B_{n,m}\right), \quad n \gg \Delta n > 0, \tag{2.166}$$

for the emission transition from level n to level m. In (2.165), the terms within the large parentheses indicate the depopulation rates for levels n and m, respectively. For a level with a large n, depopulation is almost the same for emission and absorption and, hence, we insert cofactors of 2 for the emission term of each level to account for absorption processes out of that level. At large principal quantum numbers, the lifetime of the upper level is approximately equal to the lifetime of the lower level, and we sum the n and m terms to get (2.166).

2.5.1.1 Galactic Background Radiation

We now find the mean intensity J for the ambient radiation field in our Galaxy at long wavelengths. Cane (1978) gives the brightness temperature of the isotropic nonthermal radiation to be $T_{NT} = 22.6 \times 10^3$ K at 30 MHz with a spectral index $\alpha = 2.55$. Therefore,

$$T_{NT} = 22.6 \times 10^3 \left(\frac{3 \times 10^7}{\nu}\right)^{2.55} = \frac{2.63 \times 10^{23}}{\nu^{2.55}} \text{ K}. \tag{2.167}$$

Because $h\nu = 10^{-19} \ll kT_{NT} = 10^{-12}$ ergs, we can use the Rayleigh–Jeans approximation to the Planck radiation function to calculate the mean intensity:

$$J_\nu \approx \frac{2kT_{NT}\nu^2}{c^2} \tag{2.168}$$

$$= \frac{2k\nu^2}{c^2} 22.6 \times 10^3 \left(\frac{3 \times 10^7}{\nu}\right)^{2.55} \tag{2.169}$$

$$= \frac{8.10 \times 10^{-14}}{\nu^{0.55}} \text{ erg s}^{-1} \text{ Hz}^{-1} \text{ cm}^{-2}. \tag{2.170}$$

The Einstein B terms result from their definition in terms of the A terms:[31]

$$B_{n,m} \equiv \frac{c^2}{2h\nu^3} A_{n,m} \tag{2.172}$$

$$= \left(\frac{c^2}{2h\nu^3}\right) \left(\frac{8\pi^2 e^2 \nu^2}{mc^3}\right) f_{m,n} \tag{2.173}$$

$$= 5.03 \times 10^{25} \frac{1}{\nu} \cdot \frac{0.19n}{(\Delta n)^3}. \tag{2.174}$$

Equation (2.174) uses an approximation for the absorption oscillator strength $f_{m,n}$ resulting from

$$f_{m,n} \approx nM(\Delta n) \left(1 + \frac{1.5\,\Delta n}{n}\right) \tag{2.175}$$

$$\approx \frac{0.19n}{(\Delta n)^3}, \tag{2.176}$$

after evaluating the equal-order Bessel functions in the definition given by Menzel (1968).

Combining (2.170) and (2.174) gives

$$JB_{n,m} = 0.047 \left(\frac{n}{100}\right)^{5.65} \left(\frac{1}{\Delta n}\right)^{4.65} \text{ s}^{-1}, \tag{2.177}$$

---

[31] This relationship is the specific intensity form. The relationship between $B_{n,m}$ and $A_{n,m}$ coefficients is a numerical rather than a directional one. The relationship results from the detailed balance requirement in TE that

$$n_1 J B_{1,2} = n_2 A_{2,1} + n_2 J B_{2,1} \tag{2.171}$$

such that the dimensions of $JB_{1,2}$ are the same as $A_{2,1}$, i.e., s^{-1}.

where we have used the approximation for the RRL frequencies given by
(1.21). At large values of n, this expression[32] gives sufficiently accurate frequencies.

2.5.1.2 Width of the Broadened Line

Summing the stimulated emission terms, we find

$$\Delta \nu_L = \frac{2}{\pi} \sum_{m=1}^{n-1} J_{n,m} B_{n,m} \tag{2.178}$$

$$\approx \frac{4.70 \times 10^{-13}}{\pi} \times n^{5.65} \left[\underbrace{\left(\frac{1}{1}\right)^{4.65}}_{\Delta n=1} + \underbrace{\left(\frac{1}{2}\right)^{4.65}}_{\Delta n=2} \right] \tag{2.179}$$

$$\approx \frac{4.70 \times 10^{-13}}{\pi} (1.04) \, n^{5.65} \qquad n \gg 1, \tag{2.180}$$

$$= 0.031 \left(\frac{n}{100}\right)^{5.65} \text{ Hz}, \tag{2.181}$$

for the full width at half-intensity of radiation-broadened RRL. Note that
two terms of the summation are sufficient; the term associated with $\Delta n = 2$
is nearly negligible.

Ershov et al. (1982) derived the same equation (their equation (9)) by
considering the collisions of ambient photons with highly excited atoms in
the nebular gas.

2.5.1.3 The Lowest-Frequency Detectable RRL

If we arbitrarily assume that an $n\alpha$ recombination line can be distinguished
from its neighbor when its width is less than, say, 1/3 of the interline spacing,
we can use (1.22) and (2.181) to equate the frequency difference between
adjacent lines to the spacing criterion:

$$6.58 \times 10^{15} \left[\frac{1}{n^3} - \frac{1}{(n+1)^3} \right] = 3 \, \Delta \nu_L \tag{2.182}$$

$$= 0.093 \left(\frac{n}{100}\right)^{5.65} \tag{2.183}$$

---

[32] Using the approximate expression of (2.176) for the f_{mn} reduces the numerical accuracy
to two significant figures, at most.

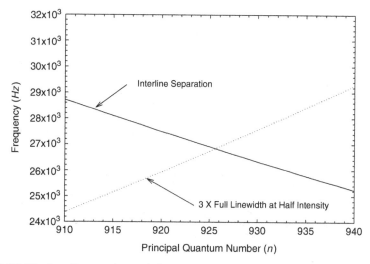

Fig. 2.37 The interline spacing and three times the line width plotted as a function of principal quantum number. The intersection gives the solution of (2.183)

and solve for n – the lower principal quantum number of the lowest-frequency, distinguishable line from the cold ISM of our Galaxy. Figure 2.37 shows the solution graphically. This equality is true for $n \approx 926$, which corresponds to a hydrogenic atom of diameter $\approx 91 \, \mu m$ – slightly larger than the thickness of this page.

This criterion for detectability is conservative, as can be seen from Fig. 2.38. If the criterion of the interline spacing is reduced to twice the line width, then the equality will give $n \approx 967$, corresponding to an even larger atom of diameter $\approx 0.1 \, mm$.

2.5.2 Existence as well as Detectability

It is possible to consider the limit from another perspective. If one considers that a quantum state will not occur when the lifetime of that state is less than the rotation period of the bound electron around the nucleus, the nonthermal background radiation would limit the quantum states to $n \leq 1{,}600$ (Shaver, 1975), corresponding to an RRL with a frequency of about 1.6 MHz. This limit is less restrictive than the broadening discussed above but the two criteria may not be in conflict. The former refers to the *existence* of the quantum states and, therefore, the existence of RRLs in *any* form rather than just being detectable. Practically speaking, though, the detectability limit is the more important one for astronomical research.

Both of these calculations suggest that atoms as quantum systems in interstellar conditions can exist up to quantum levels of $\approx 1{,}600$, at the least.

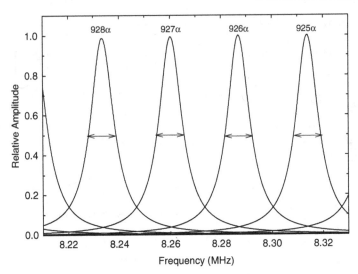

Fig. 2.38 What an observer might see: the superposition of radiation-broadened recombination lines plotted against rest frequency. The full widths at half-intensity are indicated. The separation between the H925α and H926α lines is approximately three line widths

These mean very large atoms. According to the Bohr model, such levels would correspond to atomic diameters of approximately 0.3 mm. Because ambient radiation limits their existence as well as their detectability, even larger atoms might exist and be detectable in cosmic environments with lower radiation than our Galaxy.

Despite the apparently fantastic nature of such huge atoms, their existence is real and natural. Let us imagine the rarified interstellar medium. Its most abundant element – hydrogen – is neutral. The ambient UV radiation from distant stars can only ionize atoms with ionization potentials less than that of hydrogen. Their lower – often, considerably lower – abundance means fewer electrons and ions. In some regions, the electron density is only a few cm^{-3} or so. Despite this sparseness, the free electrons and ions can recombine and produce highly excited atoms. These atoms can have significant lifetimes. The low densities and temperatures of their environment make unlikely the collisions with charge particles that can destroy them or change their state. Neutral atoms have even less influence on their existence.

Presently, astronomers seem to be approaching the excitation limits of these huge atoms. The low-frequency limit of 30 MHz for the DKR-1000 radio telescope in Pushchino allows searches for α-type RRLs up to $n \approx 600$. However, searches for higher-order transitions involving larger quantum levels are being carried out, such as the detection of the C748β line toward Cas A (Lekht, Smirnov and Sorochenko, 1989) shown in Fig. 2.39.

To date, carbon lines with the highest quantum levels have been observed at Kharkov with the UTR-2 radio telescope. Figure 2.39 shows the average of the C764α–C768α lines in the 14.7-MHz range toward Cas A, obtained by

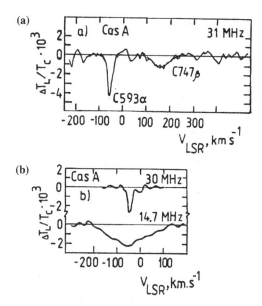

Fig. 2.39 Carbon RRLs detected toward Cas A at the highest known excitation levels of atoms. **a** The C747β line close to the frequency of the C593α line. The velocity scale is that of the C593α line. From Lekht et al. (1989). **b** At the *bottom* is the averaged spectra of the lines C764α–C768α; at the *top* is the spectrum of the C603α line shown for comparison (Konovalenko, 1990)

Konovalenko (1990). The integration time for one line was 50 h. Comparison of the C747β line with the C593α line, and of the averaged C764α–C768α lines with the C593α line, shows that the line widths increase with n due to the shortening of the lifetimes of the excited levels near the detection limit. For example, the width of the C764α–C768α averaged profile is already 20% of the separation between adjacent lines or 1,180 km s^{-1} in velocity. At $n \approx 1,000$, the widths should be so large that the adjacent lines will blend with each other into a continuous spectrum and, therefore, should disappear.

More recently, Stepkin et al. (2007) reported the detections of additional carbon lines at 26 MHz in absorption against Cas A after integrations of up to 500 h. Some of these lines are approximately 1009δ transitions and, as such, are more than 10^6 times larger than the ground-state atoms, corresponding to carbon atoms with classical diameters of 108 μm. As we have mentioned earlier, 0.1 mm is a dimension comparable to the thickness of this page.

2.6 Summary

Figure 2.40 illustrates the progress of our knowledge about atoms, the existence of excited levels, and the spectral lines from transitions between them. These data are of great interest for physics and were obtained from radio

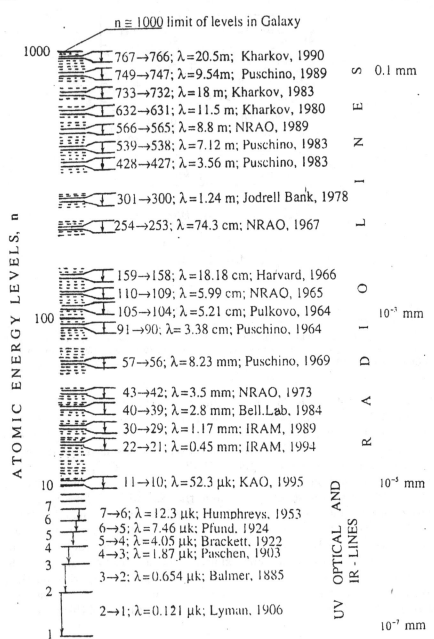

Fig. 2.40 A graphic display of the range of detected α-type recombination lines sorted by the upper principal quantum number n. The range includes the UV, IR, submillimeter, millimeter, centimeter, meter, and decameter parts of the electromagnetic spectrum. Note that interstellar atoms can exist up to excitation levels of $n = \approx 1,000$. The *right ordinate* is the equivalent diameter of the associated atoms

astronomical methods of study. Initially performed in the centimeter range at $n \approx 100$, the observations of RRLs advanced toward the longer meter and decameter wavelength ranges, where the lines occurring near $n \approx 1,000$, limited by the conditions of our Galaxy, were detected.[33] Simultaneously, observations of RRLs advanced toward the shortest wavelength region – millimeter and submillimeter waves. After observations of IR lines as far as H10α in the MWC349 hydrogen maser, this modern research became joined to the classical measurements of hydrogen lines in the IR, visible, and UV ranges which lie at the foundation of the Bohr quantum theory of atoms.

It is interesting to note that N. Bohr partially foresaw that the most highly excited atoms would be observed in space. In his classical article "About the spectrum of hydrogen" (Bohr, 1914), when explaining why the high-order lines of the Balmer series seen in celestial spectra were extremely difficult to observe in the laboratory, he wrote: "In order that the large orbits of electrons may not be disturbed by electrical forces from the neighboring atoms the pressure will have to be very low, so low, indeed, that it is impossible to obtain sufficient light from a Geissler tube of ordinary dimensions. In the stars, however, we may assume that we have to do with hydrogen which is exceedingly attenuated and distributed throughout an enormously large region of space."

Naturally, Bohr could not foresee that the most highly excited atoms would be detected by radio astronomy techniques that did not exist when he created the quantum theory of atoms. Certainly, he could not have foreseen that atoms may have up to 1,000 distinct principal quantum levels.

With these discoveries involving basic physics, one can hardly doubt that our cosmos is truly a wonderful laboratory.

---

[33] Figure 2.40 does not contain the more recent observations of Stepkin et al. (2007), which were obtained after the original edition of this book had been published.

Chapter 3
RRLs: Tools for Astronomers

Abstract This chapter describes what astronomers have learned from observations of radio recombination lines since their detection. It discusses the characteristics of gaseous nebulae, the state of ionized hydrogen and helium in these nebulae, the characteristics of ionized carbon found in the comparatively cool interstellar gas, planetary nebulae, the Sun, an unusual early star (MWC 349A) emitting time-varying masering lines, and nearby galaxies in which radio recombination lines have been detected.

Radio recombination lines (RRLs) turned out to be a powerful tool for astrophysical research. They are unique both in the number of transitions that can be detected and in the wavelength range over which they are observed. They occupy about five orders of magnitude of the wavelength scale of electromagnetic waves,[1] which is why they can be used for the study of astronomical objects that significantly differ in their physical parameters. The physical characteristics of "radio" also play a significant role here. Unlike the electromagnetic waves of the ultraviolet, optical, or infrared ranges, huge wavelength ranges of radio waves are almost unabsorbed by the interstellar medium (ISM) and therefore can be detected from very large distances.

RRLs provide us with a great deal of information about the ISM. Although its mass is only 3% of the total mass of our Galaxy ($1.5 \times 10^{11} M_\odot$), the ISM is the main component in terms of occupied volume. The study of the ISM enables us to understand the evolutionary processes that take place in the Galaxy and, by extension, in other galaxies. A continuous exchange of matter takes place between stars and the ISM. According to current thinking, stars form from the interstellar matter in regions where the physical conditions

---

[1] The etymology of "radio" refers to radiant (electromagnetic) energy used for communication or, more primitively, signaling over distances. The term was first used in the late nineteenth century, allegedly in 1898 by the French physicist Brandly in reference to a "coherer" detector. At this writing, the radio domain is considered to range from about 10 kHz to, say, 1 THz – eight orders of magnitude – but is often extended in practice, particularly toward higher frequencies. Officially, the range can be more restricted as in the definitions used within the communications industry (Emerson, 2002).

M.A. Gordon, R.L. Sorochenko, *Radio Recombination Lines*, Astrophysics and Space Science Library 282, DOI: 10.1007/978-0-387-09691-9_3,
© Springer Science+Business Media LLC 2009

in the ISM clouds – density and temperature – start the process by gravitational collapse that heats the gas that, eventually, triggers the thermonuclear fusion that causes the stars to "shine." In this way, portions of the ISM are transformed into stars.

This conversion also runs the other way. The stars return part of their mass to the ISM through stellar winds, planetary nebulae, novae, and supernova explosions. Because the thermonuclear processes within these stars have enriched their original material with new elements, this process changes the chemical composition of ISM.

At the same time, the energy radiated outward by stars in the form of UV radiation, stellar winds, and expanding shells causes fundamental changes in the structure and physical conditions of the ISM as well as in its chemical composition. That part of the ISM close to stars is ionized by the stellar UV radiation, forming H II regions with temperatures of several thousands of Kelvin. In time, these H II regions expand and the star can leave the H II region because its trajectory may differ from that of the surrounding gas. Or, in the case of a nova or supernova, the star can release its ionized shell altogether. Because of these dynamic processes, other parts of the ISM will radiate their heat energy away, allowing cold dense regions to form again, which in turn will form into new generations of stars.

In this way, a circulation between stars and the ISM takes place. This duality of cosmic processes is often called "astration," the astronomical equivalent of the biological term "symbiosis." The stars require the ISM and the ISM requires the stars. Moreover, on a large scale, the similarity of the term *astration* to the medical word "aspiration" is also appropriate, with its implication of gas moving in and out of an organism – as in breathing. It is important to note that astration is not reciprocal everywhere. That part of the ISM that forms into small and very small stars often remains there. It is the large stars that participate most vigorously in the astration process.

Hydrogen is the principal component of the ISM. It accounts for about 70% of the mass of the ISM. The remaining parts apportion by mass to 28% for helium and 2% for all other elements. Approximately, half of the hydrogen in the ISM (by mass) is in the form of molecules inside dense, cold clouds and the other half is in the form of neutral atoms (H I) and ions (H II).

Despite its importance in the astration process, molecular hydrogen does not emit spectral lines at radio wavelengths. Astronomers study it through indirect techniques or directly through spectral lines emitted in the infrared where, unfortunately, the opacity can be quite large. Hydrogen molecules collide with other molecules that can emit in the radio range, exciting quantum mechanical energy states within them, thereby allowing these molecules to radiate through rotational transitions at millimeter and centimeter wavelengths. Studying the spectra of abundant secondary molecules like CO allows astronomers to deduce the characteristics of the radio-invisible, interstellar hydrogen molecules.

On the other hand, atomic hydrogen emits the well-known, ubiquitous $\lambda = 21$ cm hyperfine line discovered in 1951 (Ewen and Purcell, 1951; Muller and

Oort, 1951). Observations of this line provided, and are still providing, the fundamental data about the distribution of neutral hydrogen in our Galaxy and in many other galaxies.

RRLs enable astronomers to study another basic component of the ISM: the interstellar ionized gas. Initially, this term referred only to discrete H II regions, which are very widespread in the Galaxy. In fact, the existence of hydrogen RRLs is the primary criterion for classifying cosmic radio sources as either thermal (H II regions) or nonthermal. Such observations revealed that the majority of discrete continuum sources at centimeter wavelengths near the plane of Galaxy are H II regions. Lockman (1989) found 462 of 500 such sources located within ±1° of the Galactic plane to emit RRLs. Moreover, among the other 38 sources, not all are nonthermal. RRLs were not detected in some of these owing to insufficient sensitivity of the survey.

Information obtained from hydrogen RRLs enables us to determine the basic physical conditions of H II regions as well as distribution of ionized hydrogen in Galaxy. Helium, the second-most abundance element of the ISM, is also ionized in the majority of H II regions. The ratio of the intensities of hydrogen and helium RRLs enables us to determine with high accuracy the relative abundance of helium, which has great significance not only for understanding the physics of the ISM, but also for understanding how the Universe formed.

Hydrogen is generally in a neutral form in the cold ISM, in H I regions, and at the surface of molecular clouds because of its high ionization potential with respect to the ambient interstellar radiation field. Figure 3.1 illustrates the

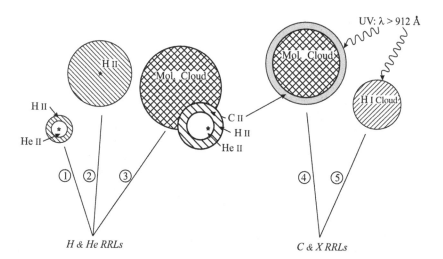

Fig. 3.1 This cartoon shows the types of ISM objects studied with RRLs. Item (*1*) represents a dense, bright H II region like the Great Nebula in Orion or a planetary nebula; (*2*) is an extended, low-density H II region; (*3*) represents C II regions at the interface between H II regions and molecular clouds; (*4*) shows the C II region boundary between molecular clouds and the diffuse ISM; and (*5*) illustrates C II regions within atomic H I clouds. "X" indicates RRLs from atoms other than carbon, helium, or hydrogen

types of ISM objects that are studied with RRLs. Note that elements with lower ionization potential can be ionized in these locations. Among these, carbon is the most abundant. Carbon RRLs are detected at many different frequencies (see Sect. 2.4.2). We shall show below that carbon lines contribute important information about intermediate ISM layers between H II regions and the parent molecular clouds from which they were formed.

In addition to these objects, RRLs have been detected from the Sun, two stellar systems, and a number of extragalactic sources.

Figure 3.2 shows the general form of this information obtained from RRLs. It shows a generalized RRL spectrum in units of antenna temperature (T_A) vs. arbitrary frequency units (ν) that could be obtained from any H II region in our Galaxy. The helium RRL is detected simultaneously with that of hydrogen, being separated by $4.078 \times 10^{-4}\nu$ owing to the difference in the Rydberg constants of the species. If the parent molecular cloud with the C II region on its border is included in beam of the radio telescope, the carbon line and the lines of heavier elements like S, Mn, Si, and Fe (see Sect. 3.3.1)

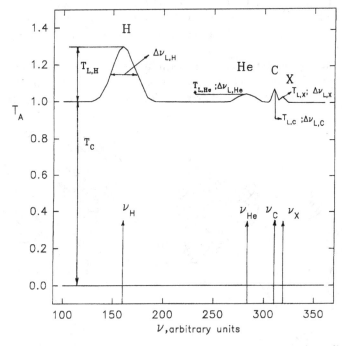

Fig. 3.2 A hypothetical spectrogram toward an H II region showing the radio recombination lines of hydrogen, helium, and carbon. The lines from heavier elements are merged into common line marked "X." The composite profile contains information about the ISM in the form of the measured antenna (brightness) temperatures of each of the constituent lines (T_L), their full widths at half-maximum $(\Delta\nu_L)$ and their line shapes, the observed (Doppler-shifted) frequencies of the line centers (ν), and the underlying continuum temperature (T_C) contributed by the free–free emission from their ionization products

would also be detected. The frequencies of the lines in this group would be blue-shifted relative to the helium and carbon lines and would overlap due to the small differences in their Rydberg constants (see Table A.2). All of this RRL emission would be superimposed on the thermal continuum contributed mainly by the H II region but also by the ionization products of the heavier elements.

A large number of parameters can be measured from the spectrogram: antenna (brightness) temperatures of each of the lines and their widths, profiles and frequencies that can determine the radial velocities of the emitting medium with respect to the local standard of rest V_{LSR}. The antenna temperature (intensity) of the underlying continuum is measured simultaneously and, when used with the line intensities, determines the excitation status of the line components. These data tell us about the physical environment of the ISM.

We now turn to what has been learned, and what can be learned, from observations of RRLs from cosmic sources.

3.1 Physical Conditions in H II Regions

3.1.1 Electron Temperature of H II Regions

RRLs of hydrogen provide the simplest and most precise method of determining the electron temperature of H II regions. Unlike observations of optical emission lines, RRLs are unaffected by reddening by interstellar dust. Further, they can be accurately measured even in weak astronomical sources.

3.1.1.1 Line-to-Continuum Observations

The possibility of such measurements stemmed from the pioneering work of Kardashev (1959). This work showed that the paramount parameter specifying the recombination and ionization processes associated with RRLs was the electron temperature T_e. This single parameter specified not only the population of the principal quantum levels of the atom, but also the degree of ionization, such that it determined the distribution of the cosmic gas between bound and unbound quantum domains. The temperature dependence of the populations of the two domains is different. As a result, the measurement of the ratio of antenna temperatures in a line and in a part of the spectrum adjacent to it allows one to determine electron temperature of H II regions.

As is almost always the case in astronomy, there are several complications in using RRLs to determine the temperature of H II regions. One is departures of the level populations from thermodynamical equilibrium values as described in Sect. 2.3.9. Another – recognized well after the initial detections of RRLs (Simpson, 1973a; Shaver, 1975) – was a systematic underestimate of the energy radiated in the line owing to Stark broadening that shifts emission from the line core to the line wings, where it easily blends with the underlying continuum. In this circumstance, fitting baselines to the profile results in mistaking wing emission for continuum, leading to the underestimate of the line emission relative to the continuum and, by (2.144), an overestimate of the electron temperature. Happily, however, there are frequencies where this diminishment of the measured line power offsets the enhancement of the line through the partial maser effect. In other words, the derived temperatures are correct.

Figure 3.3 illustrates a typical situation. The dotted line shows values of LTE L/C intensity ratios calculated from (2.124) and (2.26) for an electron temperature of 8,500 K and a turbulence broadening component of $10 \, \mathrm{km \, s^{-1}}$. These values would be typical for an H II region. Also plotted in the figure are actual observations of L/C from Hnα lines from the Orion nebula, a nearby compact H II region in the northern sky. The data cover a frequency range from 613 MHz to 135 GHz, a little over two orders of magnitude.

At low frequencies, the L/C ratios fall below the LTE values. This occurs because Stark broadening shifts radiation energy from the core of the line to the wings, where it becomes indistinguishable from the baseline of the

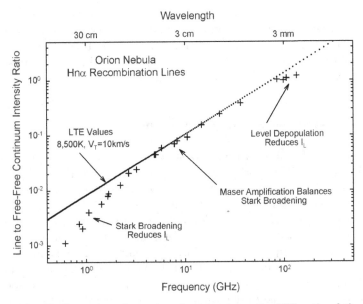

Fig. 3.3 The "line" consists of a series of points giving the LTE ratio of the line-to-continuum emission of RRLs from an H II region of 8,500 K and a turbulence broadening component of $10 \, \mathrm{km \, s^{-1}}$, calculated from (2.26) and (2.124). Each point marks the rest frequency of an RRL. Juxtaposed are observations of the L/C ratio from the Orion nebula (Lockman and Brown, 1975a; Gordon, 1989; Gordon and Walmsley, 1990)

spectra as described earlier. At high frequencies, two effects occur. First, departures from LTE lower the population of the quantum levels as described in Sect. 2.3.8, decreasing the number of atoms available to emit, and consequently decreasing the line intensity below LTE values. Second, continuum emission from dust begins to contribute significantly to the background emission as shown in Fig. 2.16, increasing the observed continuum beyond that radiated only by free–free emission. Both of these effects decrease the observed ratios of L/C below the LTE values.

There is an additional problem using observations of RRLs at low frequencies. As shown in Fig. 2.17, the opacity of the free–free emission can be significant at low frequencies. Consequently, radiation reaching the observer may not include all the matter along the line of sight. The observed values of L/C would nominally come only from regions where $\tau_C < 1$. This restriction may apply not just to a gas volume of the H II region in the foreground. Because H II regions contain density variations, the observed values of L/C could selectively reflect only the characteristics of the more tenuous gas where $\tau < 1$.

In the middle frequency range, however, it is possible to determine accurate values of electron temperature for an H II region, as discussed in detail by Shaver (1975). Stark broadening shifts energy from the line core to the line wings, thereby reducing the observed ratio L/C. Here, a fortuitous compensation occurs. The partial maser effect discussed in Sect. 2.3.8 offsets this weakening of the peak line intensity by Stark broadening such that the observed values of L/C actually fall very near the LTE values. However, owing to density inhomogeneities, the more representative observations will probably come from the upper end of this frequency range where the opacity is smaller.

Discussed in detail earlier, (2.144) provides a tool to extract an electron temperature from RRLs averaged along the line of sight in some unknown way. Equation (2.143) gives form for the enhancement of the observable quantities for the case in which $|\tau_L|$ and τ_C are much less than one. It is relatively straightforward to recast these equations into a variety of forms convenient for some particular analysis.

The problem is that, formally, one must know T_e before being able to use these equations. One must know the answer before calculating it; i.e., one must be able to determine β and b_{n_2} a priori to solve for T_e.

The solution was quantitatively given by Shaver (1980) based upon the behavior illustrated by Fig. 3.3 and an examination of the magnitude of departures from LTE. Figure 3.4 shows the calculated variation of the ratio T_e^*/T_e, i.e., the ratio of the apparent electron temperature derived from LTE considerations to the actual electron temperature, as a function of frequency and emission measure EM. The calculations assumed a filling factor for the gas of 0.1 and an excitation parameter for the central stars of $U = 100 \, \mathrm{pc \, cm^{-2}}$. The local electron density relates to the emission measure by $EM = 2 \, U \, N_e^{4/3} f^{2/3}$. Note the large area of the plot occupied by values near 1.0, which implies

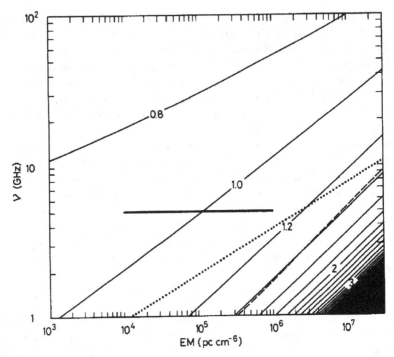

Fig. 3.4 Observing frequency vs. emission measure for various values of T_e/T_e^*. The excitation parameter $U = 100\,\mathrm{pc\,cm^{-2}}$, $T_e^* = 10{,}000\,\mathrm{K}$, and the filling factor $f = 0.1$ for all calculations. The *horizontal line* segment marks the range of H II regions observed at H109α (5 GHz) (Reifenstein et al., 1970; Wilson et al., 1970). Figure taken from Shaver (1980)

that it is difficult to make substantial errors by calculating T_e from LTE equations. To illustrate this point specifically, Shaver plotted the range of emission measures involved in two surveys of RRLs from H II regions made at 5 GHz with the H109α line (Reifenstein, Wilson, Burke and Altenhoff, 1970; Wilson, Mezger, Gardner and Milne, 1970). These observations involve the small range of $T_e/T_e^* \approx 1.0 \pm 0.1$ and, therefore, the values of T_e derived from the observations will be close to the correct values.

Shaver (1980) plotted these calculations in a different way. Figure 3.5 shows the loci of the ratio $T_e/T_e^* = 1$ for $T_e^* = 5{,}000$ and $15{,}000\,\mathrm{K}$ and for two values of the filling factors for each temperature. The narrowband between the lines would include most H II regions. The centroid of this band is

$$\nu = 0.081\,EM^{0.36}, \tag{3.1}$$

where ν is in units of GHz and EM is in units of $\mathrm{pc\,cm^{-6}}$. This equation defines a frequency at which $T_e = T_e^*$, i.e., where the LTE electron temperature calculated from (2.124) is correct.

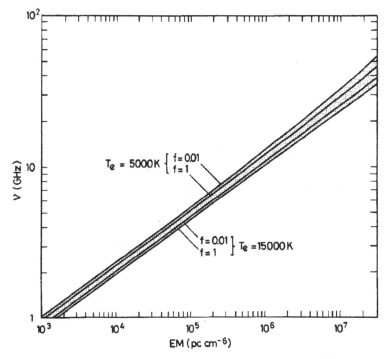

Fig. 3.5 Observing frequency vs. emission measure for $T_e/T_e^* = 1$ for $T_e^* = 5{,}000$ and 15,000 K, and the filling factors $f = 1$ and 0.01 for each temperature. Figure taken from Shaver (1980)

Still, another figure is helpful to illustrate this conclusion. Figure 3.6 shows the variation of the ratio T_e/T_e^* as a function of frequency for a range of temperatures and filling factors. Again, we see that $T_e/T_e^* \approx 1.0$ at the frequency given by (3.1).

The reason for this behavior is that, in general, there is a frequency for each H II region where the collision rates populating the levels become dominant over the radiative rates, where the line amplification induced by the slope db_n/dn of the population curve is just offset by the line weakening due to underpopulation of the quantum levels themselves, thereby obviating the effects of departures from LTE. This point is a function of the emission measure, which parameterizes the densities involved in the departure coefficients.

Using this technique requires knowing the emission measure. The free–free optical depth given by (2.95) can be rewritten in terms of the optical depth τ_C:

$$\tau_C = \frac{0.08235\, EM}{\nu^{2.1} T_e^{1.35}}, \tag{3.2}$$

where ν is in GHz and T_e is in K. As can be seen in Fig. 2.17, identifying the turnover frequency from the continuum spectrum of an H II region determines

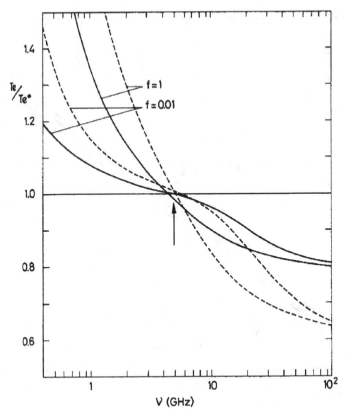

Fig. 3.6 The ratio T_e/T_e^* plotted against frequency for $EM = 10^5 \, \mathrm{pc\,cm^{-6}}$ and $U = 100 \, \mathrm{pc\,cm^{-2}}$. The *solid and dashed lines* mark $T_e = 15{,}000$ and $5{,}000 \, \mathrm{K}$, respectively, along with the indicated filling factors. The *arrow* marks the frequency given by (3.1). Figure taken from Shaver (1980)

the emission measure from (3.2) *if* the electron temperature is known. Fortunately, as Fig. 3.4 shows, the technique is tolerant to errors in the emission measure. In fact, iteration between values of T_e determined from the RRLs and subsequent re-estimates of EM from (3.2) should lead to closure, i.e., to reasonably accurate values of T_e for a given H II region.

Table 3.1 lists values of the electron temperature T_e calculated from representative millimeter and short centimeter wave RRLs from four nebulae. All observations were made with filled-aperture (single-dish) radio telescopes with beamwidths of $1'$–$2'$ at half-power. For reasons described earlier, $T_e \approx T_e^*$ for the short centimeter wave lines; i.e., the "true" electron temperature T_e is given by the LTE equation (2.124).

This situation does not apply to the millimeter wave RRLs, where underpopulation of the upper quantum levels weakens the line intensities, resulting in electron temperatures that are too large. Therefore, for the H40α and H56α lines, we corrected the temperatures (T_e^*) derived from the LTE formula by

Table 3.1 Electron temperatures derived from H II regions

RRL	T_e (K)			
	Orion nebula	NGC2024	W3	M17
H40α^{a}	$8{,}520 \pm 230$	$6{,}530 \pm 450$	$8{,}310 \pm 400$	$7{,}750 \pm 500$
	$(11{,}200 \pm 300)$	$(8{,}800 \pm 600)$	$(10{,}800 \pm 500)$	$(10{,}700 \pm 700)$
	$b_{41} = 0.73$	$b_{41} = 0.71$	$b_{41} = 0.74$	$b_{41} = 0.69$
H56α		$7{,}500 \pm 580$	$7{,}610 \pm 550$	$7{,}440 \pm 450$
		$(8{,}400 \pm 650)$	$(8{,}260 \pm 600)$	$(8{,}400 \pm 500)$
		$b_{57} = 0.88^{\mathrm{b}}$	$b_{57} = 0.91^{\mathrm{b}}$	$b_{57} = 0.87^{\mathrm{c}}$
H64α^{d}	$8{,}400 \pm 400$			
H66α^{e}	$8{,}200 \pm 300$	$7{,}200 \pm 500$		$8{,}000 \pm 300$
H76α^{f}	$8{,}600 \pm 430$	$8{,}200 \pm 400$		$7{,}300 \pm 360$
H92α^{g}			$7{,}940 \pm 140$	
Several lines	$8{,}100 \pm 100$	$7{,}400 \pm 500$		
	H40α, H56α, H66α^{h}	H41α, H63α, H90α^{i}		
Contin. 330 MHz	$7{,}865 \pm 300^{\mathrm{j}}$	$8{,}400 \pm 1{,}000^{\mathrm{k}}$		$7{,}600^{+700}_{-210}{}^{\mathrm{l}}$

Parentheses indicate T_e^*s, giving the T_es above after correction by the listed b_{n_2}s
[a]Gordon (1989), [b]Berulis et al. (1975), [c]Berulis and Sorochenko (1983), [d]Wilson and Filges (1990), [e]Wilson et al. (1979), [f]Shaver et al. (1983), [g]Adler et al. (1996), [h]Sorochenko et al. (1988), [i]Wilson et al. (1990), [j]Subramanyan (1992a), [k]Subramanyan (1992b), and [l]Subramanyah and Goss (1996)

$$T_e \approx T_e^* \, (b_{n_2})^{0.87}, \qquad h\nu \ll kT_e, \tag{3.3}$$

derived from (2.124) and (2.144). This approximation obtains because the small value of τ_C at high frequencies weakens the effect of β, the factor representing the gradient of population across the principal quantum levels of the upper and lower principal quantum levels. The departure coefficients are from Walmsley (1990), calculated from the FORTRAN program listed in Appendix E.1.

Note that the T_es shown in Table 3.1 agree with each other remarkably well for each H II region. The variation is only a few percent, indicating the accuracy of the measurements and the reliability of the theory of radiation transfer developed in Chap. 2.

The last line of the table gives the results of T_e measurements of H II regions in the continuum at 330 MHz. They were carried out at the VLA with an angular resolution of 1′.2, comparable with the resolution of the RRL observations. These data included only the brightest sources. At 330 MHz, all H II regions listed in Table 3.1 are optically thick, such that their observed brightness temperatures equal their electron temperatures, thereby providing direct measurements of T_e as described in Sect. 2.3.3.

Note the excellent correspondence between the electron temperatures derived from the RRLs and those derived from the continuum observations. Despite the higher error of the continuum measurements, they give an additional proof of the correctness of the T_e values obtained from the RRLs from H II regions. This is especially significant because the analytical techniques are very different from one another.

Determinations of electron temperatures from RRLs should be far more accurate than from the traditional methods that use the forbidden auroral and nebular emission lines[2] of nitrogen and oxygen. The problem is that the optical techniques require a high temperature to excite the auroral lines, creating a temperature threshold of about 7,000 K that applies only to certain objects such as hot, bright H II regions. These techniques cannot be used for cooler material in the interstellar medium.

No such limitation obtains for electron temperatures determined from RRLs. The I_L/I_C ratio actually increases for cooler objects. Furthermore, the radio range is relatively free of absorption and extinction such that RRLs can be observed from highly obscured objects like H II regions throughout the plane of our own Galaxy.

Observations of RRLs have provided the most accurate and extensive measurements of the electron temperature of gaseous nebulae in our Galaxy. The observations include nearly 200 H II regions seen from both the northern and southern skies, including those that are cool, faint, or obscured by dark interstellar matter. The data establish that the electron temperature falls into a range of 4–12×10^3 K. Churchwell and Walmsley (1975) detected a gradient in T_e as a function of Galactocentric radius, which was confirmed by Churchwell et al. (1978), Mezger et al. (1979), Lichten et al. (1979), Churchwell (1980), and Garay and Rodríguez (1983).

Carefully, analyzing the electron temperatures 67 H II regions observed in the southern hemisphere with observational errors of about 5%, Shaver et al. (1983) obtained the regression equation of

$$T_e = (3,150 \pm 110) + (433 \pm 40)R_G \qquad (3.4)$$

for the dependence of T_e in K as a function of the Galactocentric radius R_G measured in kpc. Figure 3.7 shows the actual data obtained from the H76α and H110α lines. The temperature gradient is easily seen.

---

[2] The terminology comes from early spectroscopy. "Auroral" transitions are named after the emission lines in the Earth's permanent aurora. They violate the LaPorte parity rule for electric dipole transitions of $\Delta \ell = \pm 1$. Instead, auroral lines involve electric quadrupole transitions where $\Delta \ell = \pm 2$ or 0 as in the oxygen line [O I] $^1D_2 \rightarrow {}^1S_0$ at $\lambda = 5,577$ Å. For this reason, they are called "forbidden" lines and are designated by enclosing the atomic symbol in square brackets. "Nebular" lines are so named because they are found in gaseous nebula as well as in the Earth's aurora. Nebular lines involve electric quadrupole *and* magnetic dipole contributions, resulting in changes in the electron configuration forbidden by the normal spectroscopic selection rules. An example is the jump from a triplet to a singlet configuration in neutral oxygen, [O I] $^3P_0 \rightarrow {}^1D_2$ at $\lambda = 6,300$ Å. In both types of transitions, the energy levels involved are called "metastable" because the transition probabilities between them are small and the lifetimes of the levels may be correspondingly long. Accordingly, the intensities of forbidden lines are sensitive to ambient temperature and density through collisions, and can be used to determine the physical conditions of their environment.

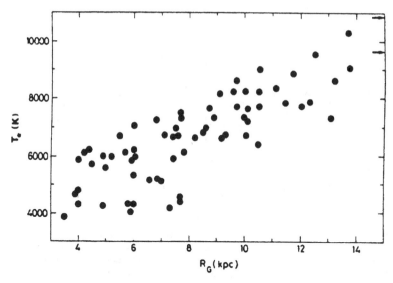

Fig. 3.7 Electron temperature of H II regions as a function of Galactocentric radius. The *horizontal arrows at upper right* indicate the N66 and 30 Doradus nebulae in the Magellanic Clouds. From Shaver et al. (1983). Reproduced with permission of Monthly Notices of the Royal Astronomical Society

Why should this gradient exist? Shaver et al. (1983) suggested that the temperature gradient could be explained by a corresponding Galactocentric gradient of metallic elements[3] in the Galaxy. Of these, the most abundant are oxygen and nitrogen that cool H II regions through radiation in spectral lines. Collisions with free electrons easily excite metastable levels associated with these elements. This kinetic energy is later released from the H II regions largely through radiation in the [O III], [O II], and [N II] forbidden lines. Consequently, H II regions with higher metallicities can cool at a faster rate than others and should have lower electron temperatures. Because the star formation rate is higher in the inner parts of the Galaxy, the metallicity should decrease – and electron temperature should increase – with increasing Galactocentric radius consistent with the trend shown in Fig. 3.7. For example, a twofold increase in the metal abundance typical of the solar vicinity will cause a decrease in the temperature of the nearby H II regions by almost 2,000 K (Afflerbach, Churchwell, Accord, Hofner, Kurtz and DePree, 1996).

Observations of optical spectral lines also indicate a decrease of metallicity with Galactocentric distance. Figure 3.8 shows the variation of the normalized abundances O/H and N/H with the Galactocentric radius of our Galaxy. The radial decrease of both constituents is obvious.

---

[3] In astronomy, "metals" refer to elements heavier than hydrogen and helium. Unlike hydrogen and helium, metals are principally produced by nucleosynthesis within the stars. The term "metallicity" refers to the relative abundance of these elements.

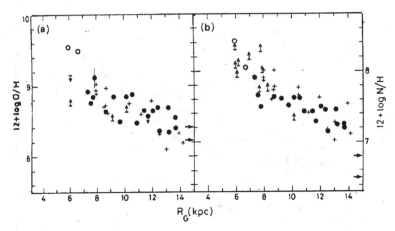

Fig. 3.8 Abundance of O and N vs. Galactocentric distance. The *open circles* denote S38 and S48 and the *horizontal arrows* represent the two Magellanic Cloud sources N66 and 30 Doradus. The *crosses* represent early data from Peimbert (1979). From Shaver et al. (1983). Reproduced with permission of Monthly Notices of the Royal Astronomical Society

At fixed distances (Galactocentric radii) from the Galactic center, Fig. 3.7 shows the electron temperatures determined from RRLs to have a large scatter. This scatter results from real differences in the electron temperatures of H II regions, in addition to measurement errors, and should be expected. Even in the same region of the Galaxy, H II regions will differ from each other in density and in the radiation field from the exciting stars. Higher densities mean more effective depopulation of metastable energy levels by collisions, reduced intensities of the associated forbidden lines, and, in turn, less effective radiative cooling and higher gas temperatures. Also, hotter exciting stars – or more of them – will increase the ambient radiation in the nebular gas and, correspondingly, increase the electron temperature of the gas.

The characteristics of ultracompact H II regions illustrate the effect of density on temperature. These are small nebulae with higher electron densities than the usual H II regions. Afflerbach et al. (1996) used H42α, H66α, H76α, and H93α lines to measure the electron temperatures of ultracompact nebulae. Despite the scatter of the observations, Fig. 3.9 shows the values of T_e for the ultracompact nebulae to exceed those of the ordinary H II regions, especially near the Galactic center. This is probably a consequence of radiation cooling being impaired by the higher nebular densities.

Together, the variation of T_e with Galactocentric distance determined from normal and ultracompact H II regions shows that the metallicity on the interstellar gas of our Galaxy decreases outward from the center. Indirectly, this gradient indicates increased star formation in the Galactic center where the nucleosynthesis enriches the heavy elements.

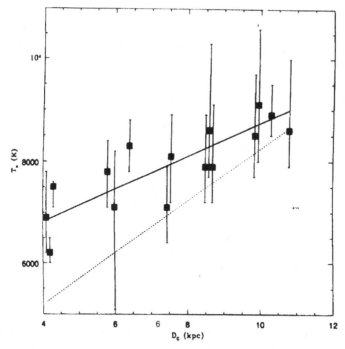

Fig. 3.9 The electron temperature of ultracompact H II regions obtained from radio recombination lines plotted against Galactocentric distance. The *solid line* is a least-squares fit to the data. The *broken line* is the gradient found by Shaver et al. (1983) for normal H II regions. From Afflerbach et al. (1996)

By providing an opportunity to measure electron temperatures through the interstellar gas of our Galaxy without concern for interstellar extinction, RRLs provide information of great value regarding evolutionary processes in spiral galaxies in general.

3.1.2 Electron Density of H II Regions

Probably, the most direct method of determining N_e for an H II regions is from the free–free continuum emission after first determining T_e, as discussed in Sect. 2.3.4. This procedure determines the optical depth of the free–free emission and, in turn, gives the emission measure of the nebula that provides the electron density once the size of the nebula has been determined.

However, it is also possible to measure electron densities from the RRLs themselves from Stark broadening. Equation (2.74) can be rewritten as

$$N_e = \frac{\Delta\nu_L}{8.2 \left(\frac{n}{100}\right)^{4.4} \left(1 + 2.2\frac{\Delta n}{n}\right)} \ \mathrm{cm}^{-3}, \qquad (3.5)$$

where $\Delta\nu_L$ is the Lorentzian component of the line width in Hz caused by Stark broadening. Determining $\Delta\nu_L$ requires observations of RRLs over a wide frequency range made with the same angular resolution or beamwidth, i.e., observations of the same volume of gas. Specifically, observations of a high-frequency (low-n) RRL would provide a purely Doppler line profile. Observations of a low-frequency (high-n) line would provide the Lorentzian profile that, with the Doppler profile, would then provide the Stark broadening component $\Delta\nu_L$ needed for evaluation of (3.5).

Presuming that the Stark broadening is well measured if it equals or exceeds the Doppler width, we can combine (2.26) and (2.74) to derive

$$n \geq \left(\frac{3.6 \times 10^{19} \Delta n}{N_e}\right)^{0.135}, \tag{3.6}$$

if $V_t = 0$ and $\Delta n/n \ll 1$. This equation determines the minimum principal quantum number that should be used to determine the Stark broadening component for any given electron density for an RRL of type Δn. For example, for α lines at $N_e = 10^4\,\mathrm{cm}^{-3}$, $n \geq 125$ whereas at $N_e = 10^3\,\mathrm{cm}^{-3}$, $n \geq 171$. Because $n \propto N_e^{-0.135}$, suitable observations of high-density H II regions like planetary nebulae could be made at modestly low values of n. Here, the line intensities would be more detectable relative to the underlying continuum because I_L/I_C varies inversely with n as can be seen in theory (2.124) and in practice (Fig. 3.3).

In practice, one implements this analytic technique by fitting a Voigt function to the higher-n line profile, using the lower-n profile to establish the purely Doppler profile. Equation (2.72) will be useful here. Rearranging this approximation, we obtain the full width of the Lorentz component to be

$$\Delta\nu_L = 7.79\,\Delta\nu - \sqrt{14.6\,\Delta\nu_G^2 + 46.1\,\Delta\nu^2}, \tag{3.7}$$

where $\Delta\nu$ and $\Delta\nu_G$ are the full widths at half-intensity of the Voigt and Doppler (Gaussian) profiles, respectively.

Excellent determinations of the Lorentz component can be made by using a series of $n\alpha$ and higher-order lines observed near a single frequency, like the series reported by Pedlar and Davies (1972) and others described in Sect. 2.2.6. Measuring the progressive changes in the pressure broadening from a series de-emphasizes the measurement errors of any single line profile and, therefore, helps define the pressure broadening accurately. Making the observations near a single frequency also ensures that the lines will originate from the same volume of ionized gas.

Values of electron density calculated from Lorentz broadening depend weakly on the temperature of the H II regions. For $5,000 \leq T_e \leq 10,000\,\mathrm{K}$, the Lorentz broadening will vary by 10% for $100 \leq n \leq 200$ as can be seen in Sect. 2.2.3.

Table 3.2 Electron densities in H II regions and planetary nebulae

Sources	From RRLs N_e ($10^4\,\mathrm{cm}^{-3}$)	From continuum $\langle N_e \rangle$ ($10^3\,\mathrm{cm}^{-3}$)	Ratio $N_e/\langle N_e \rangle$
M42 (Orion A)	1.0 ± 0.3[a]	2.3[b]	4.3
W3	1.5 ± 0.3[c]	3.7[b]	4.3
DR21	4.3 ± 0.4[c]	1.3[b]	3.3
NGC7027	6.7 ± 0.5[d]	57 ± 3[d]	1.2
IC418	1.8 ± 0.4[e]	20[e]	≈ 1

[a]Smirnov et al. (1984), [b]Berulis and Sorochenko (1983), [c]Smirnov (1985), [d]Ershov and Berulis (1989), and [e]Garay et al. (1989)

Whatever method is used to determine N_e from H II regions, it is important to note that the results may be technique dependent. For example, densities determined from the free–free continuum spectra involve the emission measure; i.e., they derive from the square root of the path-averaged emission measure, the mean electron density $\langle N_e \rangle = (EM/L)^{1/2}$. In comparison, the electron densities determined from pressure broadening result from collisions with excited atoms and give the localized actual values – a very different kind of density weighting.

We close this discussion by listing electron densities in Table 3.2 for a few H II regions and planetary nebulae determined both from RRL pressure broadening and from the free–free emission spectrum.

The rightmost column of Table 3.2 gives the ratio of the local to the mean values of N_e determined by the two methods. The RRL values – determined by the ambient conditions of the emitting atoms – exceed the continuum values by factors of 3–4 for Orion, W3, and DR21. This ratio indicates considerable fluctuations in electron density along the lines of sight through these H II regions. As one might guess, this effect is absent in the two much smaller and denser planetary nebulae NGC7027 and IC418. Not only do we learn the electron densities from the RRL line widths, but also we learn about the density structure of the sources by comparison with the continuum observations.

Odegard (1985) developed a similar method of determining the electron density of H II regions from the line profiles. He selected a high-frequency RRL to determine the gas temperature and used this information to predict the Gaussian shape and I_L/I_C ratio at line center of a lower-frequency RRL that should have Stark broadening. To correct the line intensities for departures from LTE, he used published departure coefficients b_ns and population gradients βs (Salem and Brocklehurst, 1979). Comparison of the measured line intensity with the predicted one is then a measure of the intensity of a Voigt profile and, hence, of the electron density. An important requirement for this method is that both line observations involve the same volume of gas.

Odegard's measurements show good agreement with electron densities determined from the optical forbidden lines. However, the electron densities determined directly from the line widths (see Table 3.2) are even more accurate than those he determined from the line intensities of the low-frequency lines.

3.1.3 Velocities of Turbulent Motion

In the millimeter and short centimeter wavelength ranges, the principal quantum numbers n of detectable RRLs are small. In this regime, Stark broadening is undetectable because of its steep dependence on n as seen in (2.60); i.e., the broadening $\propto n^{4.5}$. Here, however, RRL measurements can determine T_e with great accuracy and, accordingly, measure the line broadening due to turbulence within the telescope beam (microturbulence). Rewriting (2.26), we obtain an expression for the turbulence velocity V_t:

$$V_t = 0.6\sqrt{(\Delta V_G^2) - 4.55\ 10^{-2}T_e} \quad \text{km s}^{-1}, \tag{3.8}$$

from the velocity width ΔV_G of the Gaussian profile of a hydrogen RRL at half-intensity measured in km s^{-1}.

RRL observations (Berulis, Smirnov and Sorochenko, 1975; McGee and Newton, 1981; Wink, Wilson and Bieging, 1983) show that the microturbulence within H II regions lies in the range of $5 \rightarrow 25\,\text{km s}^{-1}$. Such values depend upon the angular resolution of the telescope with respect to the angular extent of these nebulae as illustrated by Fig. 2.4. The reason for this dependence of V_t upon the beamwidth is there are significant gradients of line-of-sight velocities normal to the beam as well as along the sight through the H II region as shown by Fig. 3.10.

The velocity broadening of the centimeter wave RRLs agrees well with the velocity dispersion of optical emission lines – for the Orion nebula, at least. Weedman (1966) integrated the radial velocities of Hγ measured at several thousand points at $1''.3$ intervals within a $4' \times 4'$ region of the Orion nebula centered on the Trapezium (Wilson, Münch, Flather and Coffeen, 1959). Summing the radial velocities of the optical lines and convolving the result with the average broadening of each of the components produced a composite line profile that agreed well with the profile observed for the radio H109α line made with a beam of $6''.5$ (Höglund and Mezger, 1965). The implication is that the RRLs give a faithful representation of gas dynamics internal to the nebula; i.e., of the microturbulence within the radio beam along that particular sight line.

From a survey of 82 H II regions observed in the H109α line (Reifenstein et al., 1970), the RMS turbulence[4] $\langle V_t^2 \rangle^{1/2} = 13.4\,\text{km s}^{-1}$, so that $V_t =$

---

[4] Frequently used in literature, the root-mean-square (RMS) velocity of turbulent motion $< V_t^2 >^{1/2}$ is related to V_t by the expression $< V_t^2 >^{1/2} = 1.22\,V_t$.

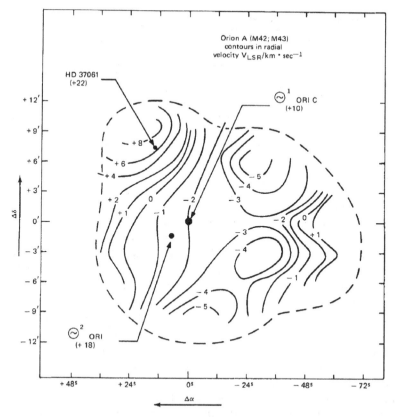

Fig. 3.10 The gradients of the radial velocity of the H109α RRL in km s⁻¹ across Orion A (M42 and M43) as a function of offsets in RA and declination from the exciting star Θ^1 Ori. The telescope beam is 6ʹ.5 at half-intensity. From Mezger and Ellis (1968)

11 km s⁻¹ for the average H II region observed with the 6ʹ.5 beam of the NRAO 140-ft. radio telescope. These values were calculated by assuming LTE values of the electron temperature T_e.

3.2 Ionized Hydrogen and Helium in the Galaxy

3.2.1 Distribution of H II Regions

3.2.1.1 Astronomical Doppler Shifts

The observed frequencies of RRL (see Fig. 3.1) usually differ somewhat from the calculated values because of Doppler shifts due to the motion of the telescope with respect to the source. The Earth rotates on its axis, the Earth

Table 3.3 Velocity components affecting observed frequencies

Component	Approximate V_{max} (km s^{-1})
Source $\rightarrow$ LSR	(Source dependent)
LSR $\rightarrow$ solar system barycenter	20
Solar system barycenter $\rightarrow$ Earth–moon barycenter	30
Earth–moon barycenter $\rightarrow$ Earth center	0.1
Earth center $\rightarrow$ telescope	0.5
Planetary perturbations upon Earth orbit	0.013

revolves around the Sun, the Sun moves within the Galaxy, and the Galaxy moves with respect to the Local Group of galaxies of which it is a member.

For convenience, astronomers adopt a kinematic reference frame in the Galaxy toward which the Sun is headed. This frame is based upon the average velocity of stellar spectral types A–G in the vicinity of the Sun without regard for luminosity class. It assumes the Sun to be moving at 20.0 km s^{-1} toward 18^m RA and 30° δ 1900.0 and is known as the "local standard of rest" (LSR). Table 3.3 lists the magnitudes of the velocity components between the telescope and the LSR. Ball (Meeks, 1976) gives a FORTRAN subroutine for calculating the velocity of a telescope with respect to the LSR.

The LSR itself moves with respect to the Galactic center. Observations (Kinman, 1959) of globular clusters show the Sun to move at 167 ± 30 km s^{-1} toward a Galactic latitude and longitude of $(\ell, b) = (90°, 0°)$. Similar observations (Humason and Wahlquist, 1955) of galaxies in the Local Group show the Sun to move at 291 ± 32 km s^{-1} toward $(106°, -6°)$. The difference between these velocity vectors is not understood. Therefore, the assumed value is taken to be 250 km s^{-1} toward $(90°, 0°)$ about the "Galactic standard of rest."

Most importantly, the rotational velocity of the Galaxy – and of most other spiral galaxies – changes as a function of Galactocentric radius as shown by Fig. 3.11. This "differential rotation" creates a variation of radial velocity with distance along a line of sight through the Galactic plane. In principal, this velocity gradient should make it possible to locate an H II region along the sight line based upon the radial velocity observed for its RRLs.

There may be a distance ambiguity, however. With the differential rotation, it is possible to relate the radial velocities V_{LSR} observed for RRLs to a distance from the center of the Galaxy. In contrast, it is not generally possible to locate the H II regions *uniquely* along sight lines passing through Galactic longitudes of $|\ell| < 90°$ on the basis of the radial velocities alone. There is no such distance ambiguity for sources located at $|\ell| \geq 90°$.

Consider the geometry shown in Fig. 3.12 for a sight line at $\ell < 90°$. Except for the locus of the subcentral points, along such sight lines are two possible locations that would have the same radial velocity with respect to

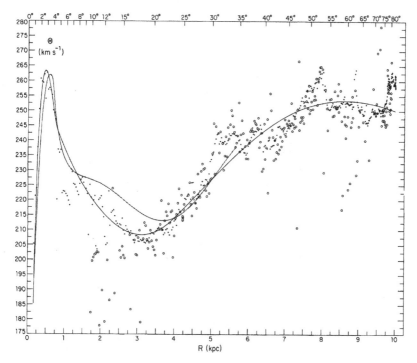

Fig. 3.11 The variation of the linear velocity of differential Galactic rotation as a function of Galactic radius. In this plot, the Sun lies at 10 kpc. *Small plus symbols* are from $\lambda =$ 21 cm H I emission and *open circles* are from CO emission. The *black line* is the fit to all observations. The *shorter, lighter line* is an earlier rotation curve proposed by Simonson and Mader (1973). From Burton and Gordon (1978)

the observer: a near and a far point. The Galactocentric radius would be the same for each location but the distance from the observer could be either of the two. Such position ambiguities can be distinguished only by additional information such as optical extinction, implied physical size, elevation above the Galactic disk, etc.

3.2.1.2 Physical Location of H II Regions in the Galaxy

Nonetheless, surveys of H II regions in hydrogen RRLs were carried out in both northern and southern hemispheres to determine the distribution of H II regions. These first surveys were initiated almost simultaneously in the northern sky in the range of Galactic longitude $\ell = 348° \rightarrow 360° \rightarrow 209°$ (Reifenstein et al., 1970) and in the southern sky for $\ell = 189° \rightarrow 36° \rightarrow$ 49.5° (Wilson et al., 1970). Both surveys searched for the H109α line toward known sources of $\lambda = 11$ cm continuum radiation located close to the Galactic equator, i.e., at $|b| \leq 1°$. Subsequently, the line was detected in 82 and 130

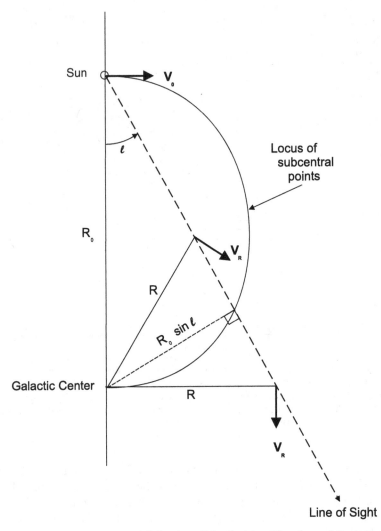

Fig. 3.12 Distance ambiguity of Galactic radial velocities. The plane of the page represents the plane of the Galaxy. R_0 is the distance of the Sun to the Galactic center. The *dashed line* indicates an arbitrary line of sight through the plane of the Galaxy at a longitude ℓ. The rotation velocity of the gas about the Galactic center varies with R. Note that two points along the sight line at same Galactocentric radius R and, hence, at the same radial velocity V_{LSR} could lie at different distances from the Sun. The *arc* marks the locus of "subcentral" points, the only positions having a unique location at a given radial velocity. After Burton (1974)

H II regions, respectively, with some sources observed in both the northern and southern surveys. The spiral structure of Galaxy was not constructed from these observations because the H II regions were located in the inner

Galaxy, where $R < R_o$ and $270° \leq \ell \leq 90°$, so that their distances from the Sun could not be determined uniquely.

Additional surveys overcame the distance problem to some extent. Downes et al. (1980) observed the H110α RRL (4,874 MHz) in 171 H II regions in the range $\ell = 357° \rightarrow 360° \rightarrow 60°$ simultaneously with observations of the formaldehyde (H_2CO) absorption line at 4,830 MHz. Of these, the H_2CO absorption resolved the distance ambiguity of 56 H II regions in the inner Galaxy on the basis of the amount of extinction along the lines of sight.

The most detailed survey of southern sky was carried out by Caswell and Haynes (1987) in the H109α and H110α lines. They detected these lines in 316 sources located close to Galactic plane within Galactic latitudes $\ell = 233° \rightarrow 360° \rightarrow 13°$. Here also, the 4,830-MHz formaldehyde line was recorded simultaneously and resolved the twofold distance ambiguity problem for 146 sources.

By enabling measurements of the distances of H II regions in the Galaxy including those located beyond the Galactic center, RRLs have provided a tool to model the large-scale structure and to locate the spiral arms. Figure 3.13 shows a composite map of H II regions plotted onto the Galactic plane. While the map was originally proposed by Georgelin and Georgelin (1976), Taylor and Cordes (1993) added observations from subsequent observations to improve the model. They also adopted the modern value of 8.5 kpc for the Galactocentric distance of the Sun.

In the figure, the spiral structure of our Galaxy is distinctly seen even though details of the arm continuity are missing. Data on the far side of the Galactic center are sparse but still adequate to suggest an asymmetry in the structure of the arms as indicated from the H II regions. Being heated by newly formed stars, the H II regions mark the location of young Galactic material, sometimes called "Population 0" material[5] as an extension to an older stellar classification scheme introduced by Walter Baade in the 1940s. Additional observations will fill in the gaps but probably not change the conclusion that our Galaxy has asymmetrical spiral arms.

The most detailed survey of H II regions in the northern sky was made by Lockman (1989). This study detected 462 sources in the H85α, H87α, and H88α lines and some in the H100α, H101α, H125α, and H127α lines as well. The study did not resolve the distance ambiguity from the Sun to those H II regions in the inner Galaxy; the observations were considered in terms of the Galactocentric distance instead.

Combining this survey with the southern survey (Caswell and Haynes, 1987) shows the distribution of 750 H II regions as a function of Galactocentric

---

[5] This newer term refers to the extremely young component (O stars and H II regions) of the interstellar gas in distinction to the older term, Population I, that includes disk stars of many ages which delineate the spiral arms of a galaxy. The term "Population II" refers to Galactic stars not associated with spiral structure, in short, to everything else. Today's revised terminology also includes subclassifications like "extreme population I," "moderate Population I," etc.

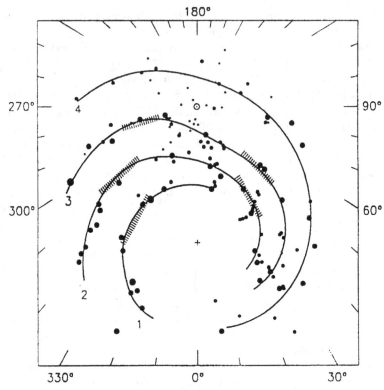

Fig. 3.13 A spiral model of the Galaxy originally proposed by Georgelin and Georgelin (1976) but now including the results of additional RRL surveys of H II regions. *Filled circles* represent the H II regions and *shaded areas* are directions of intensity maxima in the radio continuum and in neutral hydrogen. In each arm, the *lines* are cubic splines fitted to the H II region positions. A *number* marks each arm, the *cross* indicates the Galactic center, and the *open circle with dot* indicates the position of the Sun. From Taylor and Cordes (1993)

distance for our Galaxy. These distances result from the differential rotation curve proposed by Burton and Gordon (1978). The Sun is assumed to lie 8.5 kpc from the center of the Galaxy. The results shown in Fig. 3.14 confirm solidly the double-peaked, apparently toroidal distribution in the range $4 < R < 6$ kpc discovered 20 years earlier from many times fewer observations (Mezger, 1970). According to Hodge and Kennicutt (1983), with respect to surface density, the radial distributions of H II regions in spiral galaxies can be classified as (1) continuously decreasing from the center outward toward the outer boundary of the galaxy, (2) oscillating with increasing Galactocentric radius and finally decreasing toward the outer boundary, and (3) increasing from the center to a maximum and then decreasing toward the outer boundary. In this scheme, the Galaxy seems to fall into the third class.

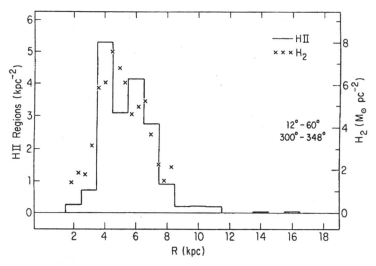

Fig. 3.14 Surface density of H II regions in the Galaxy plotted against Galactocentric distance for $R > 1.5$ kpc and $\ell = 12° \rightarrow 60°$ and $300° \rightarrow 348°$. The Sun is assumed to lie 8.5 kpc from the Galactic center and the bins are 1-kpc wide. The *crosses* mark surface densities of H_2 deduced from observations of CO emission. From Lockman (1990)

Figure 3.14 shows the radial distribution of the surface density of H II regions to follow closely that of the molecular hydrogen deduced from observations of carbon monoxide. This is to be expected because both constituents are Population 0 or I material.

For comparison, Fig. 3.15 shows the distribution of nucleons in all forms as a function of Galactocentric radius (Gordon and Burton, 1976). Subject to differences in the Sun–Center distance, the agreement of the H II distribution shown in Fig. 3.14 with the distribution of the gaseous constituents in the Galaxy is excellent. Note that the ratio of $\sigma(H_2)/\sigma(H I)$ decreases with increasing Galactocentric radius, showing that the density of gas capable of star formation decreases outward from the Galactic center. Subsequent observations (Lockman, Pisano and Howard, 1996) confirmed that the H II regions – both diffuse and compact – follow the Galactocentric radial distribution of molecular gas rather than of the H I gas. Unlike the H_2 gas, H II regions *require* newly formed stars and, therefore, indicate the presence of star-forming activity. Thus, knowing the distribution of H II regions, determined by RRLs, is essential to understanding large-scale star formation in the Galaxy.

3.2.2 Low-Density Ionized Hydrogen

RRLs have been detected not only from well-defined H II regions, but also from directions in the Galactic plane free of discrete sources. Gottesman and

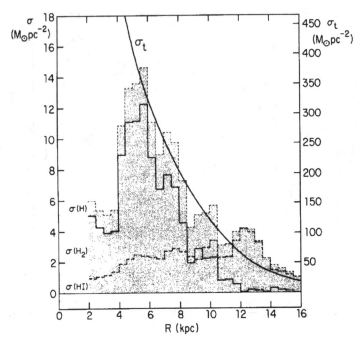

Fig. 3.15 Surface density of hydrogen plotted against Galactocentric radius for $R >$ 1.5 kpc for the Galaxy. *Lower broken line*: H I. *Solid histogram*: H_2 estimated from CO observations. *Dashed histogram*: hydrogen nucleons in all forms. *Curve* marked σ_t: total surface density predicted by dynamic models of star density (Innanen, 1973). *Note*: Here, the Sun is placed at 10 kpc from the Galactic center. From Gordon and Burton (1976)

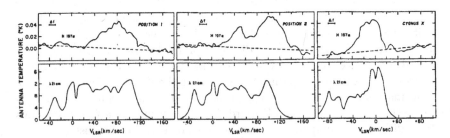

Fig. 3.16 *Top*: H157α emission observed from three directions in the Galactic plane free of discrete sources. *Bottom*: corresponding $\lambda = 21$ cm of H I in the same directions. From Gottesman and Gordon (1970)

Gordon (1970) detected rather weak ($T_L < 0.05$ K) but distinctly obvious H157α emission from the directions $\ell = 23°92$, $\ell = 25°07$, and $\ell = 80°09$ and $b < 0.9°$ free from radio sources with high surface brightness. Jackson and Kerr (1971) made similar detections almost simultaneously. Figure 3.16 shows the line spectra juxtaposed with profiles of $\lambda = 21$ cm emission from H I in nearby directions.

The qualitative agreement of the H157α line radial velocities with those of $\lambda = 21$ cm line, the much greater line widths compared with RRLs from H II regions, and the weak intensities suggested that detected lines originated in diffuse, ionized interstellar gas. Such a possibility corresponded to the steady-state two-component model of a partially ionized ISM generally accepted at that time. According to this model, interstellar gas (excluding discrete H II regions) consisted of dense, cold, $T_e = 40 \rightarrow 60$ K clouds in pressure equilibrium with an ambient hot, $T_e = 1,000 \rightarrow 10,000$ K, rarefied, intercloud medium. It was believed that low energy cosmic rays were responsible for the ionization and heating of both components (Pikelner, 1967; Field, Goldsmith and Habing, 1969; Hjellming, Gordon and Gordon, 1969).

It was impossible to calculate the electron temperature of the diffuse component from the RRL observations. The associated free–free continuum was not known well enough to use (2.124). Furthermore, the extent of departures from LTE – if any – was also unknown. All that could be safely determined from the initial observations was the quantity $\int_{\text{path}} N_e N_i T_e^{-0.35} \, dl$. Accordingly, Gottesman and Gordon (1970) calculated path-averaged electron densities for three arbitrary fractions for the free–free component of the observed continuum emission: 100, 10, and 1%. If the free–free continuum accounted for 10% of the observed continuum along the sight lines, they suggested that the value of approximately 1,000 K might be an upper limit for the electron temperature of the diffuse gas, which would imply a limit of $\langle N_e \rangle \approx 0.3 \, \text{cm}^{-3}$.

Not only subsequent observations (Gordon and Gottesman, 1971) of the H197β lines from the same positions verified the reality of the H157α line detected earlier from the diffuse gas, but also the ratios of the radiated power in the line favored a path-averaged $1,000 < \langle T_e \rangle < 10,000$ K. Unfortunately, the signal-to-noise ratio of the β line observations was insufficient to restrict the temperature further. This temperature range implied a corresponding range of the emission measure of $280 \rightarrow 5,000 \, \text{cm}^{-6}$ pc over a probable path length of 14 kpc. Although considerable fluctuations in density along the paths could be expected, these values implied $0.15 < \langle N_e \rangle < 0.6 \, \text{cm}^{-3}$, respectively, for RMS densities.

Cesarsky and Cesarsky (1971) examined other constraints on the continuum emission, noted that the unknown filling factor of the cold interstellar clouds prevented direct application of (2.124) as used to interpret RRLs from H II regions, and concluded that two models were possible with respect to the theoretical models of the ISM extant at that time. The diffuse RRL emission could come either from intercloud gas of approximately 800 K or from embedded clouds of $40 < \langle T_e \rangle < 60$ K.

To pursue their suggestion observationally, Cesarsky and Cesarsky (1973) noted that T_e has a different exponent for line and for continuum emission (see (2.95) and (2.116)). They observed RRLs in the direction of the supernova remnant 3C391 and compared the line intensity with the turnover frequency observed for the source. Their results suggested that the average electron temperature for the diffuse gas must be less than 400 K. However, the weakness of the line intensity made this a difficult experiment to perform with high accuracy. Furthermore, if the clouds along the sight line are in pressure equilibrium with hot intercloud gas, line emission from the cold clouds would dominate. Without knowing the distribution of conditions along this line of sight, one found it difficult to know what component the temperature referred to.

These models of a diffuse, partly ionized interstellar gas were consistent with observations of pulsars, discovered (Hewish, Bell, Pilkington, Scott and Collins, 1968) a short time before the detection of RRL emission from the diffuse ISM. All the pulsars exhibited a delay in the time of arrival of the pulses that varied inversely with frequency. This effect had a natural explanation of propagation dispersion caused by the ionized component (electrons) of the ISM. Fitting the frequency dependence of the delay determines the integrated electron density or "dispersion measure" along the path to the pulsar, $DM \equiv \int_{\text{path}} N_e \, dl$. The observed values of DM could not be explained by the ionization of ISM constituents with ionization potentials lower than $H\,\textsc{i}$; the resulting DM would be too small. The only possible source for the DMs was the ionization of the principal constituent of the ISM along the lines of sight – hydrogen.

The likely source of this ionization was soft cosmic rays with an energy $\approx 2\,\text{MeV}$. The UV radiation from embedded hot stars could not permeate the ISM widely enough; in fact, most $H\,\textsc{ii}$ regions are considered to be ionization bounded rather than density bounded for this very reason.[6] The rate of ionization necessary to account for observed pulsar dispersion measure was $\zeta_H = 2.5 \pm 0.5 \times 10^{-15}\,\text{s}^{-1}$ (Hjellming et al., 1969). The intensity of cosmic rays measured on Earth could only produce $\zeta_H = 6.8 \times 10^{-18}\,\text{s}^{-1}$ but these cosmic rays are strongly attenuated by the magnetic field of the Sun and solar wind, and could be significantly weaker than the ambient flux of cosmic rays in the ISM. Calculations of the possible intensity of subcosmic rays in the ISM, based on the energy and frequency of supernova explosions in the Galaxy, gave the value $\zeta_H = 1.2 \times 10^{-15}\,\text{s}^{-1}$ (Spitzer and Tomasko, 1968). This value was close to what was required to explain the observed dispersion measures.

The first radio astronomical measurement of ζ_H was even more encouraging. It agreed well with the theoretical estimate based upon supernovae and with the empirical value needed to explain the pulsar dispersion measures. Comparison of absorption in the $\lambda = 21\,\text{cm}$ line of $H\,\textsc{i}$ with free–free absorption toward three sources of nonthermal emission – 3C10, 3C123, and 3C340 – gave the value of $\zeta_H = 2.0 \times 10^{-15}\,\text{s}^{-1}$ (Hughes, Thompson and

---

[6] The term "Strömgren sphere" (Strömgren, 1939) describes a density-bounded ionization zone around a hot star in a neutral medium. The sphere results from the balance between the UV flux emitted by the star and the number of ionizations possible in the ambient medium. This is the physical principle underlying the luminous, discrete $H\,\textsc{ii}$ regions called *gaseous nebulae*. The concept stems from the recognition that ionizing UV photons cannot travel very far in the ISM.

Colvin, 1971). This result was exactly what was needed, and the existence of a diffuse component to the ISM with ionization sufficient to account for the diffuse RRLs and the pulsar dispersion appeared to be justified.

Unfortunately, subsequent observations increasingly cast doubts on the cold gas model. New, more accurate observations gave $\zeta_H \leq 2 \times 10^{-16} \, \mathrm{s}^{-1}$, one order of magnitude below the value needed to explain the DM measured by pulsars. Additional observations confirmed this upper limit (Shaver, 1976a; Sorochenko and Smirnov, 1987) (see also Sect. 3.3). As more and more data accumulated, observations of hydrogen RRLs toward directions free from discrete H II regions indicated conditions similar to the RRLs observed from normal H II regions.

An attempt was made to determine the temperature of the diffuse gas by measuring its scale height above the Galactic plane (Gordon, Brown and Gottesman, 1972). These observations implied a scale height less than 70 pc, a crudely determined value consistent with a kinetic temperature $\approx 4,000 \, \mathrm{K}$ or less.

Contrary evidence to the cold gas model lies in the measurements of the physical conditions of the medium in which the diffuse RRLs were formed. Jackson and Kerr (1975) determined electron temperatures from observations of the H110α line in nine directions in the Galactic plane from $\ell = 359° \rightarrow 0° \rightarrow 80°$ that avoided discrete H II regions. The result was $\langle T_e \rangle = 4,400 \pm 600 \, \mathrm{K}$.

The second contrary evidence came from measurements of the electron density along similar sight lines. Analysis of H110α line intensities gave $\langle N_e \rangle = 5 \rightarrow 10 \, \mathrm{cm}^{-3}$ (Shaver, 1976b). Similar values of temperature and density came from the H166α lines. Observations in 13 directions along the Galactic plane from $\ell = -1° \rightarrow 47°5$ gave $\langle T_e \rangle = 6,000 \pm 1,000 \, \mathrm{K}$ and $\langle N_e \rangle = 2 \rightarrow 10 \, \mathrm{cm}^{-3}$ (Matthews, Pedlar and Davies, 1973). The beamwidth for this survey was $\approx 30'$ where mainly extended sources contribute to the observed emission. Taken together, these results pointed to physical conditions expected in extended, low-density H II regions rather than the cold, partially ionized interstellar gas originally suggested by Cesarsky and Cesarsky (1971) for the source of the diffuse RRL emission.

The observing geometry pointed to extended H II regions. Jackson and Kerr (1975) noted that the line profile changed greatly between two directions separated by only 6′. Significant changes in the line profiles over small changes in direction were also noticed by Lockman (1980) in a survey of H166α RRLs. Figure 3.17 illustrates these changes with spectra from that survey taken along the Galactic plane over the range $\ell = 33°0 \rightarrow 37°0$.

The location of the diffuse RRL emission in the Galaxy was also a clue to its origin. Gordon and Cato (1972) observed the H157α line in nine directions along the Galactic plane from $\ell = 9°4 \rightarrow 80°6$ in directions free from discrete H II regions. Most of the line emission arose from the range of Galactocentric distance of $R = 3 \rightarrow 9 \, \mathrm{kpc}$ where the H II regions are located, as shown in Fig. 3.14. Hart and Pedlar (1976) also noted this spatial correlation from their

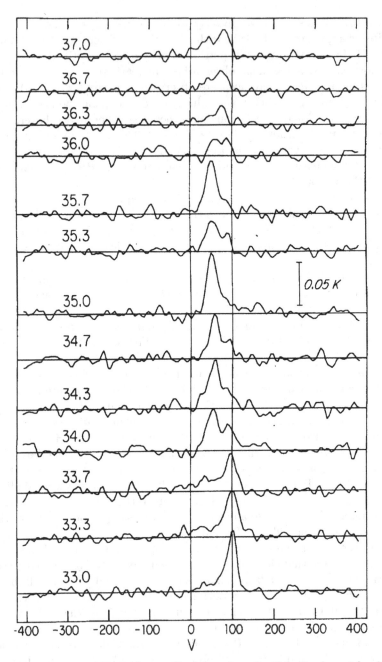

Fig. 3.17 A sequence of H166α line profiles taken along the Galactic plane at longitudes marked in each spectrum. The *vertical scale* is antenna temperature in K at the NRAO 140-ft. telescope and the *abscissae* are velocity in km s^{-1} with respect to the LSR. From Lockman (1980)

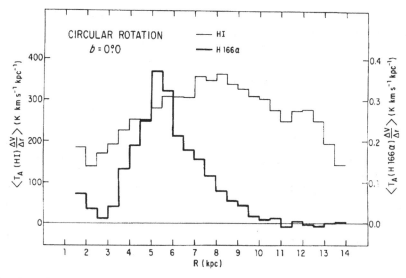

Fig. 3.18 Comparison of H166α RRLs from Galactic diffuse gas with H I emission at λ = 21 cm. *Abscissa* is Galactocentric distance with $R_o = 10$ kpc. From Lockman (1976)

H166α observations. Note that only these observations do indicate spatial correlation between the diffuse gas and the discrete H II regions but these spatial distributions differ greatly from that of H I emission as can be seen by comparing Figs. 3.15 and 3.18.

The difference between the Galactocentric distribution of the diffuse RRL emission and the H I gas is especially clear in Fig. 3.18. The data result from samplings of both lines at intervals of $\Delta\ell \approx 1°$ over a range of $358° \le \ell \le 50°$. Additional observations of the H166α emission were made at $\ell \approx 44°$ and $\ell = 100° \rightarrow 125°$. While the intensity of the λ = 21 cm line is nearly constant over the range $R = 5 \rightarrow 12$ kpc, the intensity of the H166α line increases sharply in the range $R = 3 \rightarrow 5$ kpc and rapidly decreases in the range $R > 5$ kpc. Thus, it appears certain that the RRLs from the diffuse gas are not connected with the H I component of the ISM (Lockman, 1980).

All of these observational results – electron temperature, electron density, location – suggested that the diffuse RRL emission arises in the vicinity of H II regions rather than in a cold, partially ionized component of the general ISM. The remaining question involved the origin of the low-density, ionized hydrogen with a temperature of a few thousands of Kelvins.

Examination of the location of O stars with respect to H II regions provided a surprising answer. Mezger and Smith (1975) found that only 20% of O stars were located within H II regions.[7] Mezger (1978) further suggested that the remaining 80% of the O stars formed extended regions of fully ionized,

---

[7] For this study, H II regions or radio H II regions were defined to be those sources of thermal radio emission that provided antenna temperatures at 5 GHz of $T_A \ge 1$ K at the

low-density gas that he called the "extended, low-density" (ELD) component of the ISM. Because these enormous Strömgren spheres overlapped each other, the ELD could occupy a large volume of the ISM. He estimated that the ELD would have typical values of 7,000 K and 3 cm^{-3} for the electron temperature and density, respectively.

New low-frequency observations provided confirmation. Being more sensitive to extended emission of low surface brightness, a survey (Anantharamaiah, 1986) of the Galaxy in the H272α (325 MHz) line showed that the main part of the diffuse emission indeed comes from the low-density, outer envelopes of normal H II regions. The observations indicated the electron densities to be $1 \rightarrow 10$ cm^{-3}, the electron temperatures to be $3,000 \rightarrow 8,000$ K, and the emission measures to be $500 \rightarrow 3,000$ cm^{-6} pc. The sizes of these envelopes were impressive: $30 \rightarrow 300$ pc. In the Galactic plane, a line of sight at $\ell \leq 40°$ will intersect at least one of these envelopes.

Figure 3.19 illustrates the observational results. A longitude–velocity diagram shows contours of the H166α emission, an RRL with a frequency near

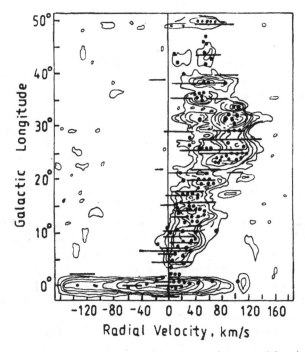

Fig. 3.19 A longitude–velocity diagram of the H272α (*horizontal lines*), H166α (*contours*), and H110α (*points*) RRLs observed for our Galaxy. The length of the *horizontal lines* marks the full width at half-intensity of the H272α line profile. From Anantharamaiah (1986)

NRAO 43-m and the CSIRO 65-m telescopes. This threshold corresponds to a minimum emission measure of $EM_{min} \approx 10^4$ cm^{-6} pc.

the H157α lines in which the diffuse, ionized gas was originally detected (Gottesman and Gordon, 1970). The diagram also shows the full widths at half-intensity of the low-frequency H272α lines that are more sensitive to low emission measure gas. Superimposed upon these are points marking detections of the H110α lines from discrete H II regions. Note that everything agrees rather well. The locations of the discrete H II regions are generally those of the ionized, diffuse gas. It is obvious that the halos of those H II regions *are* the diffuse gas, just as Mezger (1978) had suggested with his ELD.

To investigate the phenomenon even more thoroughly, Heiles et al. (1996) observed the H165α, H167α and, to a lesser extent, the H157α and H158α lines at 583 positions along the Galactic plane within the range $\ell = 0° \rightarrow 60°$. The angular resolution was 36′. RRLs were detected in 418 positions and were easily explained as emission from the extended low-density warm ionized medium (ELDWIM), a concept introduced by Petuchowski and Bennet (1993) that extended the older ELD.

Physical conditions derived from the new observations agreed well with those derived from the older ones. The average electron temperature was about 7,000 K when maser amplification was included in the analysis of the T_L/T_C ratios. The ELDWIM regions appeared to be located in the Galactic arms, occupying approximately 1% of the volume and containing an average electron density of approximately 5 cm^{-3}.

Taken together, the observations leave little doubt that the RRL emission observed along the Galactic plane outside of distinct H II regions is also emitted by more spatially extended, low-density H II gas. The situation is evidently somewhat different from the simpler situation originally modeled by Strömgren (1939) for ionization regions that "should be limited to sharply distinct bounded regions in space surrounding O-type stars." The clumpy, irregular nature of the ISM evidently allows leaks of UV photons from the immediate vicinity of the geometrically distinct H II regions to form extended, low brightness, ionized regions with indistinct borders. These regions can have diameters of tens if not hundreds of pc, average electron densities of $1 \rightarrow 10$ cm^{-3} and emission measures of $<10^4$ cm^{-6} pc.

Figure 3.20 describes the probable situation. The figure shows the locations of bright H II regions and RRL emission from the ionized diffuse gas in terms of the sizes of the observing beams. The spatial relationship between the two kinds of objects seems obvious. We expect that the increasing sensitivity of radio telescopes will facilitate the detections of additional RRL emission in the inner Galaxy that will close the spatial gaps between the discrete and diffuse sources.

What is the origin of these H II regions with low densities and surface brightness? There may not be a single mechanism. Some may have evolved by the expansion of compact H II regions (Anantharamaiah, 1986; Lockman et al., 1996), leading to tenuous H II regions with lowered electron densities and emission measures. Some of them may be formed in the evolution of

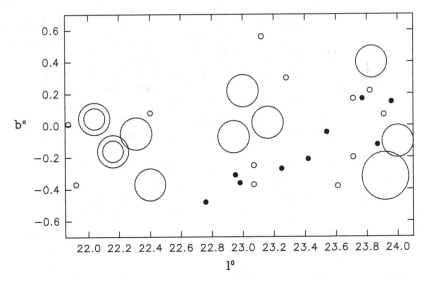

Fig. 3.20 Ionized gas observed in the inner Galaxy in a region bounded by the indicated range of b and ℓ. The *filled and open small circles* mark discrete H II regions observed in a range of RRLs from H85α to H127α by Downes et al. (1980) and Lockman (1989), respectively. The sizes of the *circles* correspond to the beams used in the observations. The *open middle-sized circles* indicate extended "diffuse" H II regions with diameters ≈12′ observed by Lockman et al. (1996). The *double circles* mark the location of two H II regions observed in the same directions but with different radial velocities. The *large circle* marks the direction and beam size in which RRLs from the diffuse medium were observed for the first time (Gottesman and Gordon, 1970)

H II regions from a giant molecular cloud (GMC) in a manner illustrated by Fig. 3.21. Massive stars spend only part of their lives in the parent clouds, about 20% or 10^6 years. There, these hot stars that form as young stellar objects disperse their cocoons with stellar winds as the star formation progresses through the giant cloud, increasingly leaving their UV emission available to ionize large regions of the ISM. In addition, isolated compact H II regions could disperse their H I mantles, so that they become density bounded rather than ionization bounded locally, also allowing their UV photons access to a much larger volume of the ISM (Churchwell, 1975). Finally, soft cosmic rays from supernovae surely ionize regions of the ISM to some extent as suggested earlier by many astronomers (Pikelner, 1967; Spitzer and Tomasko, 1968; Field et al., 1969; Hjellming et al., 1969).

Following Strömgren (1939), we examine the balance of the UV ionizing flux density to the recombinations in the ambient, lower density interstellar gas. The radius of a Strömgren sphere can be enormous in a rarified medium. For ionization-bounded media involving only hydrogen, this radius in meters is

$$R = \left(\frac{3N_L}{4\pi\alpha\, N_e\, N_{H^+}} \right)^{1/3} , \qquad (3.9)$$

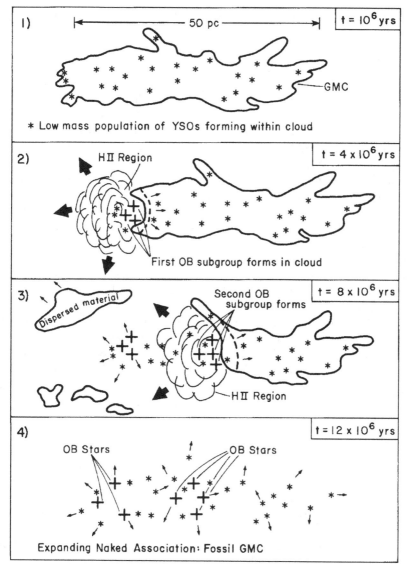

Fig. 3.21 The panels from *top to bottom* depict the formation and unveiling of an OB association from a giant molecular cloud. As the region refills with cold ISM, this process can repeat, as has happened in the Orion cloud. From Lada (1987)

where the total number of Lyman continuum photons from O stars per second $N_L \approx 10^{49}\,\mathrm{s}^{-1}$ (Spitzer, 1978) and the hydrogen recombination coefficient to all but the first level $\alpha = 2.06 \times 10^{-11}\, T_e^{-1/2}\, \phi_2$. The factor ϕ_2 is weakly dependent on temperature and is tabulated by Spitzer (1978). The electron (N_e) and ion (N_{H^+}) densities are in units of cm^{-3}. Equating these densities allows the transformation of (3.9) into a form that gives the H II mass within the sphere:

$$\frac{M_{HII}}{[M_\odot]} = \frac{8.4 \times 10^{-58} N_L}{\alpha N_e}. \tag{3.10}$$

This equation demonstrates that an O star will ionize a larger mass in a tenuous medium than that in a dense one. Physically, the recombination rate of the ions $\propto N_e^2$. Accordingly, the lifetime of the H^+ ions is higher in lower density H II regions. For the same reason, a Lyman photon flux density ionizes more hydrogen atoms in a lower density gas. Consequently, the amount of ionized mass in a Strömgren sphere varies inversely with N_e.

Numerical examples illustrate the situation. If $T_e = 7,000\,\mathrm{K}$ and $\phi_2 = 1.41$, the Strömgren sphere of an ionization-bounded H II region with $N_e = 10^4\,\mathrm{cm}^{-3}$ will have a diameter of 0.27 pc and a mass of 2.4 $M_\odot$ according to (3.9) and (3.10). Yet, with a lower density of $N_e = 1\,\mathrm{cm}^{-3}$, the sphere will have a diameter of 123 pc and a mass of $2.4 \times 10^4\,M_\odot$.

Let us apply these concepts to the diffuse, ionized component of the Galaxy. For the ELD with $\langle N_e \rangle = 3\,\mathrm{cm}^{-3}$, the ionized mass $\approx 8 \times 10^3\,M_\odot$. If the total number of O stars in the Galaxy is 2.5×10^4 (Petuchowski and Bennet, 1993) and 80% (Mezger and Smith, 1975) of them lie outside of the discrete H II regions in which they were formed, then the mass of low-density hydrogen ionized by these stars is the product of the mass per O star, the number of O stars, and the percentage available – or $\approx 1.6 \times 10^8\,M_\odot$, exactly the mass estimated by Mezger (1978) from considering the total flux of Lyman continuum from O stars in the Galaxy. Evidently, the principal mass of H II in the Galaxy is contained in these extended regions ionized by unobscured O stars, supplemented by the discrete H II regions located in the spiral arms.

This ionization also accounts for the dispersion of emission from pulsars that lie in the Galactic plane, especially when the geometry of the spiral arms is considered (see Fig. 3.13). For longitudes $\ell = 280° \rightarrow 310°$, where lines of sight go along an arm, a large dispersion measure is observed. Twenty-two pulsars have DMs in the range $260 \rightarrow 715\,\mathrm{cm}^{-3}\,\mathrm{pc}$. For the longitude range $50° < \ell < 80°$ with a wider gap between arms, only one pulsar has a DM $> 260\,\mathrm{cm}^{-3}\,\mathrm{pc}$. In fact, Taylor and Cordes (1993) believe that they have determined the distance to pulsars from a similar Galactic model, augmented with a few additions and refinements, with an accuracy of about 25%.

3.2.3 Thickness of the Ionized Hydrogen Layer

In general, astronomers observed RRLs from the diffuse, ionized component of the Galaxy to lie at low Galactic latitudes, i.e., in a very thin layer within $|b| < 1°$. In terms of scale height above the plane, these limits ranged from $h \approx 36\,\mathrm{pc}$ (Gordon et al., 1972) to $<80 \to 100\,\mathrm{pc}$ (Hart and Pedlar, 1976; Mezger, 1978; Anantharamaiah, 1986).

However, the optical recombination line $H\alpha$ was observed up to much higher Galactic latitudes. Its greater latitude extent is probably a consequence of its much greater intensity compared with RRLs *and* the much greater sensitivity of the optical instruments – by orders of magnitude. For example, with a high resolution of $0.26\,\text{Å}$ ($12\,\mathrm{km\,s^{-1}}$), Reynolds (1990) used a Fabry–Perot spectrometer to detect $H\alpha$ emission to a limiting intensity[8] of approximately $0.25\,\mathrm{R}$, i.e., corresponding to a threshold emission measure of approximately $0.5\,\mathrm{cm^{-6}\,pc}$ at a temperature of $8{,}000\,\mathrm{K}$. For comparison, the most sensitive survey (Heiles, Reach and Koo, 1996) of RRLs from the diffuse gas could only detect emission measures $\geq 120\,\mathrm{cm^{-6}\,pc}$. Therefore, the detection of diffuse $H\alpha$ emission at high Galactic latitudes does not conflict with the observations of the narrower confinement of the RRLs to the Galactic plane.

While strong interstellar absorption makes it difficult to observe in the Galactic plane, $H\alpha$ emission is an effective tool to observe ionized gas at high Galactic latitudes. Observations show that $H\alpha$ is emitted from every direction with an intensity ranging from $0.25 \to 0.8\,\mathrm{R}$ at the Galactic pole to $3 \to 12\,\mathrm{R}$ in directions approaching the Galactic plane. The implication is that the diffuse, ionized gas extends to heights above the Galactic plane of $|z| \approx 1{,}000\,\mathrm{pc}$ – considerably higher than the ionized gas observed with RRLs. Reynolds (1990) estimates the temperature of this gas to be approximately $8{,}000\,\mathrm{K}$ from the width of the $H\alpha$ profile.

Pulsar observations indicated the presence of ionized gas at high Galactic latitudes. Taylor and Manchester (1977) used the dispersion measures to determine that $|z| \approx 1{,}000\,\mathrm{pc}$, in good agreement with the value measured later with $H\alpha$ emission. Furthermore, observations of pulsars in high-latitude globular clusters at $|z| > 3\,\mathrm{kpc}$ determined the structure and electron density of the ionized halo (Reynolds, 1991) from their dispersion measures. Observations of $H\alpha$ emission in the same directions and, hence, from the same gas contributing to the signal dispersion determined an average electron density. Because the emission measure $EM \equiv \int N_e^2\,ds$ and the dispersion measure $DM \equiv \int N_e\,ds$, their ratio gives $\langle N_e \rangle$ while the ratio DM^2/EM gives the characteristic sizes of the ionized regions $\langle D \rangle$. These ratios indicated that the ionized halo is inhomogeneous. The average electron density is approximately $0.08\,\mathrm{cm^{-3}}$ in "clouds" that occupy $\geq 20\%$ of the columns along the lines of sight (Reynolds, 1991).

---

[8] The Rayleigh (R) is a unit of photon emission rate named in honor of the fourth Lord Rayleigh (R.J. Strutt) who made the first measurement of the night airglow in 1930. One R is defined as $10^6/4\pi$ photons $\mathrm{cm^{-2}\,s^{-1}\,sr^{-1}}$ (Hunten, Roach and Chamberlain, 1956). Consequently, at the wavelength of $H\alpha$ emission, $1\,\mathrm{R} = 2.4 \times 10^{-7}\,\mathrm{ergs\,cm^{-2}\,s^{-1}\,sr^{-1}}$. The emission measure of $H\alpha = 2.75\,T_4^{0.9} I_\alpha\,\mathrm{cm^{-6}\,pc}$, where T_4 is the temperature in units of $10^4\,\mathrm{K}$ and I_α is the intensity of $H\alpha$ line in Rayleigh (Reynolds, 1990). Therefore, at $T = 8{,}000\,\mathrm{K}$ and $I_\alpha = 0.25\,\mathrm{R}$, the threshold emission measure of the diffuse $H\alpha$ emission $\approx 0.5\,\mathrm{cm^{-6}\,pc}$.

We conclude that observations of RRLs, pulsar dispersion measures, and Hα emission indicate a layer of warm, tenuous ionized gas above and below the Galactic plane extending to a height of approximately $|z| \approx 1{,}000$ pc. In contrast, the discrete H II regions are generally confined to $|z| \approx 100$ pc. Both components have approximately the same temperature. The sources of this ionization are not yet known in detail but O stars are certainly capable on the basis of their collective, radiated energy. However, it is not clear how UV photons from these stars can travel hundreds of pc through the H I gas of the Galactic plane (Reynolds, 1984; Reynolds, 1993) to form the ionized layer.

Heiles et al. (1996) described a possible mechanism for the transport of the UV photons and of ionized gas from the Galactic disk to the halo domain. Based upon their observations of RRLs near 1.4 GHz, they believed that the Galaxy contains "chimney-like" structures connecting the disk constituents with the halo like the one sketched in Fig. 3.22. Similar to the closed-end "worms" seen in earlier observations of the $\lambda = 21$ cm emission of H I, the 408-MHz continuum emission, and most recently the IR (Koo, Heiles and Reach, 1992), these open-end chimneys allow UV photons and hot, ionized gas to flow freely from the disk to the halo. The chimneys themselves result from explosions of supernovae in the disk, blowing conduits to the halo. Typical dimensions of these structures could be 1,000 pc in diameter and hundreds of pc in length.

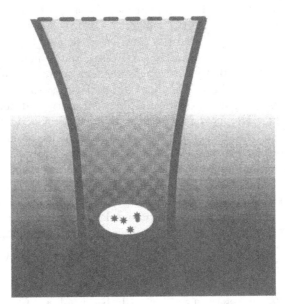

Fig. 3.22 A sketch of a "chimney" connecting the disk ISM with the gaseous Galactic halo. It would result from an older generation of massive stars in a disk star cluster that exploded as supernovae, creating the large cavity that contains the hot rarefied gas. The new O stars in the same cluster would produce ionizing photons that would travel freely through the chimney. The *white ellipse* represents the molecular cloud from which a new generation of stars will form. From Heiles et al. (1996)

Additional observations – including RRLs – will resolve the origin of the observed warm, tenuous, ionized component of the Galaxy. The worms themselves might explain all of the ionization or additional mechanisms may be found to account for the propagation of the UV flux and diffuse, ionized gas through the Galaxy.

3.2.4 Helium in the Galaxy

Helium recombination lines, i.e., from the common isotope ^{4}He, were detected soon after those of hydrogen. In many cases, both lines fell within the same spectral window of the spectrometer because the 0.04% change in the Rydberg constant (Table A.2) shifts the helium RRLs only to slightly higher frequencies (1.17) than those of hydrogen. In almost all cases, the line widths are such that the H and He lines of the same order fall side by side; i.e., the cores of the two RRLs are well separated in a spectrum and the lines do not overlap significantly. Figure 3.23 shows the original detections of the He158α (1.72 GHz) and He159α (1.62 GHz) lines from the H II region M17 with the 60-ft. telescope of Harvard College Observatory (Lilley, Palmer, Penfield and Zuckerman, 1966). That the helium recombination lines fell at the "correct"

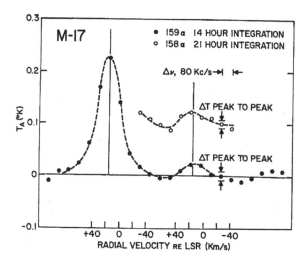

Fig. 3.23 The first detection (Lilley, Palmer, Penfield and Zuckerman, 1966) of helium recombination lines: plots of the H158α and H159α spectra (*left*) from M17 with extensions covering the corresponding helium lines (*right*), in units of antenna temperature and radial velocity with respect to the LSR. The *vertical lines* mark the theoretical separation between the two lines. Figure taken from Palmer (1968). Reproduced with permission of Nature

velocity offset[9] from the hydrogen RRLs made it certain that the detections were actually helium RRLs rather than some other emission line.

The initial detections brought no surprises. The velocity separation of the H and He RRLs was just as predicted by theory. At that time, the cosmological values of the density[10] ratio N_{He}/N_H were predicted to range from $0.08 \to 0.10$, and optical observations of extragalactic objects gave the ratio to be 0.10 (Peimbert and Spinrad, 1970). The integrated intensity ratio I_{He}/I_H of the new RRL detections was 0.10 ± 0.05, roughly consistent with the optically determined values even though the relative sizes and temperatures of the H and He ionization zones in M17 were not known in detail.

The importance of the helium RRL detection was that it offered a new tool to investigate the ionization structure within H II regions and, possibly, through better determinations of the elemental abundance ratio N_{He}/N_H an opportunity to investigate the evolution of the Galaxy and the origin of the universe.

3.2.4.1 Helium in H II Regions

The physical characteristics of the helium atom allow it to behave very differently than hydrogen within an astronomical H II region. First, its ionization energy is 24.6 eV ($\lambda = 504$ Å) compared with the 13.6 eV ($\lambda = 912$ Å) of hydrogen. This means that the size of a Strömgren sphere for He II may be smaller or larger than that of H II in the same nebula, depending primarily upon the spectrum of the UV radiation emitted by the exciting stars. Accordingly, astronomers often refer to these two spheres as being either spatially coincident or noncoincident.

Second, helium has two electrons, allowing it to exist in the He III form in addition to the more common He II. The ionization energy to produce this form is 54.4 eV ($\lambda = 228$ Å), requiring considerably more energy than that required to create the He II form.

For these reasons, the abundance of helium within an H II region is the sum of the relative number densities of the neutral, singly ionized, and doubly ionized forms of helium:

$$y = y^0 + y^+ + y^{++}. \tag{3.11}$$

---

[9] Because the rest frequencies of RRLs scale linearly with the Rydberg constant, the *radial velocity* offsets of lines of increasing mass (He, C, etc.) from the same gas will be the same relative to the H RRLs for the same values of n and Δn. Figure 3.23 is an example. Also see footnote 2.

[10] Cosmological values of hydrogen, helium, and heavier elements are often given in terms of their fractional mass X, Y, and Z, respectively. $X + Y + Z = 1$. Here, we refer to units of fractional elemental number density, expressed by the lower case letters x, y, and z, which are more closely related to the intensity of the RRL emission. *Note:* For convenience, authors sometimes use $y = N_{He}/N_H$ which, of course, is an approximation to the formal definition of y made possible by the cosmological dominance of hydrogen gas.

By definition, $x + y + z = 1$, where x and z refer to the number densities of hydrogen and heavier atoms, respectively. Each of these values refers to the integral of the constituent over the entire H II volume of the nebula. Only the components y^+ and y^{++} can be observed through RRLs. The neutral component of helium within the H II region, y^0, must be determined by other means.

Because of the enormous energy required to produce significant amounts of y^{++} with respect to y^+, this component can be ignored in (3.11). Calculations of blanketed stellar atmospheres show that the flux at $\lambda = 228\,\text{Å}$ is about two orders of magnitude less than that at $\lambda = 504\,\text{Å}$ even in the hottest stars (Kurucz, 1979). More importantly, observations of helium RRLs in H II regions gave limits for the y^{++} lines of approximately one order of magnitude below detections of y^+ RRLs (Churchwell, Mezger and Huchtmeier, 1974). Optical searches for the $\lambda = 4,686\,\text{Å}$ line of He II lower these limits by another order of magnitude (Peimbert and Goldsmith, 1972; Shaver, McGee, Newton, Danks and Pottasch, 1983). However, y^{++} lines have been detected in the much hotter environment of planetary nebulae.

In contrast, the neutral component y^0 cannot be ignored. As mentioned above, the He II zone may be smaller than the H II zone; i.e., all of the helium may not be ionized depending upon the spectral types of the exciting stars. Additional factors involve properties of the nebulae itself, manifesting themselves through radiation transfer factors like line profiles, departures from LTE, and continuum opacity including the effects of dust.

The stellar type is the most significant factor. Figure 3.24 shows the ratio γ of the helium ionizing photons to the hydrogen ionizing ones plotted

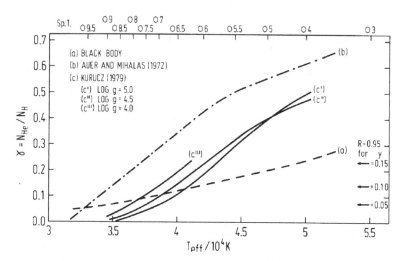

Fig. 3.24 The ratio γ of He ionizing to H ionizing photons as a function of stellar type and effective temperature. *Curves a, b, and c* refer to a black body, a non-LTE atmosphere, and three model atmospheres that include line blanketing, respectively. From Mezger (1980)

as a function of stellar type for three kinds of atmospheric models. In general, when $y = 0.10$, the diameters of the He II and H II Strömgren spheres are approximately equal (spatially coincident) for $\gamma > 0.20$ (Mathis, 1971). According to the figure, this would occur for stellar types earlier than approximately O6 based upon calculations of model atmospheres that include line blanketing (Kurucz, 1979). More recent calculations give similar results.[11] Arrows on the right side of the figure indicate where the diameter of the He II zone reaches 95% of the diameter of the H II zone for other values of y.

The compact H II region NGC2024 (Orion B) is an example of non-coincidence, where the sizes of the H II and He II ionization spheres are very different within the same nebula. Early RRL observations indicated $y^+ < 0.02$ (Gordon, 1969; Churchwell et al., 1974) compared with the value of $y^+ \approx 0.08$ observed for its nearby compact nebula, NGC1976 (Orion A). Later observations did detect weak helium RRLs in NGC2024, giving $y^+ = 0.02 \rightarrow 0.06$ depending upon where one looks (McGee and Newton, 1981; Krügel, Thum, Martín-Pintado and Pankonin, 1982).

While the exciting star of this nebula is highly obscured, some optical observations suggested that it could be a O9.5 Ib star (Becker and Fenkart, 1963). If so, the calculations shown in Fig. 3.24 indicate that the He II Strömgren sphere would be much, much smaller than that of H II even for $y \approx 0.1$, thereby accounting for the weak helium RRLs in this nebula. More recent observations in the near IR show the presence of two candidate stars but their spectral types are unknown, so that the exciting star(s) of NGC2024 remains a mystery (see Frey et al. (1979)).

Other discrete H II regions also exhibit anomalously small values of y^+. A prominent example is the giant H II region near the Galactic center, Sgr B2, for which RRLs give $y^+ \leq 0.024$ (Churchwell et al., 1974). Later, Lockman and Brown (1975b) detected $y^+ = 0.08$ from Sgr B2 with a smaller beam and higher frequency. They claimed the earlier nondetections resulted from instrumental effects. Subsequent observations by the VLA showed that the differences in observed values of y^+ were due to a combination of the beam size and a considerable variation in y^+ over this huge, obscured H II region (Roelfsema, Goss, Whiteoak, Gardner and Pankonin, 1987). In other words, $y^+ = 0.8 \rightarrow 0.10$ for Sgr B2 if one looks in the right places with the right angular resolution. This confirms the "geometric effect" originally suggested by Mezger (1980).

A few H II regions are close enough to have sufficiently large angular extents to allow imaging even by single dishes.[12] These observations indicate

---

[11] In his Chap. 2, Osterbrock (1989) calculates the ratios of the radii of the He II and H II ionization-bounded spheres as a function of effective temperature (related to stellar type) using an "on-the-spot" approximation.

[12] Single dishes capture all of the flux within their beams but cannot respond to detailed angular structure within these beams. In contrast, synthesis telescopes "see" the detailed angular structure within the primary beams of their individual antennas through their interferometric nature. However, they can miss the angularly "extended" component unless

spatial variations of He II within these nebulae. For example, Pankonin et al. (1980) used the He101α line to find a radial decrease of y^+ in the Orion nebula from 10% near the center to 6% toward the edge. Using different frequencies, Tsivilev et al. (1986) found an *increase* of y^+ from approximately 8% at the center to approximately 11% at the edge, for which they proposed a blister model for the He II component of the nebula. Similar observations of helium RRLs in the H II region W3 show variations of the observed values of $y^+ \approx 6 \to 8\%$ as the angular size of the telescope beam changes.

Higher-resolution observations reveal even more structure in He II within nebulae. Observations of y^+ for W3 made with a synthesis telescope at an angular resolution of 4″ shows values up to 9% (Roelfsema, Goss and Geballe, 1989). More recent radio interferometric observations in W3A give even greater values, with an extreme of $y^+ = 34 \pm 6\%$. Figure 3.25 shows these observations (Roelfsema, Goss and Mallik, 1992).

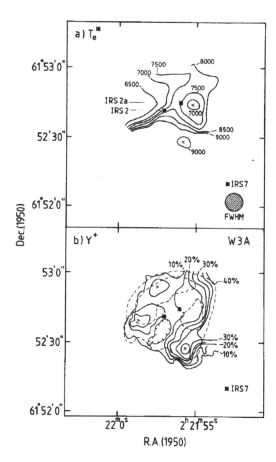

Fig. 3.25 *Top*: distribution of T_e over the H II region W3A derived from H76α RRLs (14.7 GHz) observed with the VLA at a resolution of 4″. *Bottom*: the corresponding variation of y^+ derived from the ratios of the He76α to H76α line intensities, superimposed upon the free–free emission. From Roelfsema et al. (1992)

special provision is made to obtain it. Therefore, the two kinds of radio telescopes often see different aspects of the same source.

What could be happening within these nebulae? The situation is clearly more complicated than a simple noncoincidence of the He and H Strömgren spheres. The locally observed values of y^+ can be much greater than the cosmologically predicted values of approximately 10% even though some helium is known to exist in the neutral y^0 form.

Brown and coworkers (Brown and Gómez-González, 1975; Brown and Lockman, 1975) suggested that the explanation lies in non-LTE radiation transfer effects. The basis for this idea was that measured values of y^+ seemed to be a function of the observing frequency ($n\alpha$ lines) used for the observations. However, subsequent ad hoc observations did not confirm this.

Another suggestion was that a peculiar stellar radiation field could preferentially ionize helium with respect to hydrogen such that the intensity ratio of He/H $n\alpha$ RRLs is enhanced because of a local underabundance of H II (Roelfsema et al., 1992). Unfortunately, their calculations indicated that this is rather difficult to achieve to the extent required to explain the observations of large values of y^+ observed in W3A.

Still another possibility for the large values of y^+ is a localized enrichment by nearby hydrogen-burning stars, producing helium as an "ash," and later introducing it into the ISM through a combination of convective transport and subsequent helium-enriched stellar winds as the stars evolved.

Most intriguing of all are the results of new calculations. Gulyaev et al. (1997) re-examined the possibility of the diameter of an He II ionization zone exceeding that of an H II zone. Using the best available ionization cross sections, including a 2^3s level of orthohelium, and adopting a shell model for W3A, they found that a narrow zone can exist where the He II would be enhanced relative to H II owing to a hardening of the UV radiation field, as shown in Fig. 3.26. Additional calculations show that the presence of dust would widen this zone. The implication is that the source-averaged value of $y^+ \approx 0.1$ *is* the characteristic value for W3A best representing y and that higher values of y^+ are local anomalies that depart significantly from the true helium abundance y.

3.2.4.2 Galactocentric Gradient of He II

Early observations of y^+ as a function of Galactocentric distance suggested an astrophysically important relationship: that the ionized helium abundance systematically increased with distance from the Galactic center. Here, it must be noted that the concentration of discrete H II regions in a narrow range of Galactocentric radii (see Fig. 3.14) makes it difficult to determine existence of any Galactocentric gradient of y^+. There is an unavoidable selection effect. The data are dominated by a few discrete H II regions at the Galactic center and by a clustering of most within the range $4 < R < 8$ kpc. Nonetheless, the possibility of a gradient was intriguing.

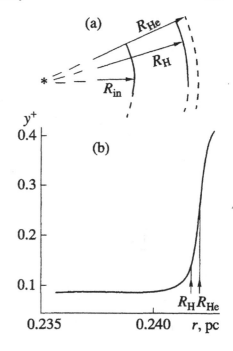

Fig. 3.26 (a) Cartoon of a shell model for W3A. R_{in} marks the inner radius of the shell. R_H and R_{He} mark the outer boundaries of the respective Strömgren spheres. (b) The calculated variation of y^+ as a function of radius from the exciting star. From Gulyaev et al. (1997)

To explain this gradient, Mezger et al. (1974) proposed a selective absorption by dust. This absorption would filter the UV radiative field from the embedded stars such that the observed ratio of He II/H II would vary with Galactocentric distance independently of y. To some extent, the observed excess IR emission of H II regions lent support to this idea. However, it proved difficult to model dust grains that could produce this effect (Mathis, 1980) and the proposal was later abandoned.

Alternatively, Panagia (1979) suggested a Galactocentric gradient in the temperature of the O stars producing the discrete H II regions. Because the star formation rate was expected to be much greater in the center of the Galaxy than that in its outer regions, the resulting gradient in metallicity could easily produce a Galactocentric gradient in stellar types of stars exciting the H II regions and, hence, in y^+.

In time, additional data made it possible to investigate the Galactocentric variation of y^+ more thoroughly. Adding to the radio He$n\alpha$ observations are the optical values determined from the $\lambda = 4,471$, $5,876$, and $6,678$ Å lines of singly ionized helium. RRLs made these new optical values possible; the analyses used the T_es accurately determined from H$n\alpha$ observations for the same nebulae (Shaver et al., 1983).

Figure 3.27 shows the results: no Galactocentric gradient of y^+ is apparent. The early, low upper limits for the Galactic center are due to a geometrical averaging within large telescope beams; newer observations give a value of ≈ 0.1 depending on resolution and position. Note how well the optical values

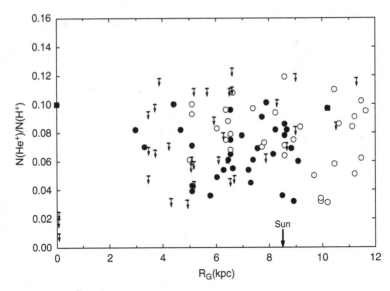

Fig. 3.27 The variation of y^+ with Galactocentric radius, determined from H II regions. The Sun lies at 8.5 kpc. *Filled circles*: measurements with RRLs. *Open circles*: measurements with optical lines. *Filled square*: average of H76α observations made with a 4″ resolution. *Upper limits* are also indicated. Data from Churchwell et al. (1974), Shaver et al. (1983), Roelfsema et al. (1987), and Gulyaev et al. (1997)

agree with the radio ones – in the range of scatter as well as in the mean value of the data. The radius-averaged value $\langle y^+ \rangle = 0.081$ and the median is 0.074.

This null result is perplexing. Figure 3.8 clearly shows a negative Galactocentric gradient of metal abundance. The higher metallicity probably results from a higher astration rate near the Galactic center, giving a higher percentage of metals in the ISM. Furthermore, Fig. 3.9 shows a corresponding positive Galactocentric gradient of the electron temperatures of compact H II regions, which can be explained by the higher cooling rates facilitated by the high metallicities of the gas nearer the Galactic center. These observations indicate that, on average, stellar types should increase from late (cool, higher metallicity) to early (hot, lower metallicity) with increasing Galactocentric distance. If the helium abundance relative to hydrogen is constant throughout the Galaxy, we would expect that y^+ should increase from the center outward, reflecting the increasing hardening of the stellar UV radiation from the Galactic center outward. Clearly, this is not seen.

The explanation must lie in the relationship of y^+ to y. It is not clear how well y^+ represents y in general. The percentage ionization of helium may well increase from the center outward but perhaps the total amount of ^{4}He is correspondingly decreasing because of a gradient in the star formation rates. Observations show that some Galactic nebulae have $y^+ \ll y$, some have $y^+ < y$, and some have $y^+ \approx y$. To understand this situation better, we must await new information on the ionization structure of helium within H II regions and, especially, on the Galactocentric distribution of helium itself in all its forms.

3.2.4.3 Cosmology

By themselves, RRLs have not yet proven useful for cosmological studies. In principle, the lines of helium and hydrogen should provide values of y that might allow separation of the primordial y_p produced by cosmological nucleosynthesis from that produced within hydrogen-burning stars. As we have seen, the best that RRLs can measure is y^+ which is not always a measure of y owing to variations in the UV radiation emitted by the exciting stars.

A more important tool for cosmological studies is the abundance of the isotope ^{3}He relative to H. Because stars might be net producers of this isotope, the lower limit to this ratio is an upper limit to the primordial value and the baryon-to-photon ratio. This limit constrains models for the chemical evolution of the Galaxy (Burles, Nollett and Turner, 2001). Furthermore, production of ^{3}He by solar-type stars ($M < 2\,M_\odot$) would lead to local enhancements in the ISM that could be detected.

Direct observations of ^{3}He$^+$ by RRLs are extremely difficult, beside the fact that only the ion can be observed with RRLs. The masses of ^{4}He and ^{3}He are about 4.0 and 3.0 amu, respectively. Equation (1.19) shows that this mass difference would give a frequency difference between their RRLs of only 0.005%. To avoid Stark broadening of the RRL profiles while still providing the most favorable line intensities, the observations would have to be made in the centimeter wavelength regime. Since the RRLs would come from H II regions, the typical turbulence broadening of approximately $20\,\mathrm{km\,s^{-1}}$ would exceed the line separation, resulting in a blend of RRLs at these frequencies that would be difficult to separate. Furthermore, the ratio of the number densities of ^{3}He$^+$/^{4}He$^+$ may be approximately 10^{-4}, which would ensure that the target line of ^{3}He$^+$ would be buried within the line profile of the ^{4}He$^+$ RRL. Finally, the superposition of ^{4}He$^+$ RRLs with those of greater mass on the high-frequency side of the lines, such as those from ^{12}C$^+$, would make a decomposition of line profiles nearly impossible to perform to an accuracy that would be useful.

Another avenue to the ^{3}He$^+$/H$^+$ ratio exists – one that depends upon RRLs in a partnership role. Following the detection of helium RRLs, Sunyaev (1966) and Goldwire and Goss (1967) suggested a search for the hyperfine transition of ^{3}He II that lies near 8.7 GHz. This line is a parallel to the $\lambda = 21\,\mathrm{cm}$ line of atomic hydrogen. Unlike the $\lambda = 21\,\mathrm{cm}$ line, the helium line is extremely weak.

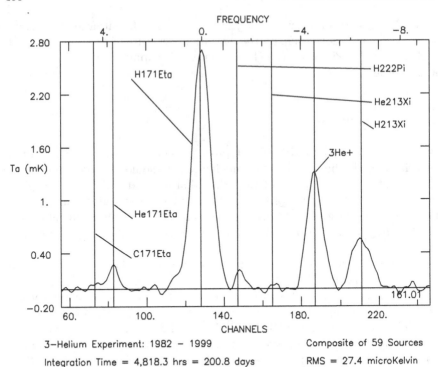

Fig. 3.28 The averaged spectra of 59 Galactic H II regions, aligned to the hyperfine line of ^{3}He II at 8.7 GHz. The effective integration time is 200.8 days and the RMS noise of the spectrum is 27 μK (1 K ≈ 3 Jy). The *ordinate* is in terms of mK of antenna temperature of the NRAO 140-ft. telescope. The *lower abscissa* is units of autocorrelator channels (78.1 kHz) and the *upper abscissa* is the offset in MHz from the H171η recombination line. From Bania (2001). See also Bania et al. (2000)

Figure 3.28 shows a spectrum (Bania, 2001) resulting from the averaged spectra of 59 H II regions and planetary nebulae, representing an integration time of about 201 days and an RMS of 27 μK. Here, an antenna temperature of 1 K ≈ 3 Jy. In addition to the detection of the ^{3}He II hyperfine-structure line, the spectrum shows the 171η ($\Delta n = 7$) RRLs of H, He, and C; the 213ξ, ($\Delta n = 14$) lines of H and He; and possibly the H222π ($\Delta n = 16$) line.

Converting observations of ^{3}He II into an abundance ratio of ^{3}He$^+$/H$^+$ is challenging. The technique consists of direct observations of the hyperfine line, of hydrogen recombination lines, of helium recombination lines, and the underlying free–free continuum. Using electron temperatures determined from the RRLs allows the extraction of the emission measure from the continuum emission. The ratio of the He/H RRLs then corrects this emission measure for electrons contributed by He II. Most difficult of all is the correction for density fluctuations within the discrete H II regions – often known as *clumping.* Emission measures involve $N_e^2 \, d\ell$; observations of clumped nebulae weight high densities much more than the low densities. In contrast, the hyperfine line is a linear function of the electron column density because it is excited by collisions; i.e., the intensity $\propto N_e \, d\ell$. Combining these measurements on the basis of a uniform sphere model for the clumped nebula would then give too small a ratio of $^3\text{He}^+/\text{H}^+$ unless detailed, complex models of the clumped nebulae were made (Balser, Bania, Rood and Wilson, 1999) to correct the ratio.

The solution was serendipitous. These studies discovered that large, diffuse nebulae tended to have little structure. Observations show that there are at least 21 of these simple H II regions in our Galaxy. Consequently, combining results for the emission measure and the hyperfine line from these nebulae would then give accurate results for the $^3\text{He}^+/\text{H}^+$ ratio.

The results are interesting (Bania, Rood and Balser, 2002). First, there does not appear to be a Galactocentric gradient of $^3\text{He}^+/\text{H}^+$ in the disk from the Galactic center to a radius of 15 kpc. Nor is there any relationship with the metallicity in the Galaxy. This result was unexpected because theory indicates that low mass stars can produce ^3He by nucleosynthesis. Possibly, the new material does not circulate widely within the ISM, or the net stellar production of ^3He is much lower than predicted either through the generation process itself or from an unknown loss mechanism. Second, the minimum value found – the upper limit on the primordial ratio – is $^3\text{He}/\text{H} = (1.1 \pm 0.2) \times 10^{-5}$. This number is consistent with an open universe, i.e., a universe that will expand forever.

3.3 Exploration of the Cold ISM by RRLs

The properties of the cold ISM differ considerably from those of the discrete H II regions surrounding hot stars. The most important and the most obvious is that it is not heated by ionizing radiation from embedded hot stars. Consequently, helium and, with a small exception, hydrogen RRLs are not detected from the cold ISM.

Nevertheless, RRLs have become an important tool to study the cold ISM. Carbon is the leading player in this drama. As described in Sect. 2.4.2, only carbon recombination lines have been detected at high principal quantum numbers. These high-n lines revealed unique information regarding the physics of highly excited Bohr atoms unobtainable in terrestrial laboratories. In addition, the carbon RRLs unveiled new secrets about our Galaxy, by teaching us about the characteristics of the cold interstellar gas that pervades the disk environment.

3.3.1 C II Regions at the Boundaries of H II Regions and Molecular Clouds

Historically, carbon RRLs were first detected in hydrogen spectra from the H II regions NGC2024 (Orion B) and IC1795 (W3) (Palmer, Zuckerman, Penfield, Lilley and Mezger, 1967). In back-to-back papers, Palmer et al. (1967) and Goldberg and Dupree (1967) announced the detection of a new RRL and offered the tentative identification that it was from interstellar carbon, respectively. Figure 3.29 shows the original spectrum from NGC2024.

Initially, there were problems with the identification. Observations of the H110α showed the same line, thereby proving it to be an RRL. However, the exact offset of the new line from the H109α in amu was not certain because of the marginal frequency sampling of only three points across the line profile. All that could be said was that the new RRL was from an element with a mass between 8 and 12 amu. There was also a problem with intensity. If the new RRL was due to interstellar carbon (12 amu) whose cosmic abundance was just below hydrogen and helium, the intensity was wrong. In the Sun, the abundance of carbon is 5×10^{-4} that of hydrogen. Figure 3.29 shows the integrated intensity of the new line to be approximately 3% of hydrogen – approximately 60 times greater than the carbon abundance. Yet, carbon seemed the best candidate on the basis of abundance and frequency.

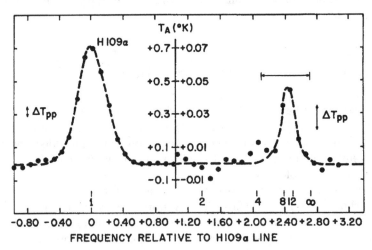

Fig. 3.29 The detection of the C109α line in NGC2024 (Orion B) with the NRAO 140-ft. telescope (Palmer et al., 1967). *Left*: the H109α line at 5 GHz. *Right*: the newly detected line later identified as the C109α line. The *ordinates* are antenna temperature in K. The *outer abscissa* is the frequency offset in MHz from the H109α line. The *inner abscissa* is in units of amu relative to the hydrogen line. Figure from Palmer (1968). Reproduced with permission of Nature

What could be happening? Goldberg and Dupree (1967) supplied the answer. They calculated that the upper quantum levels of carbon were being overpopulated through dielectronic recombination[13] and that the line intensities were correspondingly amplified above their LTE values. Therefore, they had no problem in identifying the new RRLs to be those of carbon.

Additional observations of the carbon RRLs from several nebulae gave more surprising results. Maps showed the centroid of the carbon line emission to be offset from that of the hydrogen emission, the radial velocities were often different from the hydrogen RRLs, and the line widths were always much narrower. Although the carbon lines were spatially associated, they probably did not come from the region of the H II gas itself. For these reasons, Zuckerman and Palmer (1968) suggested that the lines originated in the outer parts of a dense H I region bounding the discrete H II regions. Particularly for NGC2024, they suggested that the carbon lines arose in the same region where IR emission was observed.

Still more observations made this suggestion a certainty. Ball et al. (1970) found the width of the carbon line in NGC2024 to be only $4 \, \mathrm{km \, s^{-1}}$. This width required the line to be generated in a gas with $T_e < 1,500 \, \mathrm{K}$, which excluded origination in the H II region. Furthermore, while the radial velocities of the $\mathrm{C}n\alpha$ lines often differed from the $\mathrm{H}n\alpha$ lines, they usually agreed very well with those of the $\lambda = 21 \, \mathrm{cm}$ H I, OH absorption, and $\mathrm{H_2CO}$ lines known to be associated with the cold ISM.

All of the observations pointed to the C II regions being formed in the outer layers of molecular clouds at the boundaries with H II regions (Balick, Gammon and Hjellming, 1974; Zuckerman and Ball, 1974; Dupree, 1974). Here, UV photons with $\lambda > 912 \, \text{Å}$ leave the H II region, where they dissociate molecules and ionize atoms with ionization potentials less than hydrogen. The most abundant of these ISM constituents is carbon which has an

---

[13] This process of ordinary dielectronic recombination differs from the special one described earlier in Sect. 2.4.2 for low-frequency carbon RRLs in a cold ($\approx 100 \, \mathrm{K}$) medium. In the much warmer H II region environment where kT can be large, the process involves excitation of a bound electron of a carbon ion to a different n-state during the recombination of the singly ionized ion and a free electron, temporarily creating a neutral carbon atom with two electrons in excited states. In other words, the ion core becomes excited as well as the recombined outer electron. The newly formed atom then either autoionizes, which is of no interest to us, or stabilizes, usually by spontaneous emission of the inner electron via a resonance line, leaving the outer electron in its high quantum state. Because the rate of dielectronic recombination can be much greater than that of simple radiative recombination that occurs in the one-electron hydrogen atom, the result can be a greatly enhanced number of carbon atoms in highly excited states and, consequently, a great enhancement of the high-n level populations and their db_n/dn gradient. This produces carbon RRL lines with intensities much greater than those expected from the numerical abundance of the carbon atoms.

As described earlier in Sect. 2.4.2, in the cold environment of the low-frequency carbon spectra, the result is similar but the process is different because of the comparatively small value of kT. Stabilization occurs via collisions. To emphasize the differences between the two dielectronic processes, Watson et al. (1980) suggested the term "dielectronic captures" for the low-temperature process.

ionization potential of $V_{ion} = 11.3\,\mathrm{eV}$ for the first electron. Along with carbon, other elements with $V_{ion} < 13.6\,\mathrm{eV}$ are ionized. Among these are sulfur ($V_{ion} = 10.4\,\mathrm{eV}$), magnesium ($V_{ion} = 7.6\,\mathrm{eV}$), iron ($V_{ion} = 7.9\,\mathrm{eV}$), and silicon ($V_{ion} = 8.2\,\mathrm{eV}$) listed in terms of decreasing relative cosmic abundance. The result is the formation of partially ionized regions located between the cold, nonionized material of the parent molecular cloud and the fully ionized, discrete H II region. Such C II regions are also known as "photodissociation regions" (PDRs) because the far-UV photons from the H II region stars play a significant role in their creation.

Improved instrumentation revealed an RRL from another element. Figure 3.30 shows two spectra obtained in 1977 with the MPIfR 100-m telescope toward NGC2024 and W3. These spectra clearly show the H166α and C166α lines (Pankonin, Walmsley, Wilson and Thomasson, 1977). No He166α lines appear because of the late spectral types of the stars producing these H II regions. In addition, both spectra show a weak new line – marked "X" – at a lower velocity (higher frequency) than the carbon line. In the spectrum of NGC2024 where the C and X lines appear more crisply, their velocity

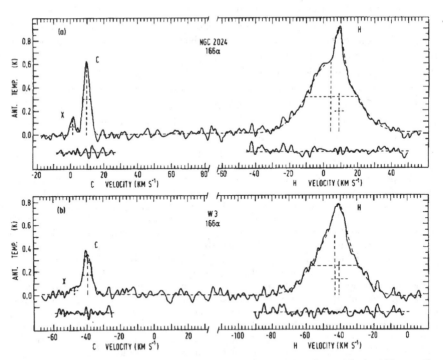

Fig. 3.30 The spectra of 166α lines from NGC2024 and W3 taken with the 100-m radio telescope of the MPIfR in Germany with a spectral resolution of $0.83\,\mathrm{km\,s^{-1}}$. The *ordinate* is antenna temperature and the *abscissae* are V_{LSR} with respect to C (*left*) and to H (*right*). *Dashed lines* indicate the Gaussian fits. The residuals to the fit are plotted below the spectra. From Pankonin et al. (1977)

separation is $8.5\,\mathrm{km\,s^{-1}}$. If both C and X have the same radial velocity, i.e., originate from the same gas, the position of the X line corresponds exactly to that of sulfur. Specifically, the difference in the ionization potentials of C and S corresponds to a velocity separation of $8.58\,\mathrm{km\,s^{-1}}$, in excellent agreement with the observations. This detection of sulfur RRLs confirmed the tentative identification made 2 years earlier in a much noisier H158α spectrum of the ρ Oph dark cloud (Chaisson, 1975).

Tielens and Hollenbach (1985) analyzed the situation theoretically to see how the composition changed as a function of depth into a cloud. They modeled the structure of a PDR with the most realistic parameters for the H II region–molecular cloud complex available. The UV radiation field was taken to be $1.6 \times 10^2\,\mathrm{erg\,cm^{-2}\,s^{-1}}$ and the density of hydrogen nuclei $n_0 = 2.3 \times 10^5\,\mathrm{cm^{-3}}$. The depth into the cloud is in terms of visual extinction; specifically, the column density $n_0 L = 2 \times 10^{21}\,\mathrm{cm^{-2}}$ at $A_V = 1\,\mathrm{mag}$.

Figure 3.31 shows their results. The geometry is a plane-parallel slab illuminated from the left by the UV radiation. The two panels are the same except for the plotted constituents. At the boundary with the H II region $(A_V = 0)$, all of the molecules are dissociated and the PDR consists only of atoms. Moreover, atoms with ionization energies $<13.6\,\mathrm{eV}$ are completely ionized here. They include C^+, S^+, and Mg^+. Because carbon is dominant, they adopted an electron abundance equal to those released from carbon, $N_e = 3 \times 10^{-4}$.

Moving deeper into the PDR, we find conditions changing further because of the attenuation of the UV radiation from the H II region by scattering, absorption, and other mechanisms. At $A_V \simeq 2$, H_2 molecules form from

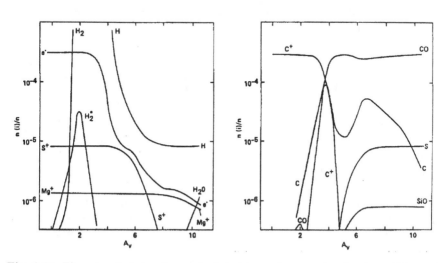

Fig. 3.31 The computed abundance of selected ions, atoms, and molecules, $n(i)/n_o$, as a function of the visual extinction, A_V into a photodissociation region. From Tielens and Hollenbach (1985)

hydrogen atoms, part of them in the vibrationally excited state H_2^*. At greater depths where $A_V > 2$, the density of excited hydrogen molecules H_2^* begins to fall because of a decrease in the UV photons necessary to excite these levels. These molecules return to the ground states through collisions and spontaneous transitions.

Note that C^+ and S^+ also vary from the edge of the PDR toward the center. At $A_V \simeq 2$, the ionized carbon atoms begin to change into neutral carbon and CO molecules. The C II layer begins to end at $A_V \simeq 3.5$ and then falls off sharply. The calculated column density for C^+ is $\simeq 2 \times 10^{18}\,\mathrm{cm}^{-2}$ with a weak dependence on the gas density (n_0) and on the incident UV radiation.

The layer of S^+ is thicker than C II, reaching a depth of $A_V \simeq 6$ because of its lower ionization potential relative to carbon. Sulfur in any form is less abundant than carbon; Tielens and Hollenbach (1985) adopted a column density for this constituent $\approx 10^{17}\,\mathrm{cm}^{-2}$.

The study of C II regions is important for understanding the complete picture of how stars form. Although the process itself takes place within dense molecular clouds, the secondary processes play a significant role in shaping the environment. The H II region created by the newly formed stars contains density gradients that facilitate expansion and fragmentation of the region's edges. This promotes mixing of enriched material into the ISM, locally at first but ultimately throughout the ISM of the host galaxy. At the same time, the ionization front and associated shock waves from the nascent H II regions create higher densities that, in turn, lead to new star formation (Elmegreen and Lada, 1977) that sequentially amplify the process. The RRLs from carbon and sulfur enable us to watch the part of this rather grand process that takes place at the boundary between H II region and the host molecular cloud, in what is now called the PDR.

3.3.2 C II Regions: Information from Carbon RRLs

Unlike H II regions where hydrogen RRLs can determine temperature and density rather accurately, C II regions are much more difficult to investigate through carbon RRLs. First, it is impossible to separate the continuum emission of electrons contributed by carbon atoms from the strong continuum background of electrons contributed by hydrogen atoms. Accordingly, the ratio T_L/T_C cannot be determined for the carbon RRLs, making the electron temperature for C II regions unavailable by this method (see Sect. 3.1.1). Second, it is not possible to obtain densities from Stark broadening of the carbon RRLs. The electron densities in the C II regions are too small to give detectable broadening except for carbon RRLs with $n > 300$, which are impossible to observe with adequate signal-to-noise ratios.

Consequently, the physical conditions within the C II regions have been determined primarily by comparing the observed intensities of the carbon RRLs, including α and β transitions, over a range of principal quantum numbers with those predicted from models made with various temperatures and densities. Specifically, the calculated brightness intensities at line center T_L result from the expression

$$T_L \approx b_n \tau_L^* T_e - b_n \beta \tau_L^* T_{BG}, \qquad (3.12)$$

where τ_L^* is the optical depth of the carbon RRL from the C II region in LTE; b_n and β are the departure coefficient and correction term for stimulated emission at the lower principal quantum number n, respectively; and T_{BG} is the brightness temperature of the continuum emission of the associated H II region at the line frequency.

Equation (3.12) is very simple. The first term is the spontaneous emission in the carbon recombination line corrected for departures from LTE. The second term corrects the first term for emission stimulated by the background free–free emission from an associated H II region when that region lies behind the C II region – which is often the case for observations of carbon RRLs. This equation is similar to (2.142), here expressed in terms of brightness temperature rather than specific intensity. The same assumptions apply: $|\tau_L|$ and τ_C are much less than 1 and $h\nu \ll kT_e$.

Table 3.4 gives the physical characteristics of a few C II regions obtained by fitting models to observed carbon RRLs. Although carbon RRLs were detected from more than 30 H II regions, only nine regions were suitable for analysis by this technique. In these sources, a few carbon RRLs were observed over a wide range of principal quantum numbers $n = 85 \to 220$. The table lists the source regions, their galactic coordinates, the observed lines, and the values of electron temperature and density resulting from the fits.

Examination of the table shows a large range of the values of temperature and density obtained from the fits, indicating fundamental difficulties with the analysis method. In a majority of the cases, the modeling assumed that the C II region lies both in front of and behind the H II region; i.e., it assumed that some of the photons from the H II region traveled through the C II region en route to the observer and some did not. This model causes large variations in the values derived for temperature and density because of radiation transfer conditions peculiar to the wavelength range ($\lambda > 5\,\text{cm}$ or $\nu < 6\,\text{GHz}$) where the carbon lines were observed. The problem is that the stimulated emission from the C II region is comparable with the spontaneous emission (see (3.12)). In these circumstances, $\beta < 0$ and the two kinds of emission add. Therefore, a unique specificity of the physical conditions of the C II region is difficult to achieve because different conditions can produce the same line intensities.

There is one other concern regarding the modeling described above. The departure coefficients (b_n) and gradients (β) were assumed to be hydrogenic, calculated from a program identical or similar to the one listed in Appendix D.

Table 3.4 Physical conditions of C II boundary regions from C RRLs

Source	ℓ (°)	b (°)	Carbon RRLs	T_e (K)	N_e (cm^{-3})	References
S64–W40	028.8	+3.5	100α, 125α	50→200	0.2→5	Vallée (1987a)
W48	035.2	−1.7	140α, 167α	30→100	10→100	Silverglate and Terzian (1978)
	035.2	−1.7	109α, 125α, 158α, 166α, 167α	100[a]	10[a]	Vallée (1987b)
S88	061.5	+0.1	140α, 167α	30→100	15→100	Silverglate and Terzian (1978)
DR-21	081.7	+0.5		10→90	300	Vallée (1987c)
S140	106.8	+5.3	142α, 166α	10	0.15	Knapp et al. (1976)
W3	137.7	+1.2	85α → 220α (12)	50→200	≈10	Hoang-Binh and Walmsley (1974)
			85α → 183α (9)	100	20→40	Dupree (1974)
			157α, 197β	≈100	≈20	Pankonin (1980)
S235	173.6	+2.8		100	1→8	Vallée (1987d)
NGC2024	206.5	−19.4	85α → 220α (12)	50	10	Dupree (1974)
			85α → 220α (12)	50→200	≈10	Hoang-Binh and Walmsley (1974)
			157α, 197β	≈100	≈20	Pankonin (1980)
Orion A	209.0	−19.4	85α → 220α (11)	10→100	> 1	Dupree (1974)
			85α → 220α (12)	50→200	≈10	Hoang-Binh and Walmsley (1974)
			85α, 109α, 137β	70→150	10→20	Jaffe and Pankonin (1978)

The number in parentheses in column 4 indicates the number of RRLs used for the modeling

[a] An error three times the value given

Subsequently, it has become clear that the low-temperature dielectronic recombination is an important process regarding the excitation of carbon in the ISM – a process not included in that program. Considering this process somewhat changes the numerical values of the departure coefficients and their gradients, and could lead to different results for the physical conditions derived for the C II regions bounding H II regions. Nevertheless, the general nature of the C II models would be about the same with regard to RRLs produced from quantum levels $n = 80 → 220$.

Despite the large range of temperatures and densities for a specific C II boundary region, the carbon RRLs indicate that these regions are cold ($T_e ≈ 100$ K) and dense ($N_e ≈ 3 → 30$ cm^{-3} and $N_H ≈ 10^4 → 10^5$ cm^{-3}).

3.3.2.1 The S II Boundary Layer

In the nine sources connected to the discrete H II regions where carbon RRLs have been detected, sulfur RRLs have also been detected. As examples, Fig. 3.32 shows spectra of the C166α and S166α lines detected toward the H II regions W48 and S87. Table 3.5 gives some numerical details.

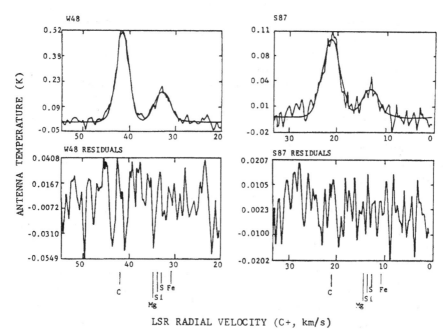

LSR RADIAL VELOCITY (C+, km/s)

Fig. 3.32 Spectra of C166α and S166α lines obtained with the 1,000-ft. (305-m) radio telescope of the US National Astronomy and Ionospheric Center at Arecibo, PR. The radial velocity is referenced to the carbon RRL frequency at the LSR. Gaussians have been fitted to and superposed on the spectra. The residuals resulting from the subtraction of the fits from the spectra are shown below the spectra. From Silverglate (1984)

Table 3.5 Observed parameters of the C166α and S166α lines

Source	RA (1950)	Declination (1950)	Full line width at half-intensity		V_{LSR}	
			C166α (km s^{-1})	S166α (km s^{-1})	C166α (km s^{-1})	S166α (km s^{-1})
W48	$18^h 59^m 14^s$	$01° 08' 29''$	3.25 ± 0.04	3.98 ± 0.09	42.52 ± 0.02	42.51 ± 0.04
S87	$19^h 44^m 16^s$	$24° 28' 28''$	4.32 ± 0.01	4.30 ± 0.23	21.43 ± 0.04	21.61 ± 0.10

From Silverglate (1984)

The high signal-to-noise ratios and spectra resolution of these spectra ($0.26\,\mathrm{km\,s^{-1}}$) ensure the identification of the sulfur recombination line. First, the situation is similar for both sources: the relative radial velocities of both carbon and sulfur RRLs are the nearly the same. The table shows their respective differences to be $0.18\,\mathrm{km\,s^{-1}}$ in S87 and $0.01\,\mathrm{km\,s^{-1}}$ in W48, i.e., nearly within the measurements errors. Second, if the line belonged to silicon or magnesium or iron rather than sulfur, these differences would be 0.71, 1.64, and $-2.19\,\mathrm{km\,s^{-1}}$, respectively – greater than the quoted measurement error. The possibility of a blended line from several elements is also excluded because it would have an asymmetrical profile, which is not seen. It is then easy to conclude that the new line is indeed that of sulfur and, further, has about the same width as the carbon line.

Why do we see RRLs of sulfur rather than those of silicon, magnesium, and iron whose solar abundances are comparable with sulfur (Cameron, 1973)? A likely answer is that the other three elements have condensed onto – or, into – interstellar dust grains, thereby depleting their abundance in the gaseous phase of the ISM (Field, 1974). From the observational limits of his spectral observations, Silverglate (1984) estimated the depletion factors for Mg, Si, and Fe to be $\approx$10 for S87 and $\approx$27 for W48.

At this writing, sulfur lines have been observed in four sources listed in Table 3.4: W48, W3, NGC2024, and Orion A. These observations facilitated modeling of the S II boundary regions similar to that done for the C II regions (Vallée, 1989). The problems were the same as for the C II models. Nevertheless, electron temperatures and densities were determined for the S II regions with an uncertainty factor of about 3. The modeling found that the S II layer extends into the host molecular cloud 2–3 times further than the C II one, that its average temperature is $\approx$40 K, and that its average density is $\approx$6 cm^{-3} (Vallée, 1989).

These experimentally determined parameters agree well with theory (Tielens and Hollenbach, 1985). Their calculations plotted in Fig. 3.31 predict that the thickness of the S II boundary layer is about twice that of the C II layer in terms of visual extinction. They also predict the temperature to decrease as a function of depth, reaching about 50 K at the middle of the S II layer.

3.3.2.2 The H^0 (H I) Layer

The spectra shown in Fig. 3.30 include one more detail of fundamental importance to our understanding of the boundary layers of discrete H II regions. Unlike most hydrogen RRLs, these profiles are asymmetrical for both sources. They consist of a narrow line blended with a broad line. This effect, first discovered by Ball et al. (1970), reveals the presence of two hydrogen recombination lines in NGC2024. One is a conventional broad hydrogen RRL from the hot H II region and the other is a narrow hydrogen RRL that must originate in the cool gas ahead of an ionization front. Separation of the two profiles is complicated, requires some judgement, and probably cannot result in unique results. Nevertheless, these spectra establish the presence of a cool hydrogen layer associated with some discrete H II regions as an astronomical fact.

In addition to NGC2024 and W3, seven other sources exhibit these narrow hydrogen lines. These include K3-50 (Roelfsema, Goss and Geballe, 1988); DR21 (Roelfsema et al., 1989); W48, S87, S88 (Onello, Phillips and Terzian, 1991); NRAO584 (Onello and Phillips, 1995); and GGD12-15 (Gómez, Lebron, Rodríguez, Garay, Lizano, Escalante and Canto, 1998). In all of these sources, the width of the "narrow" hydrogen component ranged from 3.6 to $10.5\,\mathrm{km\,s^{-1}}$, i.e., smaller than the widths of the normal hydrogen RRLs from the H II regions by a factor of 3 or more.

We refer to this gas as the H^0 component of the ISM. As mentioned earlier, the narrow widths of the lines and their spatial association with discrete H II regions require that they arise in very cool hydrogen gas linked to these emission nebulae. Because the lines are RRLs, this gas must contain an ionized component. Therefore, the term *partially ionized component* of the ISM is an appropriate description. Astronomically, it could also be called H I gas, although this specific spectroscopic term usually refers only to the widely distributed neutral component of the ISM that radiates the spin-flip $\lambda = 21\,\mathrm{cm}$ hydrogen line.

The source of this H^0 component is not known but it must be connected to some characteristic of the discrete H II region, such as the exciting stars. Two solutions have been proposed for the ionization. First, the narrow hydrogen line could be generated in the ionization front of the H II region. Calculations suggest that an H^0 line of appropriate intensity could be formed in cold, $T_e \approx 100\,\mathrm{K}$, gas at the outer side of weak D-type ionization fronts.[14] These fronts would be typical for H II regions expanding into the surrounding neutral medium (Hill, 1977).

A second possibility is that a soft X-ray flux from the vicinity of the exciting stars is ionizing the neutral gas of the boundary region. This flux could arise from a stellar wind with a velocity of $500\,\mathrm{km\,s^{-1}}$ and would easily pass through the H II region itself, creating a layer of partially ionized gas ahead of the ionization front adequate to generate RRLs of the observed intensities and widths (Krügel and Tenorio-Tagle, 1978).

The observations are not definitive enough to confirm one theory over the other at this writing. Either model can be fitted to the observed values of the narrow hydrogen RRLs (Onello et al., 1991). To complicate the situation further, VLA observations of two H^0 regions led to conflicting results. One source, W48, has a small H^0 region with characteristics consistent with the ionization front model rather than the soft X-ray model (Onello, Phillips, Benaglia, Goss and Terzian, 1994). The other, S88, has an H^0 region apparently

---

[14] "Weak D-type ionization front" means that the front propagates through a rather dense medium, causing only relatively small changes in the ambient density (Spitzer, 1978).

with an electron density of approximately $250 \, \text{cm}^{-3}$ and a depth of $0.3 \, \text{pc}$ – a few orders of magnitude larger than one might expect from the Hill's hypothesis of weak ionization fronts (Garay, Lizano, Gómez and Brown, 1998).

The data on the physical conditions within the H^0 regions are rather poor. Based on the initial observations of narrow hydrogen lines, Pankonin (1980) concluded that conditions in such regions are probably close to those of C II regions. The width of the H168α narrow line in W48 indicated that the temperature of the H^0 region in this source must be $<200 \, \text{K}$ (Onello et al., 1991). Based on the H92α, H110α, and H166α observations (Garay, Lizano, Gómez and Brown, 1998), the temperature of the H^0 region associated with S88 may be $\approx 800 \, \text{K}$. This is a large difference in temperature.

3.3.3 The Relationship Between H II, H^0, and Molecular Gas

The radial velocities of different interstellar components can tell us about the dynamics associated with each H II region, star-forming complex.

Table 3.6 lists the radial velocities observed for spectral lines from seven emission nebulae. These include normal RRLs from the H II gas, the so-called H^0 RRLs from the partially ionized gas, carbon RRLs from the C II gas, sulfur RRLs from the S II gas, and molecular lines from the parent cloud of the discrete H II region. The radial velocities are with respect to the LSR.

Examination of the table entries tells us immediately about the environment of these sources. First, the radial velocities of the partially ionized gas of the H^0, C^+, and S^+ layers *and* of the molecular lines are very nearly the same for each source region. These constituents must then be spatially associated, as considered by Tielens and Hollenbach (1985) in their analysis of PDR models.

Second, the radial velocities of the discrete H II regions themselves differ substantially from the velocities of the partially ionized gas. Depending upon the direction of motion, the hot gas moves either toward, away from, or perpendicular to the velocity of the parent cloud. From the table, we see that the sources NRAO584, S87, W3, and NGC2024 correspond to the first case, W48 and S88 correspond to the second case, and K3-50 corresponds to the third case.

These velocity differences tend to confirm the streaming of the newly ionized, hot gas from the vicinity of the stars exciting the nebulae; i.e., the hot H II gas is streaming away from the parent molecular cloud – which would have the same velocity as the newly formed stars – as described by the "champagne" evolutionary model. In the particular case of K3-50, it is also possible that the hot H II gas is not leaving the molecular cloud at all, and a champagne flow has not developed for this nebula.

Table 3.6 H II region radial velocities from various spectral lines

Source	Line	V_{LSR} (km s^{-1})					References
		H⁰	C⁺	S⁺	HII	Molecule	
NRAO584 (G34.3+0.1)	85α	58.1±0.1	58.7±0.1	60.6±0.2	53.0±0.2		Viner et al. (1976)
	168α				52.2±0.4		Onello and Phillips (1995)
W48 (G35.2−1.7)	109α		41.7±0.5		45.7±0.1		Churchwell et al. (1978)
	140α		42.8±0.2	42.5±0.1	46.7±0.1		Silvergate and Terzian (1978)
	166α		42.5±0.1				Silverglate (1984)
	167α		42.7±0.3		46.1±0.4		Silvergate and Terzian (1978)
	168α	42.3±0.3	42.8±0.1	43.1±0.1	44.9±0.5		Onello et al. (1991)
	CO					41	Zelik and Lada (1978)
S87 (G60.9−0.2)	140α				15.2±0.7		Silvergate and Terzian (1978)
	166α		21.4±0.1	21.6±0.1			Silverglate (1984)
	167α		21.2±0.2		13.9±1.1		Silvergate and Terzian (1978)
	168α	21.4±3.0	22.4±0.1	22.5±0.1	16.2±1.0		Onello et al. (1991)
	CO					22.7	Blitz et al. (1982)
S88 (G61.5+01)	140α		19.9±0.2		26.1±0.7		Silvergate and Terzian (1978)
	166α		20.8±0.1	21.0±0.1			Silverglate (1984)
	167α		20.6±0.2		26.3±0.8		Silvergate and Terzian (1978)
	168α	19.6±0.4	20.9±0.1	21.4±0.1	23.4±1.0		Onello et al. (1991)
	NH₃					≈21.7	Gómez et al. (1995)
	CO					≈22	Turner (1970)
K3-50 (G70.3+1.6)	85α	−24.2±0.4			−22.4±0.3		Viner et al. (1976)
	168α		−22.7±0.1	−22.4±0.2	−24.1±0.4		Onello and Phillips (1995)
W3 (G137.7+1.2)	56α		−41.9±0.6		−42.3±0.4		Sorochenko and Tsivilev (2000)
	158α	−41.1±0.2	−40.0±0.2	−38.3±0.5	−42.6±0.2		Pankonin et al. (1977)
	166α	−40.6±0.2	−40.1±0.1	−39.5±0.7	−43.2±0.2		Pankonin et al. (1977)
	H₂CO					−39.4	Dickel et al. (1996)
NGC2024 (G206.5−16.4)	56α		11.2±0.4		7.0±0.5		Sorochenko and Tsivilev (2000)
	76α	9.4±0.6	10.6±0.2	11.1±0.5	6.8±0.2		Krügel et al. (1982)
	109α	8.7±0.2	10.4±0.1	9.3	5.2±0.1		Churchwell et al. (1978)
	157α	9.0±0.1	9.3±0.2	9.3±0.2	4.3±0.2		Pankonin et al. (1977)
	166α		9.3±0.2		4.1±0.1		Pankonin et al. (1977)
	HCN					≈11.5	Evans et al. (1987)
	CS					10.9	Evans et al. (1987)
	CO					10.6, 11.2	Loren et al. (1981)
							and Graf et al. (1993)

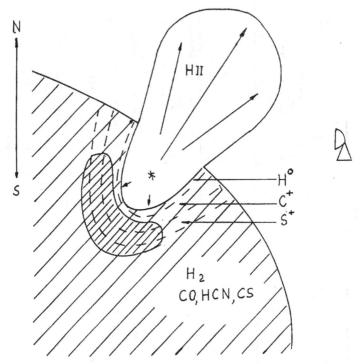

Fig. 3.33 Location and dynamics of the ionized (*bubble*), partly ionized (*dashed iso-chasms*), and molecular gas (*hatched*) components of the NGC2024 star-forming complex. Closely spaced lines indicate the dense core in the molecular cloud where protostars may be forming. An *asterisk* marks the location of the exciting star. A radio telescope shows the direction of the observer

Figure 3.33 illustrates the probable situation for NGC2024, based upon the observed radial velocities. The model is a cut through the nebula at constant right ascension. The partially ionized layers of H^0, C^+, and S^+ lie near the boundary of the H II region and the parent molecular cloud. The outflow of H II gas leaves the nebula from the side near the observer at an angle to the line of sight, as the maps of hydrogen and carbon RRLs suggest (Krügel et al., 1982).

The decrease in the radial velocities of the $Hn\alpha$ lines as n increases establishes the kinematics. At lower frequencies (longer wavelengths), the line emission comes from the outer regions of the H II "fountain" because of the greater opacities. At higher frequencies where the opacity is small, the emission comes from the core region of the fountain – the nebula itself.

In principle, the radial velocities of the carbon and sulfur lines can be interpreted similarly. However, the detailed characteristics of these lines are a little different. Figure 3.33 shows these lines to originate from two layers: a far one and a near one with respect to the observer. The lines from the near layer include emission stimulated by the background H II emission from the fountain. The brightness temperature of this continuum emission decreases with frequency (increases with wavelength), correspondingly increasing the stimulated emission in the carbon and sulfur RRLs. For this reason, the velocity decrease of the carbon and sulfur RRLs occurs at higher quantum numbers. Line emission from the layers nearest the observer plays a more important role in the line profiles.

At the highest frequency lines detected for carbon, spontaneous – not stimulated – emission dominates the line profiles. The radial velocities of these lines, C56α and C76α, are approximately 11 km s^{-1} and represent gas deep in the molecular cloud. These velocities also correlate well with those of the HCN, CS, and CO molecular lines because they originate in the same general region.

Using all of these radial velocities, we can determine the general dynamics of NGC2024. Expansion of its H II fountain is primarily toward the north, into the less dense ISM. The expansion of the H II region itself within the molecular cloud is either absent or very small, i.e., with the scatter of the velocity measurements of approximately 1 km s^{-1}.

There is a second possibility for interpreting the dynamics of NGC2024. Observations of the continuum emission at $\lambda = 1.3$ mm with an angular resolution of 10$''$ reveal the presence of six condensations of cold gas and dust with a number density of $N_H \approx 10^8 \rightarrow 10^9$ cm^{-3}. These lie in the compressed core of the molecular cloud close to the discrete H II region. Mezger (1988) suggested that these condensations have masses of about 60 $M_\odot$ and could be protostars. Subsequently, the Infrared Space Observatory (ISO) confirmed the presence of these condensations by detecting in them a large number of CO molecular lines in the FIR range of 45 $\rightarrow$ 200 μm (Giannani, Nisino, Lorenzetti, DiGiorgio, Spinoglio, Benedettini, Saraceno, Smith and White, 2000). It seems possible that sequential star formation is taking place in the host molecular cloud, perhaps triggered by the star-forming event of the NGC2024 H II region.

There are similar stories to be deduced from the other sources listed in Table 3.6. The positive or negative sign of $V_{HII} - V_{H^0}$ (or $V_{HII} - V_{C+}$) tells us whether a listed H II region lies at the far side of its parent molecular cloud and is moving away from us, or lies on the near side and is moving toward us. On the other hand, if $V_{HII} - V_{C+} > 0$ and $V_{C+} - V_{mol} < 0$, or if $V_{HII} - V_{C+} < 0$ and $V_{C+} - V_{mol} > 0$, then the expansion of the H II region is taking place deep within the molecular cloud where the exciting stars formed.

All of this information regarding partially ionized regions came from RRL observations with high angular resolution. These observations led to our understanding the location and characteristics of the C II layers with respect to the H II region ionization fronts and to the host molecular gas (Wyrowski, Schilke, Hofner and Walmsley, 1997). These relationships are particularly apparent in the Orion nebula complex. There, the discrete H II region (Orion A) was formed in the near side of the molecular cloud, having a cup-like shape with its symmetry axis slightly inclined to the direction of the observer. It penetrates about 0.6 pc into the molecular cloud (Hogerheijde, Jansen and van Dishoeck, 1995). At about 2′ southeast of the star[15] θ^1C Ori in the region of the Orion Bar, the plane of the C II layer surrounding the discrete H II region is almost parallel to the line of sight. The C II layer, the ionization front, and the boundary with the molecular cloud are seen almost edge on. This geometry is ideal for observations that can test the model.

Figure 3.34 shows just such observations. This image juxtaposes the $\lambda =$ 3.5 cm C91α and free–free continuum emission against the ^{13}CO(3 → 2) emission from the molecular gas. The $\lambda = 3.5$ cm emission has an angular resolution of 10″; the CO emission, 20″.

There is a distinct spatial separation between the ionized H II gas (the continuum), the partially ionized C II gas (the C91α emission), and the molecular gas (CO emission). The distance between the ionization front and the center of the C II emission is about 20″ (0.05 pc), which is approximately the same distance between the C II region and the molecular gas. The thickness of the C II layer is about 28″ (0.07 pc).

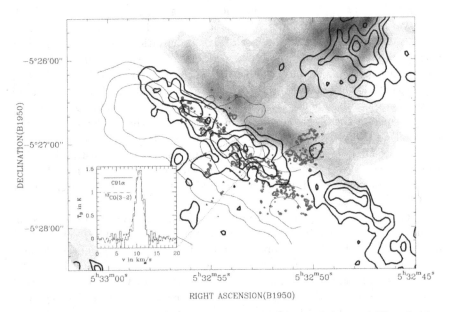

Fig. 3.34 A comparison of continuum emission, carbon RRL emission, and CO molecular emission in the vicinity of the "Orion Bar." The *gray scale* marks the $\lambda = 3.5$ cm emission observed with the VLA. The *thick lines* mark the 30, 50, 70, and 90% contours of the peak intensity of 5.5 K km s^{-1} of the C91α line. The *thin lines* mark the ^{13}CO(3 → 2) emission in the immediate vicinity of the bar (Lis et al., 1997). The *gray lines* mark contours (small, busy ones) of H$_2$(1 → 0 S(1)) emission (van der Werf et al., 1996). The *inset* at the bottom left compares a C91α line (smoothed to 20″) with ^{13}CO(peak 29 K) emission toward the position in the Orion Bar marked with *dotted lines*. From Wyrowski et al. (1997)

---

[15] The stellar coordinates $(\alpha_{1950}, \delta_{1950})$ are $(5^h32^m49^s, -5°25′16″)$.

From these observations, we can calculate the column density of the C II region. If the number density of hydrogen ranges from 5×10^4 to 2.5×10^5 cm^{-3} (Wyrowski et al., 1997), the corresponding column density of the C II layer, $N_H L$, has to range from 10^{22} to 5×10^{23} cm^{-2}. Such values correspond to an opacity range of $A_V = 5 \rightarrow 25$, which exceeds the range calculated in the PDR model shown in Fig. 3.31.

What could be the problem? The answer lies in the model used for the calculations. The real PDR medium is inhomogeneous, whereas the theoretical calculations assumed a homogeneous one (Stutzki, Stacey, Harris, Jaffe and Lugten, 1988; Howe, Jaffe, Genzel and Stacey, 1991). The inhomogeneities allow the stellar UV radiation, causing the molecular dissociation and the ionization of carbon to penetrate considerably further into the molecular cloud. The inset within Fig. 3.34 shows a nearly perfect correspondence between the profiles of the C91α and ^{13}CO line emission, implying that these constituents are well mixed in the PDR (Wyrowski et al., 1997).

VLA observations of hydrogen, carbon, and sulfur RRLs from S88 provided some interesting results. At angular resolutions ranging from $3''$ to $10''$, these observations showed the C II source region to consist of two components of dimensions $6''$ and $16''$, sandwiched between the molecular cloud and two ionized hydrogen regions. Evidently, two discrete H II regions were formed nearly simultaneously in S88 and are now in the champagne phase in which ionized gas is leaving the host molecular cloud (Garay, Gómez, Lisano and Brown, 1998; Garay, Lizano, Gómez and Brown, 1998).

Similar VLA observations of NGC2024 but with a larger angular resolution of about $50''$ have produced equally interesting results. These showed that the C II and H^0 regions overlap spatially. The intensity peaks of the C166α and the narrow H166α lines coincide with the location of maximum free–free continuum emission, thereby confirming the importance of stimulated emission in the line formation process (Anantharamaiah, Goss and Dewdney, 1990).

To summarize, RRLs – mainly from carbon – have provided information crucial to our understanding the interfaces (PDRs) between discrete H II regions and their host molecular clouds. The characteristics of these PDRs correspond well to the theoretical model produced by Tielens and Hollenbach (1985), modified to account for inhomogeneities in the ambient ISM. Taken together, the observations and the model have provided us with a well-grounded picture of the structure and interconnection of these transition regions.

However, the observations still need to improve to allow refinement of the models for the gas interfaces. Table 3.4 cites values of temperature and density for C II regions that have uncertainties of multiples of the observed values.

Is it possible to measure these parameters more precisely and more accurately? Which parameters have the dominant effect upon making more detailed models of the PDRs? The answers to these questions appeared only recently, and we discuss them in the next section.

3.3.4 Physical Conditions from Carbon RRLs, IR Fine-Structure Lines of C^+, and $O\,\textsc{i}$ Lines

The difficulty in analytically determining the physical conditions in C II regions from carbon RRLs alone rests upon one important restriction. Their intensities are functions of *three* unknown quantities: the temperature, local volume density, and column density of C^+ ions along the lines of sight. This multiparameter dependence is the reason astronomers build models which they fit to imperfect observations to derive the physical characteristics of the PDRs.

Incorporating data from other observables could simplify the process and greatly increase the accuracy of the results. For example, we could include data derived from the fine-structure line of carbon ions, $^2P_{3/2} \rightarrow \ ^2P_{1/2}$, at a wavelength of 158 μm in the IR. The probable existence of this line toward some H II regions became evident after the detection of the carbon RRL (Palmer et al., 1967), and it was detected soon after the construction of an appropriate high-resolution IR spectrometer (Russell, Melnick, Gull and Harwit, 1980). Both line types involve the same carbon ions. After recombination into highly excited levels, the subsequent cascades result in RRLs while transitions between fine-structure levels emit the 158 μm line. The importance here is that, while both lines originate from the same region, their intensities have different dependencies on the ambient physical conditions.

The physical conditions transmitted by these lines can be easily determined. In the optically thin case, the intensity of the 158 μm line is proportional to the *first* power of the density, whereas the intensity of the RRL is proportional to the *square* of the density. Consequently, the ratio of the line intensities enables us to determine the local hydrogen density as a function of temperature in the emission region (Natta, Walmsley and Tielens, 1994; Smirnov, Sorochenko and Walmsley, 1995; Wyrowski et al., 1997).

In the general case with no assumptions as to the optical depth, the mathematics are somewhat complicated. Smirnov et al. (1995) give the hydrogen density as a function of temperature to be

$$N_H = \frac{2.33 \times 10^5 \Delta\nu_L T_L \alpha_{1/2} \beta_{158} T^{1.5} \exp(-1.58 \times 10^5/n^2 T)}{b_n \Delta\nu_{158} \left(1 - \frac{\beta T_{bg}}{T}\right) \ln \left[\frac{\exp(91.2/T_{158})-1}{\exp(91.2/T_{158})-\exp(91.2/T_{ex,158})}\right]}, \qquad (3.13)$$

where the parameters $\Delta\nu_L$ and T_L are the full width at half-intensity and the brightness temperature of the carbon RRL in Hz and K, respectively, and $\Delta\nu_{158}$ and T_{158} are similar values for the carbon 158 μm fine-structure line. The background temperature $T_{bg} = 3.55 + T_C$ includes the continuum emission from the H II emission behind the C II region. The b_n and β factors correct the quantum population levels of carbon for departures from LTE, including the effects of dielectronic recombination (Ponomarev and Sorochenko, 1992) extended to higher values of T and N_H.

In addition, the equation contains parameters defined as

$$\alpha_{1/2} = \frac{1 + R_{158}}{(1 + R_{158}) + 2\exp(-91.2/T)},$$ (3.14)

$$\beta_{158} = 1 - \frac{\exp(-91.2/T)}{(1 + R_{158})},$$ (3.15)

and

$$T_{ex,158} = \frac{91.2T}{91.2 + T\ln(1 + R_{158})},$$ (3.16)

which, respectively, determine the fraction of carbon atoms in the lower level, $^2P_{1/2}$, relative to their total number density N_{C+}; the correction for stimulated emission; and the excitation temperature of the fine-structure levels.

All of the parameters above depend upon R_{158}, which is defined as the ratio of the radiative transition rate to the deactivation rate of the $^2P_{3/2}$ level by collisions with electrons, hydrogen atoms, and molecules. Considering that collisions with hydrogen atoms are the dominant deactivation process, we can write

$$R_{158} = \frac{A_{3/2-1/2}}{N_H \gamma_H} = \frac{4.1 \times 10^3}{N_H T^{0.02}},$$ (3.17)

where $A_{3/2} = 2.4 \times 10^{-5}\,\text{s}^{-1}$ is the probability of spontaneous transitions between the $^2P_{3/2}$ and $^2P_{1/2}$ levels, and $\gamma_H = 5.8 \times 10^{-10}T^{0.02}$ is the deactivation rate of the $^2P_{3/2}$ level by collisions with hydrogen atoms (Tielens and Hollenbach, 1985).

The derivation of (3.13) assumes that $N_C = 3 \times 10^{-4}N_H$, that all the carbon in the C II region is completely (but singly) ionized, and that all electrons result from this ionization, i.e., $N_C = N_{C+} = N_e$.

There was one small observational obstacle that had to be overcome. The spectral resolution of the IR spectrograph was too small to resolve the 158 μm fine-structure line adequately. Only the integral of the line profile could be measured. Assuming the line width to be identical to that of the carbon RRL, Sorochenko and Tsivilev (2000) derived the brightness temperature of the fine-structure line in K from

$$T_{158} = \frac{91.2}{\ln(1 + 1.08 \times 10^{-10}\Delta\nu_{158}/I_{158})},$$ (3.18)

where they applied a slight correction for broadening as a consequence of the optical depth τ_{158}.

There, remaining parameter required for evaluation of (3.13) is the gas temperature T. This can be obtained from our third source, the O I lines. In the PDR, oxygen exists as neutral atoms because its ionization potential of 13.62 eV slightly exceeds that of hydrogen, i.e., most of the photons below that energy ionize hydrogen. Figure 3.31 shows that oxygen atoms begin to bind into CO molecules only when the visual extinction $A_V > 4$. The ground state of atomic oxygen is split into three fine-structure levels: $^3P_2$ (lower), $^3P_1$, and $^3P_0$ (upper). Therefore, oxygen atoms radiate the fine-structure lines $^3P_1 \rightarrow {}^3P_2$ and $^3P_0 \rightarrow {}^3P_1$ at 63 and 146 μm, respectively, which can be observed in the IR.

The fine-structure lines of carbon and oxygen have greatly different dependencies on density, which makes them useful for determining the density of the PDR. The cross section for collisional excitation of the C^+ fine-structure levels by hydrogen atoms (Launay and Roueff, 1977a) is one to two orders of magnitude greater than those of O I (Launay and Roueff, 1977b). Consequently, for PDR environments where the density $N_H \geq 3 \times 10^4 \, \mathrm{cm}^{-3}$, the fine-structure levels of carbon are already thermalized and the intensity of the carbon 158 μm line is independent of density. In contrast, there is no thermalization of the O I IR lines under such conditions; their intensities are direct functions of the hydrogen density (Tielens and Hollenbach, 1985). Therefore, the ratios I(C II 158 μm)/I(O I 63 μm) and I(C II 158 μm)/I(O I 146 μm) depend only on the density and can be used to resolve ambiguities regarding the physical conditions within the C II layers of the PDR.

For comparison with the fine-structure lines, it is best to use carbon RRLs at frequencies above 20 GHz ($\lambda < 1.5$ cm). In this range, the opacity is small and stimulated emission (the $\beta T_{bg}/T$ term in (3.13)) can be neglected. This approximation substantially simplifies the interpretation of the observations. Figure 3.35 shows the RRL spectra used for the analysis. These include H, He, and C56α lines observed toward NGC2024, W3, and the Orion Bar region of the Orion nebula (Sorochenko and Tsivilev, 2000).

The corresponding fine-structure lines came from observations made from the Kuiper Airborne Observatory (KAO). These are the C^+ 158 μm line from NGC2024 (Jaffe, Zhou, Howe and Stacey, 1994), the O I 63 μm line from NGC2024 (Luhman, Jaffe, Sternberg, Herrmann and Poglitsch, 1997), and the O I lines at 158 and 146 μm from W3 (Howe, 1999) and from the Orion Bar (Herrmann, Madden, Nikola, Poglitsch, Timmermann, Geis, Townes and Stacey, 1997).

Results. Figure 3.36 shows the intensity ratios plotted as a function of volume density and temperature for the nebulae NGC2024 and W3. The ratios for C56α/C II(158) as a function of N_H and T resulted from (3.13). The other ratio, C II(158)/OI(63) for the case of NGC2024, was calculated iteratively from the observations of I(O II 158 μm)/I(O I 63 μm) = 0.4. A similar calculation was performed for the ratio C II(158)/OI(146) for W3. Sorochenko and Tsivilev (2000) give details of these calculations.

To summarize: The essence of this analysis is that these spectral lines have different dependencies upon temperature and density.

RRLs are based upon an ion–electron recombination process, for which the line intensities (related to the populations of the bound levels) are proportional to the square of the density – as noted above. Furthermore, since the partition of electrons between bound and unbound states is a function of temperature, the level populations and the line intensities decrease as

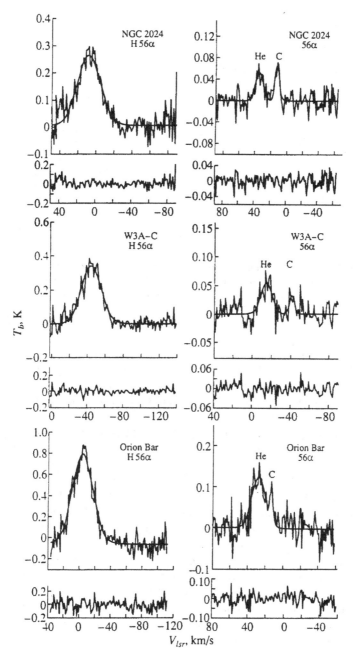

Fig. 3.35 56α lines of hydrogen, helium, and carbon observed with the 22-m telescope at Pushchino, Russia. *Smooth lines* indicate the Gaussians fitted the lines. The spectra below the lines gives the residuals from the fits. From Sorochenko and Tsivilev (2000)

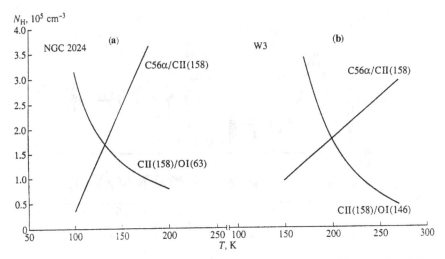

Fig. 3.36 Ratios formed from the observed intensities of C56α RRLs and the IR fine-structure lines of carbon 158 μm, O I 146 μm, and O I 63 μm plotted against volume density (N_H) and temperature. Their *intersections* mark the solutions for the temperature and density of the C II PDRs in the nebulae NGC2024 and W3

temperature increases, actually $\propto T^{1.5}$. These RRL dependencies on N_e and T are shown by the Saha–Boltzmann relationship of (2.113). On the other hand, the intensities of the fine-structure lines *increase* with temperature. Therefore, the intensity ratio C56α/C II(158) has a positive slope in Fig. 3.36; higher densities correspond to higher temperatures.

The intensity ratio of the IR fine-structure lines of carbon and oxygen has a different dependence on density and temperature. Both IR lines have their levels populated by collisions with hydrogen atoms. For this reason, their intensities increase with temperature (more frequent collisions) and density (more colliders). However, the different physical "sizes" of C and O atoms mean different excitation rates: slower for carbon and faster for oxygen. Consequently, a I(C)/I(O) ratio will have a negative slope in Fig. 3.36.

Because of these different behaviors, the two ratio curves of Fig. 3.36 intersect, enabling the fixing of the density and temperature of the C II region within a PDR. Table 3.7 gives the results (Sorochenko and Tsivilev, 2000). The table also gives results for the source S140/L1204, where IR observations were compared with the C165α (C166α) lines. The first three columns on the left give the characteristics of the discrete H II regions and the exciting star, and the columns on the right give the lines used and the results derived from them for the abutting C II regions.

The physical conditions for the C II regions derived in Table 3.7 generally agree with those obtained earlier in Table 3.4, obtained exclusively from RRLs. However, the parameters obtained by comparing RRLs and IR

Table 3.7 Characteristics of C II regions

Source	H II region		C II region				
	Stellar type	N_e (cm^{-3})	RRL	IR (μm)	T (K)	N_e (cm^{-3})	N_H (cm^{-3})
Orion Bar	O6	10^{4}[a]	C56α	158, 146	215	39	1.3×10^5
W3	O5–O6	1.7×10^4[b]	C56α	158, 146	200	54	1.8×10^5
NGC2024	O9.5	1.9×10^3[c]	C56α	158, 63	132	51	1.7×10^5
S140/L1204	B0 V	10^d	C165α	158, 63	67–85	3	$\approx 10^4$

[a]Smirnov et al. (1984), [b]Colley (1980), [c]Berulis and Sorochenko (1973), and [d]Smirnov et al. (1995)

fine-structure lines are much more accurate. Sorochenko and Tsivilev (2000) concluded that the uncertainties from the newer method are about 20–30% for temperature and 30–50% for density.

Even with only four regions, the data in Table 3.7 reveal new information regarding the C II regions. Higher temperatures should be expected for C II regions lying at the boundaries of H II regions excited by earlier-type (hotter) stars. This result seems reasonable. The early stars would have a harder UV emission that, in turn, would provide more energetic photons entering the surrounding PDR.

Finally, these data generally agree with the theory of PDRs (Tielens and Hollenbach, 1985). According to this theory, the temperature of the C II layer should decrease from about 1,000 K at the boundary with the H II region to ≤100 K at $A_V \approx 4$ near the outer boundaries of the layer. The carbon RRLs are actually averaged over a range of depths of the C II layers and, therefore, give average temperatures. The only discrepancy with theory seems to be the linear dimensions – and that is probably the result of the simplifying but unrealistic assumption of a homogeneous medium within the PDR.

3.3.5 Carbon RRLs from Atomic and Molecular Clouds

C II regions occur not only in complexes of discrete H II regions but also in cold clouds in the ISM that are exposed to the general interstellar UV radiation field. There, the flux density of the UV is much less and the correspondingly weaker intensity of the carbon RRLs in these regions makes them more difficult to observe. Nonetheless, astronomers have been able to detect the carbon RRLs toward a few strong background sources.

3.3.5.1 Cold Interstellar Clouds Observed Toward Cassiopeia A

The radio source Cassiopeia A or, simply, Cas A is an interesting object.
Lying about 2.8 kpc from the Sun in the plane of our Galaxy, it is a remnant
of a supernova that occurred in the latter part of the seventeenth century. It
is an intense source of radio waves.

For us, however, Cas A is an important tool for determining the characteristics of the interstellar gas. Serendipitously, cold interstellar clouds lie
along the line of sight to this object. In fact, the majority of information
about C II regions in cold interstellar clouds was obtained by observing this
cosmic radio beacon, much like determining the characteristics of fog by view-
ing a distant street light. Carbon RRLs have been detected toward this object over a wide range of frequencies and principal quantum numbers, from
$n = 766 \rightarrow 166$ and from 15 MHz to 1.5 GHz – a frequency range of two
decades (Konovalenko, 1990; Sorochenko and Walmsley, 1991).

Figure 3.37 shows two examples of carbon spectra obtained toward Cas A.
The top spectra are two components of the C221α line ($\nu \approx 600$ MHz) in

Fig. 3.37 *Top*: two components of the C221α line
($\nu \approx 600$ MHz) in emission
obtained toward Cas A
with the RT-22 radio telescope at Pushchino, Russia
(Sorochenko et al., 1991).
The integration time was
about 170 h. Residuals
from the Gaussian fits are
shown immediately below
the spectrum. *Bottom*: the
average of eight transi-
tions (C571$\alpha \rightarrow$ C578α at
$\nu \approx 34.5$ MHz) in absorp-
tion toward Cas A obtained
with the E–W arm of the
T-shaped radio telescope
at Gauribidanur, India
(Kantharia et al., 1998).
The total integration time
for one transition was about
400 h. The *broken line*
shows the best-fitting Voigt
profile and the *dashed line*
shows the residuals after
fitting. The *abscissae* are
V_{LSR}

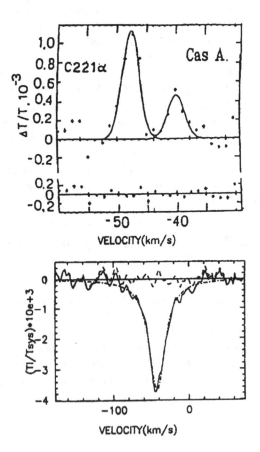

emission at the radial velocities -40 and $-48\,\mathrm{km\,s^{-1}}$. The bottom spectrum is the average of eight transitions (C571α →C578α at $\nu \approx 34.5\,\mathrm{MHz}$) seen in absorption. The population of the quantum levels of carbon determines whether the lines appear in emission or absorption, as described in Sect. 2.4.2. At low frequencies, the two velocity components blend into a single Voigt absorption profile. In general, the line width of the carbon α-type RRLs toward Cas A increases with n. Figure 3.38 clearly shows this relationship in which all known observations have been plotted.

One has to be careful when comparing a two-component emission spectrum with a single-component absorption spectrum. In Fig. 3.38, the widths of these carbon RRLs were referenced to the width of the emission component at $V_{LSR} = -48\,\mathrm{km\,s^{-1}}$ seen in Fig. 3.37 and measured to be $3.7\,\mathrm{km\,s^{-1}}$. Because the absorption spectra are a blend of the two components seen in emission, the widths of these absorption blends were decreased by the separation of the emission components, $8\,\mathrm{km\,s^{-1}}$ to facilitate comparison with the emission profiles. The final data set includes spectra collected Sorochenko and

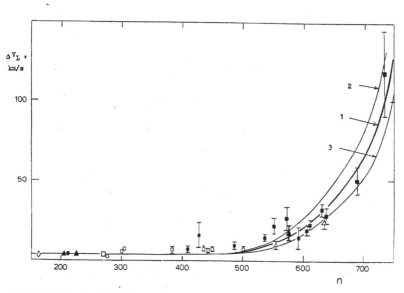

Fig. 3.38 Width of carbon RRLs toward Cas A plotted against principal quantum number. *Filled squares and circles* show respective data from Kharkov (Konovalenko, 1984) and Pushchino (Ershov et al., 1984; Ershov et al., 1987; Lekht et al., 1989; Sorochenko et al., 1991; Kitaev et al., 1994). *Open triangles* (Anantharamaiah et al., 1985) and *open circles* (Payne et al., 1989) represent observations from Green Bank, *open squares* represent observations from the VLA (Anantharamaiah et al., 1994), and *diamonds* from Effelsberg (Sorochenko and Walmsley, 1991). The data on the C640α line have been corrected (Konovalenko, 1995). *Curve 1* shows the widths $\Delta V_L(n)$ calculated for $N_e(T_e/100)^{0.62} = 0.1$, and *curves 2 and 3* show the same but with $N_e(T_e/100)^{0.62}$ increased and decreased by a factor of 1.5, respectively. From Sorochenko (1996) (recent observational data in three frequency ranges (Kantharia et al., 1998) have been added to this figure as *filled triangles*)

Smirnov (1990) augmented by more recent spectra on the lines C165α and C166α (Sorochenko and Walmsley, 1991), C220α (Sorochenko et al., 1991), C270α (Anantharamaiah et al., 1994), C537α →C540α (Kitaev et al., 1994), C201α →C206α, C223α →C229, and C571α →C578α (Kantharia et al., 1998).

Following (2.72), we can describe the full velocity width of the Voigt line profile at half-intensity as

$$\Delta V_V = 0.53L + \sqrt{0.22L^2 + D^2} \quad \text{km s}^{-1}, \tag{3.19}$$

with an accuracy better than 1%. Here, L and D are the full widths of the Lorentz and Doppler components of the profile at half-intensity. The Lorentz component itself consists of two parameters:

$$L = (\delta\nu_{col} + \delta\nu_{em})\frac{3 \times 10^5}{\nu} \quad \text{km s}^{-1}, \tag{3.20}$$

where $\delta\nu_{col}$ and $\delta\nu_{em}$ represent the broadening contributed by collisions and emission, respectively. In the conditions of the cold ISM, (2.160) and (2.181) give these parameters as

$$\delta\nu_{col} = 1.16 \left(\frac{n}{100}\right)^{5.1} \left(\frac{T_e}{100}\right)^{0.62} N_e, \quad \text{Hz} \tag{3.21}$$

and

$$\delta\nu_{em} = 3.1 \times 10^{-2} \left(\frac{n}{100}\right)^{5.65}, \quad \text{Hz}, \tag{3.22}$$

where n, T_e, and Ne are the lower principal quantum number of the line, the electron temperature in K, and the electron density in cm^{-3}, respectively.

The best agreement of the measured and observed values occurs at

$$N_e \left(\frac{T_e}{100}\right)^{0.62} = (0.1 \pm 0.02) \quad \text{cm}^{-3}\,\text{K}^{0.62}. \tag{3.23}$$

Figure 3.38 shows these values as a thick line.

Observations of the width of the carbon RRLs over a broad range of frequency give important information about the nature of C II regions in the general ISM. Unfortunately, they do not give sufficient information to determine the physical conditions within these regions. Equation (3.23) provides only a functional relationship between electron temperature and density of the C II regions.

One way to specify conditions in the C II regions is to make use of the intensities of the carbon RRLs observed toward Cas A. Two extreme models have been considered: "cold" ones with $T_e = 16 \to 20$ K and $N_e = 0.27 \to 0.4$ cm^{-3} and "warm" ones with $T_e = 50 \to 100$ K and $N_e = 0.05 \to 0.15$ cm^{-3} (Walmsley and Watson, 1982; Ershov et al., 1982; Ershov et al., 1984; Ershov et al., 1987; Konovalenko, 1984; Anantharamaiah et al., 1985; Payne et al., 1989). A "cold" model with $T_e = 18\,\text{K}$ and $N_e = 0.3\,\text{cm}^{-3}$ and a "hot" model with $T_e = 50\,\text{K}$ and $N_e = 0.15\,\text{cm}^{-3}$ both gave satisfactory agreement between the calculated values and the observational data.

The details of these models are somewhat different. The "cold" model assumed hydrogen-like carbon atoms with the usual hydrogen-like mechanisms of populating the quantum levels. In contrast, the "hot" model included dielectronic recombination of carbon atoms, described in Sect. 2.4.2 (Sorochenko and Smirnov, 1990).

Distinguishing these models required the expansion of the carbon RRL observations into the decimeter wavelength range. There, the intensity of the lines is a strong function of the temperature of the C II regions.

Figure 3.39 compares the calculated and observed values of the integrated line-to-continuum ratios for four models. The same data shown in Fig. 3.38 have been used but with corrections to the intensities of the C603α, C611α, C621α, and C640α lines (Konovalenko, 1984; Payne, Anantharamaiah and Erickson, 1994). In the formation of carbon RRLs along the line of sight to a strong background source like Cas A, stimulated transitions play a major role in the observed intensity of the lines. Specifically, the rightmost term of (3.12) becomes important. The ratio of the integrated line to the underlying continuum emission is

$$\int \frac{T_L}{T_C}d\nu = -\int \tau_L d\nu = -\frac{2.0 \times 10^6\, EM_{CII} b_n \beta}{T_e^{5/2}} \quad \text{Hz}, \qquad (3.24)$$

where EM_{CII} is the emission measure of the C II region and $\tau_L = \tau_L^* b_n \beta = k_L^* \ell\, b_n \beta$ is the line optical depth, where k_L^* is the LTE line absorption coefficient derived from 2.116 and ℓ is the thickness of the C II region.

The solid curves of Fig. 3.39 show curves for the line ratio calculated for temperatures of 25, 50, and 75 K and corresponding densities derived from (3.23). The values of b_n and β include dielectronic recombination (Ponomarev and Sorochenko, 1992). All curves are referenced to the observations of the C537α–C540α lines, where the measurement error is small owing to an integration time of 1,224 h (Kitaev et al., 1994).

The calculations give a solution for the characteristics of the C II regions along the sight line to Cas A. The best agreement of observations and calculations occurs for $T_e = 50\,\text{K}$ and $N_e = 0.15\,\text{cm}^{-3}$ – thereby excluding the "cold" models. With these conditions, the emission measure for the C II region, EM_{CII}, is $1.7 \times 10^{-2}\,\text{cm}^{-6}\,\text{pc}$. The critical observations determining the most appropriate model were the high-frequency carbon RRLs: C165α and C166α (Sorochenko and Walmsley, 1991); C220α (Sorochenko et al., 1991); C270α (Anantharamaiah et al., 1994); and C300α–C303α, C308α, and C310α lines (Payne et al., 1994). As before, the additional observations involving $n \approx 205$, 225, and 575 (Kantharia et al., 1998) have been added to the figure and agree with the solution.

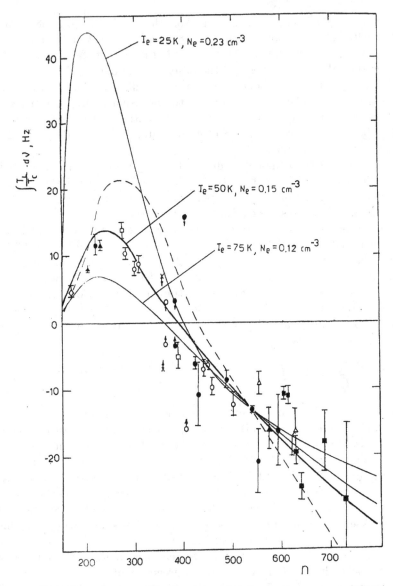

Fig. 3.39 The observed ratios of the line integral to the continuum toward Cas A compared with values calculated from different models. The emission profiles of two components at $V_{LSR} \approx -40\,\mathrm{km\,s^{-1}}$ and $V_{LSR} \approx 48\,\mathrm{km\,s^{-1}}$ have been summed. The *solid curves* indicate models which do not consider the influence of Cas A radiation on the width of the carbon lines. The *dashed curve* indicates the model where $T_e = 50\,\mathrm{K}$ and $N_e = 0.05\,\mathrm{cm^{-3}}$, and $I_{Cas} = 0.83 I_{bgr}$

Despite this great success in fitting the observations of the $Cn\alpha$ RRLs from the general ISM, the theory could still be improved. In the calculations, the only radiation field considered for line broadening was the nonthermal background radiation, I_{bgr}, involved in the derivation of broadening parameter given by (3.22). At 100 MHz, this isotropic radiation has a brightness temperature $\approx 1,000$ K, varying $\propto \nu^{-2.55}$. However, along the line of sight toward Cas A, clouds lying close to it would be exposed to the additional nonthermal radiation from that supernova remnant, which would be significant at low frequencies. Low-frequency carbon RRLs from these clouds would have additional radiation broadening, the required broadening by electron collisions would be less, and the analysis described above would overestimate the derived value of the electron density (Payne et al., 1994; Kantharia et al., 1998).

Kantharia et al. (1998) examined this effect in more detail. At the temperature $T_e = 75$ K, they considered three combinations of radiation intensity (I) and density: $N_e = 0.15$ cm^{-3} and $I_{Cas} = 0$, $N_e = 0.11$ cm^{-3} and $I_{Cas} = I_{bgr}$, and $N_e = 0.02$ cm^{-3} and $I_{Cas} = 3I_{bgr}$. Despite better agreement of the first combination with the observations, they preferred the third combination because the gas pressure within the C II regions better conformed to the pressure of the surrounding ISM. For example, the first combination would give a gas pressure[16] of $P/k = N_H \cdot T_e = 3.75 \times 10^4$ K cm^{-3}, which about one order of magnitude larger than the value of 3,700 K cm^{-3} proposed for the ISM by McKee and Ostriker (1977). In contrast, the third combination produced a gas pressure of 5,000 K cm^{-3}. To resolve this disagreement, Kantharia et al. (1998) suggested that additional refinements to the cloud models might produce a pressure that agreed better with the 3,700-K cm^{-3} value.

To achieve the 3,700-K cm^{-3} pressure, one might imagine that it would be possible to increase the contribution of the Cas A radiation to the carbon line broadening to allow a corresponding reduction of the density of the C II region models and, correspondingly, of the pressure (see $\Delta V_L = f(n)$ in Fig. 3.38).

But, it is not possible. Collision and radiation broadening have similar dependencies on quantum number and cannot be separated given the measurement errors of the observations, but the intensity of the carbon RRL is a tight function of the electron density. Decreasing N_e results in increasing the β_n coefficient and its dependence on the principal quantum number n, which changes the carbon line intensity in the wrong sense.

One curve in Fig. 3.39 does include emission from Cas A in the radiation broadening of the carbon lines. This model is not a particularly good fit to the data. Its radiation component was taken to be $I_{Cas} = 0.83 I_{bgr}$, which corresponds to locating the C II cloud about 150 pc from the supernova remnant. The cloud characteristics were assumed to be $T_e = 50$ K and $N_e = 0.05$ cm^{-3}. Inspection shows that these conditions cause this model to overestimate the line emission at quantum numbers near $n \approx 200 \rightarrow 300$ and, although less strikingly, to underestimate the emission for $n \geq 650$. Furthermore, the curve crosses the 0 point on the ordinate (the transition from emission to absorption) at a value of n greater than that suggested by the observations.

---

[16] The assumption is that $N_e = N_{c+} = 3 \times 10^{-4} N_H$ cm^{-3} and that all of the carbon is in the gas phase.

There is also a practical restriction to the density – the observations. At most, the observations would allow a minimum density of $0.1\,cm^{-3}$ within the observational errors, as shown by Fig. 3.39. This density is equivalent to setting the minimum distance of the C II region toward Cas A to about 250 pc, compared with the known 2.8-kpc distance of Cas A from the Sun. The value $N_e' = 0.05\,cm^{-3}$ is inconsistent with these observations.

At the temperature of 50 K and electron density of $0.15\,cm^{-3}$, the thermal pressure of the C II region is $P/k = T_e N_H = 2.5 \times 10^3\,K\,cm^{-3}$, significantly higher than that of the intercloud gas of the ISM. Because we want the model to agree with observations, we will re-examine it to try to explain why the pressure of the C II region exceeds that of the general ISM.

The above analysis shows that the carbon lines observed toward Cas A originate in rather dense regions. At $N_e = 0.15\,cm^{-3}$, the hydrogen density has to be about $500\,cm^{-3}$ even without considering the partial depletion of carbon. Observations of CO lines give approximately the same number ($N_{H_2} = 300\,cm^{-3}$) for the density of molecular clouds (Goldreich and Kwan, 1974).

3.3.5.2 The Nature of C II Regions Toward Cas A

Another method of investigating the characteristics of the interstellar C II regions is to compare the $Cn\alpha$ lines with the profiles of molecular spectra seen in the same direction, in this case toward Cas A. Figure 3.40 compares the C221α recombination line (Sorochenko et al., 1991) with observations of the $1_{10} \rightarrow 1_{11}$ formaldehyde H_2CO line (Goss, Kalberla and Dickel, 1984) and CO. The observations were taken of 16 molecular clouds in the direction of the Perseus arm of our Galaxy – which contains Cas A – at an angular resolution of $10''$. The H_2CO is an average over all of the clouds to facilitate comparison with the carbon RRL.

The velocity structure of the C221α and H_2CO lines agrees well even though the recombination line appears in emission and the other, in absorption. There are two velocity components in each profile: $V_{LSR} = -49$ and $-46\,km\,s^{-1}$ and $V_{LSR} = -42$ and $-36\,km\,s^{-1}$, where we list the subcomponents visible in the H_2CO profile but which are blended together in the C221α profile. The H_2CO is intrinsic to the dense material in the molecular clouds, because only there can collisions invert its level populations so that the molecule can absorb the background radiation. In the C II regions, the carbon line is enhanced by emission stimulated by radiation from Cas A. Thus, the similarity of the two velocity profiles tells us that the C II regions correspond spatially to the dense molecular clouds.

Also shown in Fig. 3.40 are contour maps of ^{13}CO emission toward Cas A obtained with the 30-m IRAM radio telescope in southern Spain (Wilson, Mauersberger, Muders, Przewodnik and Olano, 1993). The angular resolution was $21''$. These velocity slices through the CO lines tell us where the cloud

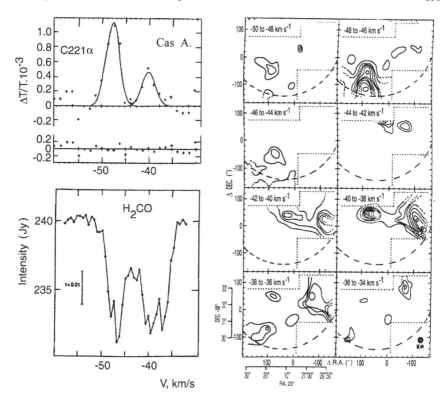

Fig. 3.40 Comparison of the C221α and molecular emission spectra toward Cas A. *Left*: spectra of the C221α line (*above*) and the $1_{10} - 1_{11}$ H_2CO line (*below*). *Right*: a contour map of the ^{13}CO line for eight V_{LSR} intervals summed over $2\,km\,s^{-1}$. The *thin dashed line* indicates the map borders and the *thick dashed line* indicates the boundary of Cas A. The *text* gives the references

components lie. Toward the south lies, a cloud with a radial velocity in the interval $V_{LSR} = -48 \rightarrow -46\,km\,s^{-1}$, exactly the velocity range of the most intense component of the C221α profile. Furthermore, the central and western parts of the maps indicate intense CO emission in the velocity interval $V_{LSR} = -42 \rightarrow -36\,km\,s^{-1}$, agreeing well with the other velocity component of the C221α profile. Finally, in the velocity ranges $V_{LSR} < -48\,km\,s^{-1}$ and $V_{LSR} > -36\,km\,s^{-1}$ where there is no C221α emission, there is also very little CO emission.

This spatial correspondence between the C II region and the molecular clouds is certainly not accidental. We expect the carbon RRLs to arise in the interfaces of the molecular clouds and the general ISM, regions illuminated by the ambient UV radiation field of the Galaxy (Ershov et al., 1984; Ershov et al., 1987). Calculations predict that the dense molecular clouds are surrounded by a layer of atomic hydrogen that protects the H_2 molecules of the cloud from dissociation by the interstellar UV radiation. This envelope also contains ionized carbon. The length scale or thickness of the $H\,I/H_2$ transition region is

$$L_{tr} = \frac{9.5 \times 10^{-5}\epsilon^{-1.4}}{\langle N_H \rangle} \quad \text{pc,} \qquad (3.25)$$

where the dimensionless parameter ϵ ranges from 6×10^{-5} to 2×10^{-4} and is defined by the ratio of the formation and destruction rates of the H_2 molecules and by the column density of hydrogen in the transition region (Federman, Glassgold and Kwan, 1979). The units of the space-averaged hydrogen volume density, $\langle N_H \rangle$, are cm^{-3}.

In fact, observations generally confirm the presence of $H\,I$ shells surrounding molecular clouds. Comparison of $\lambda = 21\,cm$ and formaldehyde maps in 12 out of 16 clouds detected with high probability toward Cas A suggests the presence of $H\,I$ envelopes. The average thickness of these shells (transition regions) was $0.19\,pc$, the average hydrogen number density was $300\,cm^{-3}$, and the average hydrogen column density was $3.4 \times 10^{20}\,cm^{-2}$ (Goss et al., 1984). Wilson et al. (1993) found similar characteristics by comparing observations of $\lambda = 21\,cm$ and CO emission lines in ten molecular clouds toward Cas A. These $H\,I$ transition layers surrounding the cloud cores had hydrogen column densities $\geq 10^{20}\,cm^{-2}$.

The temperature and volume density of the molecular clouds themselves can be determined from CO emission lines (Troland, Crutcher and Heiles, 1985; Wilson et al., 1993) and from NH_3 (ammonia) absorption lines (Gaume, Wilson and Johnston, 1994). For ^{12}CO, collisions with H_2 excite these optically thick lines, thereby providing a gas temperature from the line intensities. The optically thin lines from ^{13}CO give the column densities and, by dividing by a cloud length, the corresponding hydrogen volume densities. For the NH_3 observations, the temperatures are derived from the relative population of the rotational states when multiple rotational transitions are observed. Taken together, these observations give $T \approx 20\,K$ and $N_{H_2} = 1 \rightarrow 4 \times 10^3\,cm^{-3}$. These values are averaged over the clouds; in fact, generally the temperature increases and the density decreases outward from the cloud centers (Sorochenko and Walmsley, 1991). Only a small portion of $H\,I$ lies within the shells. The column density of atomic hydrogen in the Perseus arm toward Cas A, summed over both velocity features, exceeds $3.5 \times 10^{21}\,cm^{-2}$ (Troland et al., 1985).

Carbon RRLs originate mainly in the narrow transition layer of molecular clouds, i.e., their contiguous outer envelopes. In general, these ISM atoms are ionized by the background (ambient) UV radiation escaping from discrete $H\,II$ regions.[17] The integrated optical depth of the line emitting region $\propto N_e N_{C+}/T^{2.5}$. Therefore, the contributions from the outer regions $H\,I$ layer fall off rapidly as the density decreases and the temperature increases. For

---

[17] Carbon is also ionized in the surface layer of the molecular cloud core because the borders of $C\,II$ region and those of transition layer between $H\,I$ and H_2 overlap somewhat (see Sect. 3.3.1).

this reason, the carbon RRLs are restricted to only the small envelope volume in which the ionized medium is both dense and cool.

For this reason, the two velocity components observed toward Cas A in the C221α profile of Fig. 3.40 tell us something about the associated molecular clouds. The separate radial velocities indicate that there must be two distinct C II regions: one surrounding a cloud at $V_{LSR} \approx -48 \, \mathrm{km \, s^{-1}}$ and another surrounding a cloud at $V_{LSR} \approx -40 \, \mathrm{km \, s^{-1}}$. Furthermore, the narrowness of the components indicates fairly small dimensions for the host molecular clouds, because the C221α is not expected to exhibit either collision or radiative broadening because of its (high) frequency. The respective line widths are $\Delta V_{-48} \approx 3.7 \, \mathrm{km \, s^{-1}}$ and $\Delta V_{-40} = 5 \rightarrow 8 \, \mathrm{km \, s^{-1}}$.

With these velocity characteristics, it is possible that these clouds indicate the passage of interstellar gas through a spiral density wave associated with the Perseus arm. Spiral density waves provide a mechanism for compressing Galactic gas to stimulate star formation, thereby creating the "arms" of spiral-type galaxies like our Milky Way (see the review by Roberts (1975)). First proposed by Lindblad and Langebartel (1953), revised by Lin and Shu (1964), and further revised in a series of papers largely but not exclusively by C.C. Lin and his students, the density-wave theory envisions a quasistationary wave pattern rotating around the center of a galaxy at a fixed angular velocity. The differentially rotating Galactic gas[18] would flow through this pattern, being compressed at passage by a factor of 4 or less, thereby inducing gravity forces to create dense molecular clouds. In turn, regions within these clouds would form massive early-type stars, ultimately creating the luminous spiral arms characteristic of spiral galaxies. Of all the improvements to the theory, the most significant was the nonlinear mathematical treatment. This revision predicted the density increase and corresponding shock to occur in the ambient gas over smaller distances, stimulating star formation over equally small distances, and creating a crisper (more contrasty) spiral arm.

There has been an interesting refinement to the density-wave theory that seems to explain the two velocity components of the C221α profile of Fig. 3.40 observed toward Cas A. Roberts (1972) has examined the observational data – both optical and radio – in this direction in terms of a nonlinear, two-armed spiral shock (TASS). He envisions the Perseus arm to result from this Galactic shock wave embedded in the background density wave. Unlike standard density waves, the TASS can compress density by as much as a factor of 15, very probably accelerating star formation of that expected from the nonlinear density-wave theory.

Specifically, this TASS model predicts that the two (and sometimes more) distinct velocity components observed in the Cas A carbon RRLs and H I emission (and in optical objects) could arise in very nearly the same place within the Perseus arm, i.e., from molecular clouds that spatially lie very close to each other.

---

[18] The Galactic gas is not rigidly bound together and rotates differentially (see Fig. 3.11).

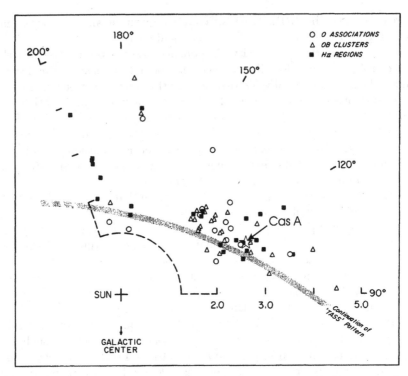

Fig. 3.41 The space distribution of young astronomical objects in the Perseus arm, based upon radial velocities and the Schmidt rotation curve (H II regions and O associations), and color-magnitude diagrams (Young Open (OB) Clusters). A *star* indicates the position of Cas A. The Galactic gas flows clockwise through the two-arm spiral shock marked by *hatching*. After Roberts (1972)

Figure 3.41 illustrates the situation. Plotted as a function of linear coordinates are the locations of Population I objects in the Perseus arm of the Galaxy. The data include O star associations with distances determined from the radial velocities of absorption spectra; OB clusters, from their color-magnitude characteristics (reddening); and H II regions, from the radial velocities of their RRLs. The Schmidt (1965) rotation model was used to convert radial velocity into distance from the Sun. An important result of this plot is the large spread of distances determined from radial velocities.

According to Roberts (1972), determining Galactic distances from a simple rotation curve, say, from the Schmidt, can give incorrect results. Much of the large distance spread shown in Fig. 3.41 could arise naturally from the local velocity perturbations imparted by the TASS wave. Figure 3.42 gives the details. The passage of the TASS through the Perseus arm creates a large velocity dispersion immediately following the leading edge of the shock. Gas located in this vicinity would have at least a 20-km s^{-1} range depending

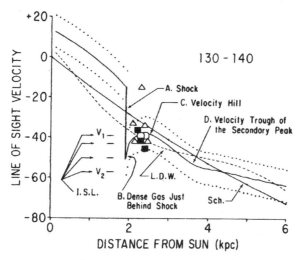

Fig. 3.42 The line-of-sight velocities in km s^{-1} created by a TASS wave within Galactic longitudes $130 \leq \ell \leq 140$. In this plot, the Sun lies 10 kpc from the Galactic center. *Curve A* marks the radial velocities created by the TASS wave, as are the velocity "hill" and "trough" following the shock front. The Schmidt (1965) and linear density-wave (L.D.W) rotation curves are also shown. On the *left dashes* indicate the velocities of Ca II and Na I interstellar absorption lines (I.S.L) observed toward O star associations in the Perseus arm (Münch, 1957). The *symbols* mark probable locations of the Population I objects considered in Fig. 3.41. From Roberts (1972)

upon its exact location with respect to the front. A similar picture[19] obtains for the longitude range containing Cas A. This model easily explains the large velocity dispersion of interstellar absorption lines observed against more distance O star associations.

More importantly for us, the model also explains the two-component velocity structure seen in the carbon RRLs and in the H I absorption observed toward Cas A. Interstellar gas (with its molecular cloud) piled up just following the shock front would contribute the emission at $V_{LSR} \simeq -48$ km s^{-1}. As would be expected from the large compression of the TASS, this H I gas is observed to have a large optical depth of $\tau > 8$ (Greisen, 1973). Furthermore, this model also explains the large longitude–velocity gradient of 0.7 km s^{-1} pc^{-1} observed for this gas with the VLA: 20 times greater than the velocity gradient expected from Galactic rotation alone (Bieging, Goss and Wilcots, 1991). The second gas component, at $V_{LSR} \simeq -40$ km s^{-1}, would lie close by but in the relaxed, more diffuse gas slightly downstream of the shock. The distance separation might be ≈ 500 pc but it could also be much smaller. Because of the unknowns regarding the detailed characteristics of the interstellar gas, the

---

[19] We reproduce the $\ell = 130 - 140°$ figure of Roberts (1972) rather than the one dealing specifically with the Cas A range, $\ell = 110 - 115°$, because of its better annotation.

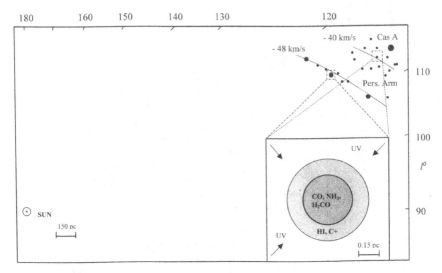

Fig. 3.43 Perseus arm based upon spiral density waves (Roberts, 1972). The *small filled circles* indicate the locations of atomic and molecular clouds formed in the possible two branches of the Perseus arm. Both kinds of clouds would exist within a tenuous medium of H I not shown in the figure. The *inset* shows a model of a molecular cloud surrounded by an H I shell containing the C II layer where the carbon RRLs originate

TASS model is more illustrative than quantitatively rigorous.[20] The molecular clouds formed in the two gas components would then contribute the carbon RRLs from their C II envelopes.

For our purpose, the density-wave theory augmented by the TASS is able to explain the multiple velocity components seen in carbon RRLs and H I spectra observed toward Cas A. It is also able to explain the high pressure in the C II regions seen in the same direction. Figure 3.43 shows a possible model for the situation, although the size scale is uncertain for reasons discussed above. The gas clouds emitting these velocity components lie at a distance of approximately 3 kpc within the Perseus arm of our Galaxy, the higher velocity one being less dense and lying somewhat further from the Sun than the lower velocity, denser one. The carbon RRLs irrefutably identify the sources as molecular clouds because they would be enveloped in C II envelopes produced by the ambient interstellar UV radiation field. The theory indicates that these clouds would have been formed as a consequence of the passage of the interstellar gas through the shock front. In time, both of these molecular clouds will likely serve as parents to OB stars and others.

Analysis indicates that the masses of all the molecular clouds detected toward Cas A fall in the range of one to tens of $M_\odot$ and, moreover, are less than the virial masses calculated from their velocity structures (Goss

---

[20] Roberts (2001) notes that adding density inhomogeneities to the TASS model would reduce the distance dimensions of the compression wave.

et al., 1984; Wilson et al., 1993). Such clouds may be in a dynamic evolutionary process, in which they are re-expanding after being created in the denser regions immediately following passage through the shock front. Such expansion would extend to the H I and C II envelopes surrounding them. This expansion can explain the higher pressure of the C II gas relative to the ambient ISM.

3.3.5.3 Carbon RRLs from Other Cold Clouds in the ISM

The detection of low-frequency carbon RRLs in cold clouds toward Cas A suggested that such lines might be found elsewhere in the Galaxy as well. Soon, the C443α and C447α lines near 75-MHz range were detected toward the Galactic center with the 140-ft. (43-m) NRAO telescope in Green Bank (Anantharamaiah, Payne and Erickson, 1988). In the same direction but at still lower frequencies, the C537α, C538α, and C539α lines were detected near 42 MHz with the E–W arm of the synthesis telescope DKR-1000 in Pushchino, Russia (Smirnov, Kitaev, Sorochenko and Schegolev, 1996). Both sets of spectra shown in Fig. 3.44 appear in absorption against the nonthermal emission of Sgr A.

Additional observations of carbon RRLs indicated that the C II emission could be spatially extended. A survey made of the C441α line (76.5 MHz) with the 64-m Parkes telescope (HPBW $= 4°$) in Australia showed emission over the range of Galactic longitude of $\ell = \pm 20°$ at $b = 0°$ (Erickson, McConnel and Anantharamaiah, 1995). The line intensity decreased with distance away from the Galactic center and, because of the large beamwidth, was not detected reliably along the Galactic equator except at

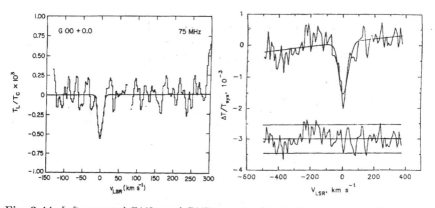

Fig. 3.44 *Left*: averaged C443α and C447α spectra fitted with a Gaussian. *Points to the left* of 75 km s^{-1} represent all observations and to the *right* represent only observations free of radio interference (Anantharamaiah et al., 1988). *Right*: averaged C537α, C538α, and C539α spectra fitted to a Gaussian. The residuals lie below (Smirnov et al., 1996)

the longitude 312°. However, at $\ell = 352°$, 358°, 0°, 2°, and 14°, the survey detected C441α emission at $b = \pm 2°$. These observations showed that C II regions in the central part of the Galaxy can extend as much as 4° in Galactic latitude.

In subsequent observations near 34.5 MHz made with the T-shaped telescope at Gauribidanur, India, Kantharia and Anantharamaiah (2000) detected eight carbon RRLs with $n \approx 575$ toward the Galactic center – a direction similar to the 76-MHz observations noted above. The angular resolution of this telescope was $21' \times 25'$. The lines clearly appeared in absorption in six directions at $\ell < 17°$.

In these same directions, searches for carbon lines were carried out near 328 MHz with the radio telescope at Ooty, India (Anantharamaiah and Kantharia, 1999). Four carbon RRLs with $n \approx 271$ were detected from all six directions. Figure 3.45 shows 328- and 34.5-MHz spectra observed from India.

As a group, these lines tell us about the C II regions in which they arise. First, all the line widths are about the same; in other words, Stark broadening is absent. This absence means that the upper limit on the electron density in the C II regions must be $N_e \leq 0.1\,\mathrm{cm}^{-3}$. Furthermore, comparison of the

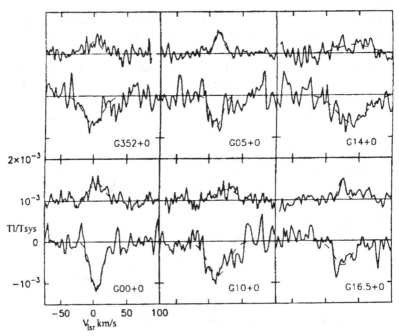

Fig. 3.45 *Above*: spectra near 328 MHz averaged over four transitions with $n \approx 271$. Velocity resolution of $1.8\,\mathrm{km\,s}^{-1}$ and angular resolution of $2° \times 2°$. *Below*: spectra near 34.5 MHz averaged over eight transitions with $n \approx 575$. Velocity resolution of $4.5\,\mathrm{km\,s}^{-1}$ and angular resolution of $21' \times 25'$. From Anantharamaiah and Kantharia (1999). Reproduced with permission of The Astronomical Society of the Pacific

line intensities at the three frequencies (34.5, 75, and 328 MHz) limits the temperature of the C II regions to $20 < T_e < 300$ K. The radial velocities of these lines place the C II regions at Galactocentric distances of $4 < R_G < 8$ kpc, in the Scutum or Sagittarius arms of the Galaxy, shown in the spiral model of Fig. 3.13 (Erickson et al., 1995; Kantharia and Anantharamaiah, 2000).

What are the detailed characteristics of these C II regions? Is ionized carbon located only in the central regions of the Galaxy? Or, is it present everywhere but requires highly sensitive telescopes for its detection?

To answer these questions, we turn to observations of the fine-structure C^+ line at 158 μm. Section 3.3.1 described how the formation of the fine-structure line and carbon RRLs takes place in the same regions. Therefore, studies of the $\lambda = 158$ μm line may give us information about the formation of carbon RRLs in the Galaxy and allow us to predict the results of additional searches for these RRLs.

Figure 3.46 shows a contour map of the carbon fine-structure line, $\lambda = 158$ μm, toward the Galactic center. The observations were made with the Balloon-borne Infrared Carbon Explorer (BICE) with a resolution of 15′. The data show bright sources within an extended, diffuse component. Many of these sources correspond to well-known bright H II regions, so the carbon emission probably comes from the PDRs enveloping these regions and molecular clouds as discussed earlier in Sect. 3.3.1.

The separation of one of these discrete sources, NGC 6334, from the underlying background is shown clearly in the $\lambda = 158$ μm C^+ emission of Fig. 3.47. These spectral data are a cut across the Galactic plane made with the Balloon-borne Infrared Telescope (BIRT), which has a higher angular resolution (3.′4) than BICE. The peak of the C^+ emission coincides with a maximum of the 5-GHz continuum emission. We note that McGee and Newton (1981) detected the C76α RRL from NGC 6334 a decade earlier.

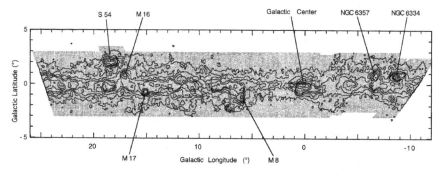

Fig. 3.46 A map of the $\lambda = 158$ μm fine-structure line of C^+ obtained by the Balloon-borne Infrared Carbon Explorer (BICE) with a resolution of 15′. The contour levels are 0.3, 0.6, 1, 1.5, 2, 3, 4, 5, 6, and 9×10^{-4} ergs s^{-1} cm^{-2} sr^{-1}. From Nakagawa et al. (1998)

Fig. 3.47 (a) The line intensity profile of the $\lambda =$ 158 μm fine-structure line of C^+ across NGC6334. The angular resolution is $4'$. (b) The 5-GHz radio continuum emission from the vicinity of the spectral scan. From Shibai et al. (1991)

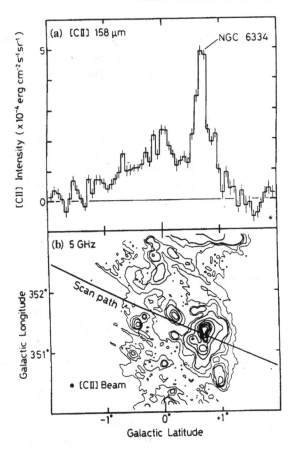

If one were to remove the $\lambda = 158$ μm emission associated with NGC6334, a strong extended component of C^+ emission from the Galactic plane over $b \approx \pm 1°$ would remain. Such an extended component of C II emission would agree with the observations of the low-frequency carbon RRL data toward this region. Unlike the carbon RRLs whose emission is confined to $\ell < 20°$, the $\lambda = 158$ μm emission extends to greater longitudes, from $\ell = 30°$ to $\ell = 51°$ at $|b| = 1 \rightarrow 2°$ (Shibai, Okuda, Nakagawa, Matsuhara, Maihara, Mizutani, Kobayashi, Hiromoto, Nishimura and Low, 1991). On the other hand, the C^+ emission in this region is weak at these extended longitudes, which might explain the uncertainty or failure of the searches for the corresponding carbon RRLs at $\ell > 20°$ along the Galactic plane. For example, Makuti et al. (1996) measured a differential of $I_{158} = (2 \rightarrow 3) \times 10^{-5}\,\mathrm{erg\,s^{-1}\,cm^{-2}\,sr^{-1}}$ in this region, which is a few times less than the brightness observed in directions closer to the Galactic center.

Figure 3.48 shows the distribution of the C^+ $\lambda = 158$ μm emission along the Galactic plane with respect to other kinds of emission. The C^+ observa-

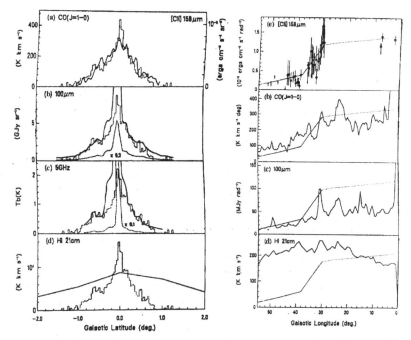

Fig. 3.48 The intensity distribution of C^+ $\lambda = 158\,\mu m$ emission in Galactic plane compared with other types of radiation. *Left*: (**a**) The C^+ profile across W43 superimposed on the latitudinal profile of the integrated ^{12}CO ($J = 1 \rightarrow 0$) intensity. (**b**) The C^+ scan superimposed on the $\lambda = 100\,\mu m$ continuum emission observed by IRAS. (**c**) The same scan superimposed on the 5-GHz continuum emission. (**d**) The same scan superimposed on the integrated $\lambda = 21\,cm$ emission of H I. *Right*: (**a**) The longitude distribution of the integrated C^+ emission. The *solid and dashed lines* represent a best-fit linear longitudinal profile, where the *dashed line* indicates interpolation between the data obtained with BIRT and the data obtained with the NASA Lear Jet (Stacey et al., 1985). (**b**) The same data plotted with the ^{12}CO distribution. (**c**) The same data and the $\lambda = 100\,\mu m$ continuum emission observed by IRAS. (**d**) The same data and $\lambda = 21\,cm$ emission of H I. From Shibai et al. (1991)

tions, made by BIRT, show the line intensity along a cut passing through the region of the H II region W43 ($\ell = 30^\circ.8$). Latitude profiles of CO, of H I and of the $\lambda = 100\,\mu m$ and 5-GHz continuum emission are superimposed upon the C^+ emission for comparison. Longitude profiles of the same emission are shown on the right-hand side of the figure.

Both similarities and dissimilarities are striking. The latitude profiles of the C^+ and CO emission almost coincide except for the narrow peak of the $\lambda = 158\,\mu m$ emission at $b = 0^\circ$ toward the W43 source. This agreement contrasts sharply with the large discrepancy between the C^+ and H I line profiles. The latitude profile of the carbon emission is much narrower than the H I profile. The longitude profile exhibits a sharp decrease in intensity at $\ell > 30^\circ$ that is not seen in the corresponding H I profile.

These observational data – the C^+ and CO emission – testify to the spatial association of the C II regions and the molecular clouds. On this basis, one might conclude that the PDRs form on the surfaces of the molecular clouds. The ionizing agent must be the interstellar background UV radiation that, in turn, would cause the $\lambda = 158\,\mu m$ emission from singly ionized carbon atoms. Note that these circumstances may be very different from the C II emission observed from the PDRs surrounding discrete H II regions. For one thing, the angular extent of the carbon emission is more diffuse than one would expect from the distribution of the discrete nebulae (Shibai et al., 1991). Here, the ionizing UV radiation may leak from within the H II regions.

Heiles (1994) offered an alternative explanation for the origin of the diffuse carbon emission in the inner part of the Galaxy. He suggested that the ELDWIM, which contains the major part of Galactic ionized hydrogen, may be connected with the $\lambda = 158\,\mu m$ line emission (see Sect. 3.2.2). The efficiency of excitation of the upper level of the carbon fine-structure line is higher in H II regions than in H I regions. The collision cross section for electron excitation of the $C^+\ ^2P_{3/2}$ level is larger than that for hydrogen atoms. That is why the $\lambda = 158\,\mu m$ brightness per nucleon is higher in H II regions than in the H I gas (Spitzer, 1978). As a result, even though the total mass of H II is much less than H I in the ISM and, moreover, the main part of the C^+ gas lies in the neutral component of the ISM, most of the carbon $\lambda = 158\,\mu m$ emission comes from regions of ionized hydrogen.

Support for this suggestion lies in the excellent correlation between the $\lambda = 158\,\mu m$ emission and the 5-GHz continuum shown in Fig. 3.48 on the left. However, the spatial resolution of the carbon line is insufficient for examination of the detailed correspondence between the C II regions and ELDWIM.

And so, from these data, we have two possible models for the origin of the C^+ fine-structure emission. In the first one, the source of the $\lambda = 158\,\mu m$ emission is a cold, neutral medium where envelopes of atomic carbon surround molecular clouds, where they are subsequently ionized by the ambient UV background radiation. In the second model, the $\lambda = 158\,\mu m$ emission originates in a warm, ionized medium; i.e., in the extended, low-density H II gas known as ELDWIM.

Can observations of carbon RRLs contribute to the viability of either of these models? These lines occur in cold regions only. The absence of Stark broadening in the carbon RRLs requires a low electron density that would be characteristic of neutral hydrogen regions. The intensity of the carbon RRLs is strongly dependent upon transitions stimulated by the background radiation field and has complex dependence upon principal quantum number. High-frequency (low quantum number) lines are observed in emission and low-frequency (high quantum number) lines are observed in absorption as described by Sects. 2.4.1 and 2.4.2. This dependence upon principal quantum number was observed for carbon RRLs detected toward Cas A. From observations of these RRLs at 328, 75, and 34.5 MHz, Kantharia and Ananthara-maiah (2000) found the "Cas A" C II regions to have temperatures ranging from 20 to 300 K.

The "warm" model in which carbon lines form in H II regions is a different situation. At the approximately $3 \, cm^{-3}$ electron densities of ELDWIM, the low-frequency carbon lines would exhibit strong Stark broadening. At these densities, (2.74) gives a line width of $\Delta V_L = 435 \, km \, s^{-1}$ for $n \approx 565$, which is an order of magnitude larger than 20–$54 \, km \, s^{-1}$ observed (Kantharia and Anantharamaiah, 2000) for the widths of Galactic carbon lines.

The line intensities of the "warm" model would also conflict with observations. Because of the recombination nature of the populations of the highly excited levels of carbon, the optical depth of these lines decreases sharply with temperature; i.e., τ varies approximately as $T_e^{-2.5}$. At conditions typical for ELDWIM where $T_e \approx 7,000 \, K$, $N_e \approx 3 \, cm^{-3}$, $N_{C^+} \approx 10^{-3} \, cm^{-3}$, and a path length of approximately $100 \, pc$, the optical depth of carbon RRLs would be determined primarily by spontaneous transitions and would be very small. For example, at a frequency of $75 \, MHz$, the LTE line optical depth τ_L^* would be approximately 10^{-5} for the carbon lines and maser effects would not be important. The carbon line would be observed in emission with a brightness temperature $(T_e \tau_L^*)$ less than $0.1 \, K$. Yet, observations at $75 \, MHz$ show the C441α RRL to be in absorption with an intensity of approximately $10 \, K$ (Erickson et al., 1995).

Because the spatial association between the C II $\lambda = 158 \, \mu m$ emission and the carbon RRL has not been clearly established, one might suppose that these lines are formed in different regions. The carbon RRL might be formed in the H I part of the ISM, whereas the $\lambda = 158 \, \mu m$ carbon line might be formed in H II regions.

Mochizuki and Nakagawa (2000) revisited the question of the origin of PDRs in molecular clouds under the influence of the UV background radiation. The earlier analysis (Tielens and Hollenbach, 1985) considered only plane-parallel geometry, whereas the new analysis considered a spherical model of a molecular cloud located within an isotropic UV radiation field. In addition to the UV radiation ($6 \, eV < h\nu < 13.6 \, eV$), this model also considered the IR radiation ($0.15 \, eV < h\nu < 6 \, eV$) associated with the interstellar dust.

The spherical geometry revealed that a much weaker UV radiation field could account for C II component of the PDRs. New calculations showed that a UV flux density which was an order of magnitude less than that of the plane-parallel model could account for the $\lambda = 158 \, \mu m$ carbon line emission in the Galaxy. This removed the problem in recognizing the PDRs as the fundamental source of this emission. If the Galactic UV flux density[21] $G_0 \approx 30$ – only a few times higher than believed earlier – then the observations of the fine-structure line would be consistent with the theoretical model.

In addition to examining the radiative energies involved, Mochizuki and Nakagawa (2000) also considered one more important factor that pointed toward the "cold" model. They noted that the survey (Nakagawa, Yui, Doi,

---

[21] G_0 is the incident far-UV flux density relative to that of the solar neighborhood value of $1.6 \times 10^{-3} \, ergs \, cm^{-2} \, s^{-1}$.

Okuda, Shibai, Mochizuki, Nishimura and Low, 1998) of the C II $\lambda = 158\,\mu$m line emission spatially correlated well with observations of the IR continuum emission. Shibai et al. (1991) mentioned such a correlation, which can be seen in the left panel of Fig. 3.48.

Observations indicate that the majority of the IR continuum is emitted by the neutral rather than the ionized medium. Therefore, Mochizuki and Nakagawa (2000) concluded that the main part of the Galactic C II $\lambda = 158\,\mu$m is formed in the neutral ISM, in PDRs associated with molecular clouds and H I gas rather than with H II regions.

Moreover, the temperatures calculated for the C II regions within the modeled PDRs agreed with those derived from the carbon RRL observations. The spherical-model calculations indicated that the gas temperatures within the PDRs decreased from 39 K at the outer surface of the C^+ envelope to as low as 20 K at the border of the PDR and the spherical, inner molecular cloud. This temperature range generally agrees with those measured from carbon RRLs emitted by C II regions; i.e., they fall within the range of 20–300 K derived from the carbon recombination lines.

The entire complex of data on both types of carbon lines indicates that the C II regions that emit or absorb carbon RRLs *and* simultaneously emit the $\lambda = 158\,\mu$m fine-structure line lie in the inner region of the Galaxy within $|b| \approx 1°$ of the Galactic plane. At the angular resolution of the observations, these regions are unresolved and contour maps of their emission indicate the presence of a spatially diffuse component. The source of the carbon ionization in these clouds is the interstellar UV radiation with an energy range of $6\,\text{eV} < h\nu < 13.6\,\text{eV}$. Under its influence, carbon is ionized in atomic clouds and on the surfaces of molecular clouds in PDRs. In both cases, most of the carbon RRL and fine-structure radiation comes from the atomic medium but some also comes from isolated clouds of H I as well as from PDR envelopes around molecular clouds. It is worth noting that H I envelopes, with thicknesses from 0.5 to a few pc, were detected with $\lambda = 21$ cm and CO emission from a number of specific molecular clouds. In particular, these included L1599, S255, Per OB2, Mon OB1 (Wannier, Lichten and Morris, 1983), and L134 (van der Werf, Goss and Vanden Bout, 1988).

The dependence of the line intensities on electron density determines where the emission is found. Carbon RRLs are proportional to N_e^2. They are more apt to be detected toward dense atomic clouds or near the surface of the PDRs surrounding molecular clouds. On the other hand, the C II $\lambda = 158\,\mu$m fine-structure line is proportional to N_e and may be expected to be found over a wider area of the ISM where atomic density can be lower.

The location and morphology of the discrete C II regions in the Galaxy is a good subject for future research. The most reliable and detailed data for such research might be obtained from the superposition of observations of carbon RRLs and of the fine-structure line.

3.3.6 Estimates of the Galactic Cosmic Ray Intensity

In the cold ISM, soft cosmic rays (CR) partially ionize the neutral hydrogen. Protons from Type 1 supernovae[22] spread through the Galaxy with typical energies of about 2 MeV, corresponding to their velocities of up to 20,000 km s^{-1} (Spitzer and Tomasko, 1968). The ionization cross section of hydrogen by protons, σ_{ion}, increases with decreasing energy. In the energy range of 1–10 MeV, $\sigma_{ion} \propto 1/E$. At 2 MeV, $\sigma_{ion} \approx 10^{-17}$ cm^{-2} (Fowler, Reeves and Silk, 1970).

These 2-MeV cosmic rays can penetrate deep into the ISM. They are absorbed by ionization losses only when the hydrogen column density along their trajectory reaches 9×10^{20} cm^{-2} (Bochkarev, 1988). Because of this penetration, it was once proposed that they caused an ionization rate of $\zeta_H = 1.17 \times 10^{-15}$ s^{-1} in the ISM, which was sufficient to account for the dispersion of pulsar signals outside of H II regions. However, further study did not confirm this hypothesis (see Sect. 3.2.2).

It became apparent that RRLs could provide measurements of the hydrogen ionization rate in the ISM and, correspondingly, the intensities of cosmic rays in the Galaxy. The ionization equilibrium can be described to be

$$\zeta_H N_H = \alpha^{(2)} N_e N_{H^+},\tag{3.26}$$

where N_e, N_H and N_{H^+} are the volume densities of electrons, neutral atoms and ions of hydrogen, respectively. The hydrogen recombination coefficient $\alpha^{(2)} = 2.06 \times 10^{-11}\Phi_2/T^{1/2}$ and includes recombination to all levels except the first one. The factor Φ_2 has weak dependence on the temperature and is tabulated by Spitzer (1978).

Combining (3.26) with the definition of the optical depth of an RRL, Shaver (1976a) derived a simple expression for the hydrogen ionization rate:

$$\zeta_H = 5.7 \times 10^{-14}\, \Phi_2 \left(\frac{\tau_H}{\tau_{HI}}\right) \cdot \left(\frac{\nu_{m,n}}{100\,\text{MHz}}\right) \cdot \left(\frac{T_e}{T_s}\right) \cdot \frac{T_e}{(b_n\beta)_H},\tag{3.27}$$

where τ_H and τ_{HI} are the optical depths of the hydrogen RRL and $\lambda = 21$ cm line, respectively; T_s is the hydrogen spin temperature in K; and $\nu_{m,n}$ is the frequency of the hydrogen RRL in MHz emitted from the transition from the upper principal quantum number m to n. As usual, the factor $(b_n\beta)_H$ corrects for deviations of the hydrogen-level populations from LTE and for the contribution of stimulated emission to the line intensity.

---

[22] Supernovae are the cataclysmic explosions of massive stars that occur with a frequency of about 1 every 30 years in large spiral galaxies like our Milky Way. Type 1 refers to those whose optical spectra show no hydrogen emission lines during the "maximum light" phase of their light curves. In contrast, the spectra of Type 2 supernovae include broad, intense hydrogen emission lines. There are also subclassifications. See Reynolds (1988).

The results were astonishing. Using (3.27) and the upper limit for the optical depth of the H352α line observed toward Cas A of $\tau_H \leq 1.7 \times 10^{-4}$, Shaver et al. (1976) calculated a hydrogen ionization rate of $\zeta_H < 3.3 \times 10^{-17}\,\mathrm{s}^{-1}$. They used a value of $\tau_{HI} > 5$ for the optical depth of the $\lambda = 21\,\mathrm{cm}$ line (Clark, 1965). The departure coefficients were calculated by Shaver (1975) for $T_e = 50\,\mathrm{K}$ and $N_e = 0.05\,\mathrm{cm}^{-3}$, values accepted for the ISM toward Cas A at that time. Later, a similar value of $\zeta_H < 2 \times 10^{-17}\,\mathrm{s}^{-1}$ was obtained by the same method but for the H300α line (Casse and Shaver, 1977). These limits were about two orders of magnitude smaller than estimated above from the soft cosmic rays.

To refine these results, it seemed worthwhile to re-examine the assumptions of average temperature and density of the ISM in the direction of Cas A. These parameters are essential to calculate appropriate values of b_n and β, which are factors in (3.27) used to determine ζ_H. Trying different values of temperature and density, Payne et al. (1989) analyzed new observations of the H308α lines toward Cas A to obtain $\zeta_H < 3.9 \times 10^{-17}\,\mathrm{s}^{-1}$ to $\zeta_H < 2.6 \times 10^{-16}\,\mathrm{s}^{-1}$. The range of these calculated results is about an order of magnitude, much larger than desired.

Fortunately, this range of uncertainty for ζ_H can be reduced further by using another method of analysis proposed by Sorochenko and Smirnov (1987) based upon simultaneous observations of low-frequency carbon and hydrogen RRLs in cold clouds within the ISM.

Consider the model of the C II shells surrounding molecular clouds that was discussed in Sect. 3.3.5. Interstellar UV background radiation with $\lambda > 912\,\text{Å}$ ionizes the carbon in these shells while keeping the hydrogen largely neutral. However, some of the hydrogen in these shells must be ionized by soft interstellar cosmic rays. As noted above, the penetration of 2-MeV cosmic rays through the ISM corresponds to a column density of about $9 \times 10^{-20}\,\mathrm{cm}^{-2}$, which corresponds approximately to the thickness of the C II–H I shells surrounding the molecular clouds. The intensity ratio of the hydrogen and carbon RRLs will facilitate the determination of ζ_H with a smaller dependence upon the local temperature and local electron volume density.

First, we determine the ionization ratio of hydrogen in the shells of the molecular clouds. If all carbon in the shells is ionized, $N_{C+} = N_C$. Using (2.116), we factor this ratio as

$$\frac{N_{H+}}{N_H} = \frac{N_C}{N_H}\frac{N_{H+}}{N_{C+}} = 3.3 \times 10^{-4} \left(\frac{\tau_H}{\tau_C}\right)_{LTE} \frac{\phi_C(\nu)}{\phi_H(\nu)}, \qquad (3.28)$$

where τ_H and τ_C are the LTE optical depths, and $\phi_H(\nu)$ and $\phi_C(\nu)$ are the RRL profiles for hydrogen and carbon, respectively.

Second, we determine the ratio of the line optical depths required by (3.28). In C II regions, the populations of the excited levels are not in LTE and the line optical depths have to be corrected accordingly. Using (2.132), we write

$$\left(\frac{\tau_H}{\tau_C}\right)_{LTE} = \left(\frac{\tau_H}{\tau_C}\right)_{obs} \frac{(b_n\beta)_C}{(b_n\beta)_H}. \tag{3.29}$$

Determining the remaining factor in (3.28) is easy. At low frequencies, the line widths are determined by Stark broadening rather than by the atomic mass of the emitters. For this reason, the line profiles $\phi(\nu)$ are identical for both hydrogen and carbon lines and $\phi_C(\nu)/\phi_H(\nu) = 1$.

Substituting these into (3.26), Sorochenko and Smirnov (1990) obtained

$$\zeta_H = \frac{6.8 \times 10^{-15} \Phi_2 N_e}{T^{0.5}} \left(\frac{\tau_H}{\tau_C}\right)_{obs} \frac{(b_n\beta)_C}{(b_n\beta)_H}, \tag{3.30}$$

where the values of optical depths may be both positive and negative. At low frequencies, these RRLs can be observed in absorption and emission.

We can now calculate the ionization rate. Observations of the C486α line toward Cas A give $(\tau_C)_{obs} = (2.35 \pm 0.22) \times 10^{-3}$; and, of the H486$\alpha$ line, $(\tau_H)_{obs} < 6.3 \times 10^{-4}$ (Ershov et al., 1987). In the C II regions, $T_e = 50\,\mathrm{K}$ and $N_e = 0.15\,\mathrm{cm}^{-3}$. For these conditions, $(b_n\beta)_H = -1.05$, $(b_n\beta)_C = 5.3$ (Ponomarev and Sorochenko, 1992), and $\Phi_2 = 3.6$. Substituting these values into (3.30) gives an upper limit of $\zeta_H = 7 \times 10^{-16}\,\mathrm{s}^{-1}$. This limit lies near the middle of the range found by Payne et al. (1989).

An even lower estimate resulted from a program of observing low-frequency RRLs especially designed to determine the interstellar hydrogen ionization rate by cosmic rays (Kitaev et al., 1994). This program used the DKR-1000 array telescope in Pushchino, configured into two parts to facilitate fast switching between Cas A and a reference region. The 42-MHz observing frequency allowed observations of both carbon and hydrogen recombination lines involving the lower principal quantum numbers 537, 538, 539, and 540. The total integration time was 1,224 h.

Figure 3.49 shows the average of the four recombination lines resulting from these 42-MHz observations of Cas A. The carbon line profile has two absorption components corresponding to velocity features from the Perseus and Orion spiral arms of the Galaxy. The Gaussian fit to the Perseus component gives $\tau_C = (-3.18 \pm 0.05) \times 10^{-3}$ and $\Delta V = 22.1 \pm 0.3\,\mathrm{km\,s}^{-1}$. On the figure is marked the expected position for the corresponding hydrogen RRL. Because the detection is unreliable, the optical depth of the hydrogen line can be only an upper limit, $\tau_H \leq 1.1 \times 10^{-4}$.

Substituting these parameters as well as values of $b_n\beta$ calculated for $T_e = 50\,\mathrm{K}$ and $N_e = 0.15\,\mathrm{cm}^{-3}$ gives an upper limit for the hydrogen ionization rate of $\zeta_H = 2.75 \times 10^{-16}\,\mathrm{s}^{-1}$ – very close to the lower limit of the range of rates found by Payne et al. (1989).

There is a convergence of the values found for the hydrogen ionization rate ζ_H from the two different techniques. If one accepts the temperatures and volume densities obtained from the carbon RRLs as valid for the calculations of appropriate departure coefficients, then the limits on the ionization rate obtained from the hydrogen optical depth ratios τ_H/τ_{HI} in (3.27) agree well

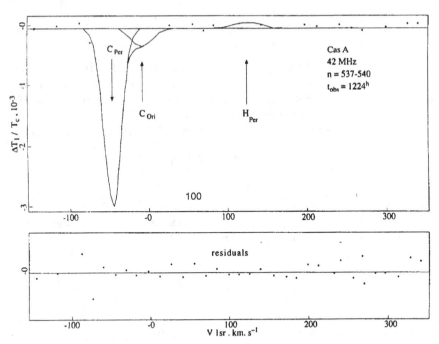

Fig. 3.49 The spectrum of Cas A at 42 MHz with a velocity resolution of $14\,\mathrm{km\,s^{-1}}$. This spectrum is the average of four recombination lines. The *ordinate* is the ratio of the line to continuum emission with negative numbers indicating absorption, and the *abscissa* is the velocity with respect to the LSR. The *lower plot* shows the residuals remaining after fitting the line components. The integration time is $1{,}224\,\mathrm{h}$. From Kitaev et al. (1994)

with the inherently more accurate result obtained from the observed ratio of τ_H/τ_C given by (3.30). The hydrogen-based limits of the ionization rates are then $\zeta_H < 4.5 \times 10^{-16}\,\mathrm{s^{-1}}$ from the H300α–$\lambda = 21$ cm combination, $< 3.8 \times 10^{-16}\,\mathrm{s^{-1}}$ from the H308α–$\lambda = 21$ cm combination, and $< 1.2 \times 10^{-16}\,\mathrm{s^{-1}}$ from the H352α–$\lambda = 21$ cm combination.

To date, the attempts to use RRLs to determine a fixed value for the hydrogen ionization rate in the cold ISM have not yet been successful. Observations have determined only an upper limit; i.e., $\zeta_H \approx 3 \times 10^{-16}\,\mathrm{s^{-1}}$ or less.

It is interesting to compare this limit to the ionization rates obtained from observations of interstellar molecules. Many of these molecules form from reactions involving hydrogen ions either as intermediate or as fundamental building blocks. For example, the dependence of the OH abundance on the hydrogen ionization rate enables independent determinations of ζ_H (Black and Dalgarno, 1973). Comparing theoretical calculations with observations of OH abundances in the diffuse molecular clouds surrounding ζ Per, ζ Oph, and o Per, van Dishoeck and Black (1986) found the value of the hydrogen ionization rate, $\zeta_H = (7 \pm 3) \times 10^{-17}\,\mathrm{s^{-1}}$.

This value of ζ_H is within a factor of 5 of the upper limit obtained from RRLs detected from the PDR shells surrounding dense molecular clouds. Someday, the sensitivity of radio telescopes will increase sufficiently to detect hydrogen RRLs from the general ISM and, thereby, determine a value of ζ_H that can be directly compared to the one found through interstellar chemistry. Comparison of the two results should then pin down this parameter that is essential to understanding the nature of the ISM within the Galaxy.

3.4 RRLs from Stars and Stellar Envelopes

RRLs offer physical insight to stars and their environments, in addition to their contributions regarding the nature of H II and C II regions and the broad interstellar medium. In particular, the stellar topics include the hot stellar envelopes known as planetary nebulae, our own Sun, and a peculiar and as of this writing a unique stellar system known as MWC349.

3.4.1 Planetary Nebulae

As the name implies, planetary nebulae are ionized regions immediately surrounding individual stars. There are no planets, however. Typically, their masses range from approximately 1 to 8 $M_\odot$, including the mass of their central star. As a class, they are much smaller than the discrete H II regions and molecular clouds that have been discussed earlier in this book. These masses range from a few to more than 10,000 $M_\odot$, depending upon where one chooses to draw the line between entity and association.

The shapes, sizes, color temperatures, and luminosities of planetary nebulae are varied. Despite this heterogeneity, most are easily recognizable as belonging to a distinct astronomical class. The hallmark is a void separating a thin nebular halo(s) from a centrally located bright star. In general, the "typical" morphology is that of a spherical (annular) envelope or shell surrounding a star, separated by a visual void between the star and the envelope. However, multiple concentric rings or shells can often be seen, as well as noncircular shapes like dumbbells. They are believed to represent a late stage in stellar evolution, connecting the red giant branch of the Hertzsprung–Russell diagram with the white dwarf branch by an unobserved process in which an older, evolved star separates into a hot ionized envelope and a core which, in turn, quickly becomes a white dwarf.

In studying them, the most important parameter is their distance (see the monographs by Pottasch (1984) and Kwok (1999)). It not only locates them with respect to other constituents of the Galaxy, but also allows conversion of their observed intensities into physically useful units at the nebulae.

Unfortunately, this is difficult to obtain. There is no direct kinematical method of determining their distances accurately. Trigonometric parallax does not work because, in general, the planetary nebulae are too far from Earth. Spectroscopic parallax does not work because the emission spectra of their central stars are so different that it is impossible to establish a standard "candle" for calibration. As a result, distances listed for planetary nebulae come from a variety of techniques, including interstellar extinction, angular expansion rates of their shells, associations with nearby normal stars, etc. The most commonly used method combined spectral estimates of electron density with a presumption that all planetaries have about the same mass – the "Shklovsky" method (Shklovsky, 1956a) – to allow angular sizes to be converted into distances. This turned out to be a flawed assumption but the method with refinements was used for more than 25 years. Consequently, one must consider quoted distances as the best available but having significant uncertainties.

Spatially, planetary nebulae are distributed within a sphere surrounding the Galaxy rather than confined to the Galactic plane. Because of this, they are often called Population II objects to distinguish them from the younger, earlier-type stars (Population I) confined mainly to the Galactic plane. Based upon a convergence of theory and observation, the stars associated with planetary nebulae are considered to mark an early phase of the development of a galaxy.

The physical characteristics of planetary nebulae vary widely. Figure 3.50 shows the continuum spectra for three compact planetary nebulae. The rightmost part of the spectra arises from heated dust in the envelope with a temperature of about one or two hundreds of Kelvins, and the leftmost arises from free–free emission from hot electrons of roughly 10,000 K in the same general location. Note the wide variation in the turnover frequency, from 2 to 200 GHz, which indicates a correspondingly wide range of large emission measures (see (2.95) and (2.105)).

Because the envelopes are composed of hot, tenuous ionized gas, they emit also recombination lines in the radio wavelength regime. However, the angular sizes of the ionized shells are small and, consequently, the RRLs are weak and difficult to detect with single element radio telescopes. For example, Terzian (1990) wrote "Although the number of identified planetary nebulae in the Galaxy is more than one thousand, only about ten have shown detectable radio recombination lines." At this writing, the number of catalogued planetary nebulae has grown to about 1,500 (Terzian, 2002) and the RRL detections have grown proportionally.

Table 3.8 lists most of the planetary nebulae for which RRLs have been detected up to the year 2001. The table includes full line widths at half-intensity (ΔV), LTE electron temperatures (T_e^*), non-LTE electron temperatures (T_e) where available, and the derived emission measures (EM).

Although not all RRL observations are listed, the listings are sufficient to identify distinctive characteristics of planetary nebulae. First, the derived

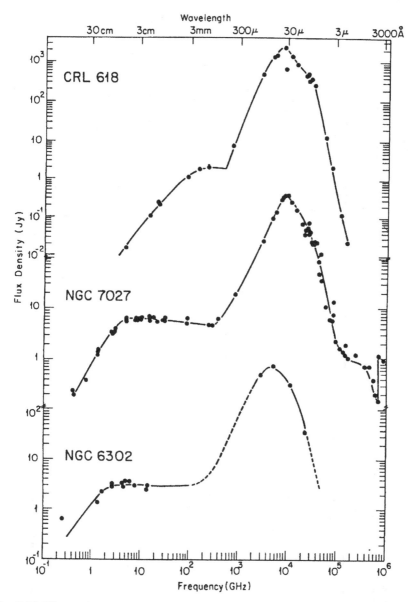

Fig. 3.50 The continuum spectra of the compact planetary nebulae CRL618, NGC7027, and NGC302. The lines have been fitted to the observations. *Broken segments* indicate uncertainty. From Terzian (1990)

LTE temperatures tend to be significantly greater than those observed for discrete H II regions. These temperatures agree well with those determined optically (Terzian, 1990). Second, the emission measures are often one or two orders of magnitude larger than those derived for most discrete H II regions.

Table 3.8 Some planetary nebulae with detected RRLs

Nebula	Dist. (kpc)	Diam. (″)	RRL	ΔV (km s^{-1})	T_e^* (K)	T_e (K)	EM (cm^{-6} pc)	References
IC418	1.8	12	H76α	26±2	8,400	9,800	2×10^7	Walmsley et al. (1981)
			H76α	27±1	9,600	9,500	6.0×10^6	Garay et al. (1989)
CRL618	1.7	≈ 0.7	H41α	30±9	17,800		$4 - 10 \times 10^{10}$	Martín-Pintado et al. (1988)
			H35α	32±1	12,400		$4 - 10 \times 10^{10}$	Martín-Pintado et al. (1988)
			H30α	48±8	13,500		$4 - 10 \times 10^{10}$	Martín-Pintado et al. (1988)
NGC1514	0.2	100	H140α	43±1	(14,000)		$4 - 10 \times 10^{10}$	Terzian (1990)
NGC2440	1.	11	H92α	60±5	16,000		8×10^5	Vázquez et al. (1999)
NGC6302	2.2	25	H76α	40±3	18,000		(7.5×10^7)	Gómez et al. (1989)
			H110α	56±2	21,000		(7.5×10^7)	Gómez et al. (1989)
NGC6369	2.0	12	H76α	43±6	12,100		1.6×10^6	Terzian et al. (1974)
NGC6572	1.2	12	H76α	32±3	12,900	15,000	3×10^7	Walmsley et al. (1981) and Kawamura and Masson (1996)
BD30+3639	1.5	11	H76α	50±5	5,900	7,600	3×10^7	Walmsley et al. (1981) and Kawamura and Masson (1996)
NGC7009	0.1	29	H76α	44±6	9,700	12,000	1.6×10^6	Garay et al. (1989)
NGC7027	1.0	14	H76α	48±2	11,700	14,000	7×10^7	Walmsley et al. (1981)
M1-78	5.0	12	H76α	37±5	16,200	17,500	5×10^7	Walmsley et al. (1981)
			H85α	54±12	12,800			Terzian et al. (1974)
			H109α	65±12	8,200		2.9×10^7	Churchwell et al. (1976)
IC5117	3	1.2	H92α	49	11,000		1×10^8	Miranda et al. (1995)
NGC7662	1.7	26	H76α	27±5	(19,000)	(19,000)	4×10^6	Walmsley et al. (1981)
Hb12	2.1	1	H92α	35	14,000		3×10^8	Miranda et al. (1995)

Note: Parentheses indicate approximate results. This table may not list all detections

Such large values of EM also imply high electron densities and, in fact, Stark broadening has been detected for the low-frequency RRLs from planetary nebulae. Walmsley et al. (1981) analyzed a series of recombination lines observed from NCC7027. While the width data had smaller signal-to-noise ratios than desired, the corresponding variation of the T_L/T_C ratios with frequency also implied a systematically increasing line width as a function of principal quantum number. On this basis, Walmsley et al. cautiously implied the presence of Stark broadening consistent with $N_e \approx 10^5 \, \text{cm}^{-3}$ for NGC7027. More recently, Ershov and Berulis (1989) found an electron density of $(6.7 \pm 0.5) \times 10^4 \, \text{cm}^{-3}$ for NGC7027 using six RRLs from H56α to H110α. VLA observations of IC418 showed the width of the H110α (5 GHz) line to be significantly larger than that of the H76α (15 GHz) line (Garay, Gathier and Rodríguez, 1989). These widths implied $N_e = 1.8 \times 10^4 \, \text{cm}^{-3}$ when assumed to involve Stark broadening.

The presence of Stark broadening has been used to determine the distance for at least three planetary nebulae. VLA observations of the H76α and H110α lines from NGC6302 also showed dissimilar line widths likely due to the presence of Stark broadening (Gómez, Moran, Rodríguez and Garay, 1989). Determining the Doppler profile and electron temperature from the H76α profile, Gómez et al. fitted a Voigt profile to the H110α line and derived an increase in the width of $20 \pm 4 \, \text{km s}^{-1}$ due to electron impacts. This width implied a mean electron density for the planetary nebula of $(2.5 \pm 0.5) \times 10^4 \, \text{cm}^{-3}$. Further, it agreed well with the density of $2 \times 10^4 \, \text{cm}^{-3}$ derived for the center of the nebula from optical data (Meaburn and Walsh, 1980). Using this density and temperature, Gómez et al. adjusted the assumed distance to the planetary nebula until the model flux density matched the observations, thereby determining a distance to NGC6302 of $2.2 \pm 1.1 \, \text{kpc}$. Applying this technique to similar observations resulted in distances of 1.0 and 0.2 kpc for NGC7027 and IC418, respectively, with comparably large uncertainties.

While none of these distances have as high precision as might be desired, nevertheless the technique points the way to a way of determining distances for planetary nebulae from RRLs. What is needed are higher accuracy line profiles to better establish N_e from Stark broadening as well as sophisticated models with which to calculate the radio flux densities.

One especially interesting way of using RRLs to determine the characteristics of planetary nebulae is to consider the effects of optical depth upon the observed radial velocities. Ershov and Berulis (1989) noted that the radial velocities of RRLs emitted by NGC7027 should vary as a function of frequency. The idea is that the line opacity would vary inversely with frequency and, because of the expansion gradient of the planetary nebula, would allow the expansion velocity to be determined from observations of RRLs over a large range of principal quantum numbers. The dependence of the velocity variation would also tell whether the nebula was expanding or contracting. An increase of radial velocity with frequency would imply expansion. Combining observations of the H56α, H66α, H76α, H85α, H90/94α, and H110α RRLs, Ershov and Berulis fitted a model to NGC7027 with $N_e \approx 6 \times 10^4 \, \text{cm}^{-3}$, $T_e \approx 15,000 \, \text{K}$, $V_{LSR} \approx 26 \, \text{km s}^{-1}$, and $V_{Exp} \approx 21 \, \text{km s}^{-1}$. Combining this linear expansion rate with the measured angular expansion of $0\rlap{.}''0047 \, \text{yr}^{-1}$ gives the distance to NGC7027 of 940 pc $\pm 20\%$ (Masson, 1986).[23]

At present, the parameters derived from RRL observations have shown excellent agreement with the physical conditions and distances determined by other techniques. This is a good foundation from which to grow as the sensitivities of radio telescopes increase. In particular, unlike optical lines, radio recombination lines are unaffected by interstellar extinction and would have an advantage for determining the physical characteristics of obscured planetary nebula lying in the Galactic plane.

3.4.2 The Sun

In principle, RRLs can be detected from the Sun. Dupree (1968) noted that highly stripped ions in the solar atmosphere have an overpopulation of their high quantum states. This overpopulation can lead to enhanced intensities – hopefully, detectable – of RRLs emitted by the Sun.

The n-dependencies of the rates describe the situation. According to Dupree, radiative recombination from the continuum is the principal process populating the upper principal quantum levels of hydrogenic atoms in the solar corona with a rate $\propto n^{-3}$. On the other hand, dielectronic recombination is the principal means of forming complex ions, with a rate $\propto n^{-1}$ for principal quantum numbers $20 \leq n \leq 200$. Because the depopulation rate by spontaneous emission is $\propto n^{-5}$, the level population of the complex ions in this range of quantum numbers can vary as n^4, giving a significant overpopulation of the principal quantum levels populated $\propto n^2$ in LTE. This gives a corresponding enhancement of the intensities of RRLs from the complex ions over the LTE values.

The initial calculations were made for Fe XV because of the availability of these rates for ordinary dielectronic recombination. Added to these were the contributions of cascade transitions between levels. Figure 3.51 shows the results for $T_e = 3 \times 10^6 \, \text{K}$ and $N_e = 10^8 \, \text{cm}^{-3}$. As the quantum levels increase beyond 10, the departure coefficients increase because of dielectronic recombination. Note that the coefficients can reach values of several hundred. When the ion becomes so large that collisional de-excitation becomes important, the departure coefficients begin to decrease sharply. The quantum number at

---

[23] The method of combining linear and angular expansion rates is a well-known method of determining distances to planetary nebula. In fact, Masson combined older determinations of the linear expansion velocity with new, precise VLA observations of the angular expansion rate. Later, Ershov and Berulis independently determined the linear expansion rate from RRL observations, which confirmed the assumed expansion rate adopted by Masson.

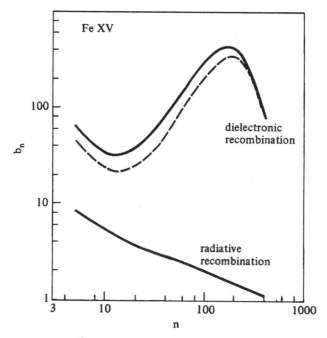

Fig. 3.51 The departure coefficients for principal quantum numbers of Fe XV in the solar corona. *Upper curves* include dielectronic calculation with (*solid line*) and without (*broken line*) cascades between levels. The *lower curve* results from populating only by radiative recombination from the continuum. From Dupree (1968)

which this inflection takes place depends upon the electron density. Higher densities move the inflection toward lower quantum numbers, because of the increased frequency of collisional de-excitation.

Similar calculations can be made for other ions in the Sun. Dupree considered recombination lines from O VI good possibilities for detection in the Sun because of its abundance. Using the departure coefficients calculated for sublevels of Fe XV, she found – very roughly – that the central intensity of its 100α recombination line ($\lambda = 1.28$ mm or 235 GHz) could occur in emission with an intensity of about 2% of the background continuum at the center of the solar disk. On the other hand, the 200α line ($\lambda = 10.2$ mm or 29 GHz) could occur in absorption because of the negative gradient of the departure coefficients of the principal quantum numbers. For large quantum numbers like this one, Stark broadening could smear out the line profile, rendering the RRL unobservable.

Such calculations were quite approximate because, except possibly for the less abundant Fe XV ion, many details regarding the population rates for ions were not known. Nevertheless, the possibility of a new tool for the study of the solar atmosphere generated a great deal of interest.

The initial searches proved disappointing. Dravskikh and Dravskikh (1969) looked unsuccessfully for solar RRLs near $\lambda = 6\,\text{cm}$ (5 GHz). Berger and Simon (1972) tried to detect them at frequencies from 85 to 92 GHz for ions from $Z = 1 \rightarrow 15$ with the 36-ft. (11-m) telescope of the National Radio Astronomy Observatory on Kitt Peak, AZ. No lines were detected. Revisiting the theory, they recalculated the departure coefficients as a function of principal quantum number and, based partly on relative abundance, estimated that solar RRLs from C III, O IV, O V, Ne VII, and Si XI were good candidates for detection. Furthermore, barring unexpectedly large Stark broadening, they showed that the best wavelength range for the searches would be $3\,\text{mm} < \lambda < 3\,\text{cm}$ based upon the calculated line intensities.

Additional searches also failed to detect RRLs from the Sun. Shimabukuro and Wilson (1973) searched at 110–115 GHz for lines from ions with $Z = 3$–14, more or less following the recommendations of Berger and Simon (1972) to avoid principal quantum numbers that might be subject to Stark broadening. The lines just were not there.

It was again time to scrutinize the theory. Possibly, Zeeman splitting from intense solar magnetic fields could broaden the lines, thereby reducing their intensities (Greve, 1975). Finally, carefully reconsidering the details of radiation transfer and including new estimates of Stark broadening, Greve (1977) calculated that the previous searches for solar recombination lines were at least a factor of 10^2–10^3 too insensitive to detect the lines because of pressure broadening by electrons. He suggested that observers would need to establish a limit of at least 10^{-5} for the line to continuum ratio to detect solar RRLs in the millimeter wavelength range. In other words, the previous observing limits were insufficient for detection by a very large factor.

Because of the difficulties in detecting RRLs from coronal ions, Khersonskii and Varshalovich (1980) examined the detection prospects for H$n\alpha$ lines for $n = 24$–36 (1,747–135 GHz or $\lambda = 0.63$–2.13 mm) originating in the solar chromosphere. In this range of n, the collision rates would determine the level populations such that the line intensities would be near LTE and, as such, might be only marginally detectable. Within the range considered, they suggested that the best candidates would lie near H24α where $T_L/T_C \approx 10^{-2}$, with the intensities of higher-n lines sharply decreasing because of increasing Stark broadening. For example, they predicted $T_L/T_C \approx 10^{-4}$ for the H36α line. They concluded that the lower-n lines would be detectable and, in special circumstances such as a flare, might actually be enhanced.

Similarly, Hoang-Binh (1982) also looked at the possibility of detecting normal hydrogen RRLs from the solar chromosphere but in the far infrared and submillimeter wavelength regimes. His calculations indicated that the H$n\alpha$ RRLs in the range $5 \leq n \leq 20$ should be much stronger than RRLs from coronal ions, making them better candidates for detection. Like Khersonskii and Varshalovich, he found pressure broadening to be an important consideration. The high electron densities of the chromosphere, $N_e \approx 6 \times 10^{10}$

to $10^{11}\,cm^{-3}$, would manifest themselves in detectable pressure broadening even at the chromospheric temperatures of approximately 6,000 K.

These two wavelength ranges proved successful. Solar emission lines from ions were first detected in the mid-infrared (Murcray, Goldman, Murcray, Bradford, Murcray, Coffey and Mankin, 1981; Brault and Noyes, 1983) and later identified as hydrogenic ($n = 7$) transitions from Mg I, Al I, and possibly Ca I (Chang and Noyes, 1983). A year later, Chang (1984) identified the remaining solar lines as due to hydrogenic transitions from Si I. Since then, astronomers have observed the H15α (1,770 GHz or 59.0 cm^{-1}) and H13α (2,680 GHz or 89.4 cm^{-1}) (Boreiko and Clark, 1986), H19α (888 GHz or 29.6 cm^{-1}) (Clark, Naylor and Davis, 2000a), and H21α (662 GHz or 22.1 cm^{-1}) (Clark, Naylor and Davis, 2000b) lines in the solar chromosphere. Figure 3.52 shows probably the highest frequency *radio* recombination line detected in the Sun.

Although challenging to observe, these lines not only are useful for probing the thermodynamic conditions within the chromosphere, but also provide another cosmic laboratory for investigating Stark broadening and other physics associated with Rydberg atoms.

Although many of these solar recombination lines lie well outside the radio range, the search for them surely originated with the theoretical work of Dupree (1968), stimulated by the detection of centimeter wave RRLs from discrete H II regions about 20 years earlier. They begin a new avenue of astronomical research involving the solar chromosphere.

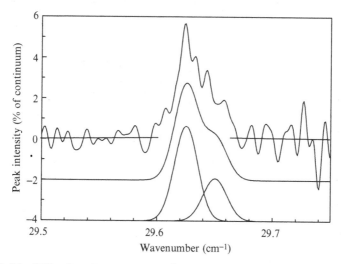

Fig. 3.52 The H19α (888 GHz or 29.6 cm^{-1}) radio recombination line detected in the solar limb with a polarizing interferometer and the James Clerk Maxwell radio telescope on Mauna Kea, HI. The two fitted profiles and their sum are shown below the spectrum. The weaker profile is probably the Mg19α line. From Clark et al. (2000a)

3.4.3 MWC349

One of the most intriguing sources of RRLs is MWC349. This is a binary system lying about 1.2 kpc from the Sun. One member, MWC349A, has been classified as a Be star (Merrill and Burwell, 1933); the other, MWC349B, a B0 III star. The optical and infrared spectra from MWC349A are complicated (Cohen, Bieging, Dreher and Welch, 1985). The luminosity of its Lyman continuum is about 10^{48} s^{-1}, comparable with that of some O stars. Many of its IR emission lines have double peaks as might be expected from a circumstellar disk viewed edge on. In the discussion below, we will refer to MWC349A as MWC349 for simplicity.

The search for RRLs from MWC349 was based upon its unusual nature. Its continuum spectra had long been known to be peculiar. Figure 3.53 shows the spectrum determined by combining radio and infrared observations (Harvey, Thronson and Gatley, 1979). First, note the $\lambda = 100\,\mu$m flux density is more than an order of magnitude lower than the peak emission, indicating that the source of the infrared emission – presumably dust – is extremely hot. Second, note the slowly varying flux density ($\propto \nu^{0.6}$) in the radio wavelength region, unlike the $\nu^{\approx 2}$ dependence seen in the optically thick region of spectra from standard discrete H II regions, such as those of Figs. 2.16 and 2.17. On the basis of its continuum spectrum, the accepted model for MWC349 was a hot star with a stellar wind of about 50 km s^{-1} (Olnon, 1975).

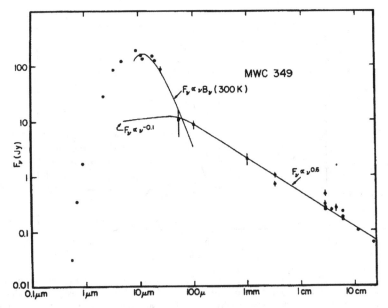

Fig. 3.53 The continuum spectrum of MWC349 fitted with a two-component model. Spectral flux density is plotted against wavelength. From Harvey et al. (1979)

Consequently, groups searching for RRLs from stars placed MWC349 high on their list of candidates. The source had the necessary characteristics: high-intensity radio emission for easy detectability, and indications of copious amounts of moderate-density ionized gas to generate RRLs with line widths less than the spectrometer bandwidths. And, in fact, Altenhoff et al. (1981) had been able to detect H76α (15 GHz) and H66α (23 GHz) RRLs from this system with the 100-m radio telescope at Effelsberg but they were very weak.

Using the 30-m IRAM telescope on Pico Veleta, Spain, Martín-Pintado et al. (1989) pressed the search for stellar RRLs to much higher frequencies. They were able to detect the H29α (256 GHz), H30α (232 GHz), H31α (211 GHz), and H41α (92 GHz) RRLs from MWC349 with high signal-to-noise ratios.

These new lines have unusual characteristics, which the more recent spectra shown in Fig. 3.54 illustrate. This figure compares the high-frequency line profiles from MWC349 with those from the normal H II region DR21. The higher-frequency lines of MWC349 have two principal components and the lower-frequency line has one. Below these, the DR21 spectra exhibit single Gaussian profiles with full widths at half-intensity of approximately $40 \, \mathrm{km \, s^{-1}}$. Of these widths, about half is due to thermal broadening from the 9×10^3 K gas and the remainder is due to microturbulence within the beamwidth (see (2.18), (2.22), and (2.26)). In contrast, the widths of each of the high-frequency MWC349 components are approximately $10 \, \mathrm{km \, s^{-1}}$– a situation one would expect from, say, a 2,000 K gas with a quarter of the microturbulence normally observed in H II regions. Furthermore, both components sit on a broad, weak pedestal of approximate width $50 \, \mathrm{km \, s^{-1}}$.

In addition, the T_L/T_C ratios differ from RRLs observed from normal H II regions. For example, those from the H II region DR21 increase with increasing frequency, just as expected (see Fig. 3.3). This behavior is very different for MWC349; the variation of T_L/T_C increases much more sharply with frequency for the high-frequency, double-peaked lines. In comparison, the lower-frequency H40α line appears almost normal.

These are the characteristics one would expect from masering RRLs. Accordingly, Martín-Pintado et al. (1989) suggested this explanation in the detection announcement. Subsequent observations of the intensity ratios of H$n\alpha$ to H$n\beta$ lines – interlocking lines – have confirmed that the level populations associated with the peculiar RRLs are far from LTE in a way expected for masering transitions (Gordon, 1994; Thum, Strelnitski, Martín-Pintado, Matthews and Smith, 1995).

Following the detection, observations revealed another interesting characteristics of the RRLs from MWC349: time variation of the line intensities and radial velocities. Figure 3.55 shows the intensity variations for the components of the H30α line over 3 years. The line areas are referenced to the continuum emission of the calibrator, DR21, because of its greater intensity and intensity stability. Note that there seems to be a decrease in the relative line emission of the pedestal component but the signal-to-noise ratio is

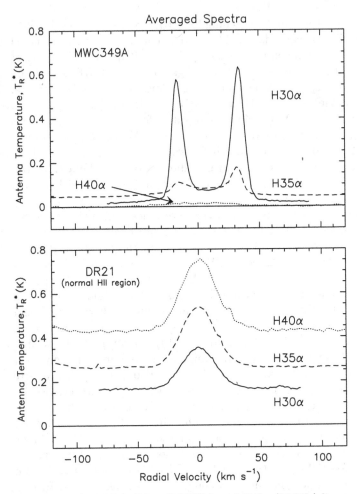

Fig. 3.54 *Top*: H30α (232 GHz), H35α (147 GHz), and H40α (99 GHz) line profiles averaged over 3 years of observations of MWC349. *Bottom*: the same lines from the normal H II region DR21. From Gordon et al. (2001)

marginal. The time variation of the red-shifted component is clearly significant in terms of the observational uncertainties. The intensity varies over a factor of 3 with no monotonic pattern. The same situation obtains for the blue-shifted component.

Figure 3.56 shows the time variations in the radial velocities of Gaussians fitted to the components of the H30α line. There are significant variations in the radial velocities over the 3-year observing period. It appears that the two red- and blue-shifted components move away each other and then toward each other, as can be seen in the two upper panels and in the lower panel. Meanwhile, the average of the two radial velocities – the "system" velocity – changes very little.

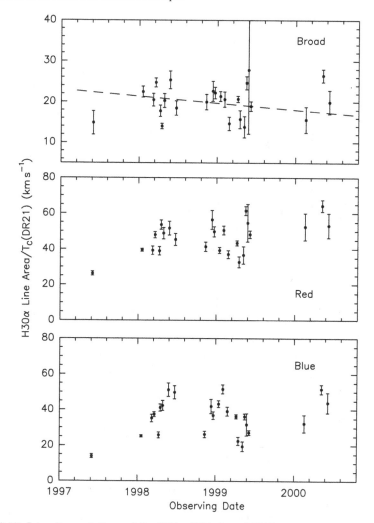

Fig. 3.55 Intensity variations of the H30α RRL from MWC349 over 3 years. *Top*: the area-to-(DR21)continuum ratio of the underlying pedestal line with a fitted regression line. *Middle*: same, for the red-shifted line component. *Bottom*: same, for the blue-shifted line component. From Gordon et al. (2001)

These characteristics all point to radial movements of the maser emission within a circumstellar disk viewed edge on (Gordon, 1992; Thum et al., 1992), the model originally suggested by Hamann and Simon (1986; 1988) to account for the two-component line profiles observed for IR and FIR lines from MWC349. The red-shifted component would be located on the side of the stellar disk moving away and the blue-shifted component would be located on the side moving toward us.

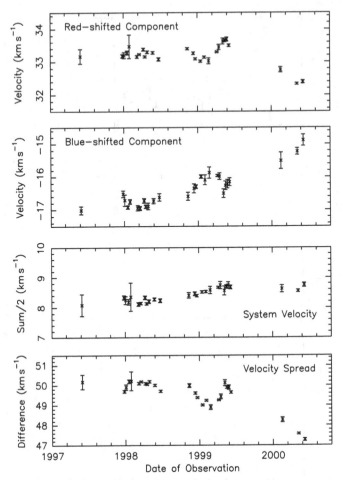

Fig. 3.56 Velocity variations of the H30α RRL from MWC349 over 3 years. *Top*: the radial velocity of the Gaussian fitted to the red-shifted component. *Upper middle*: same, for the blue-shifted line component. *Lower middle*: one-half of the sum of the velocities; i.e., the radial velocity of the MWC349 RRL "system." *Bottom*: the difference between the red- and blue-shifted radial velocities. From Gordon et al. (2001)

Observations by high angular resolution synthesis telescopes support the disk model. Planesas et al. (1992) imaged the H30α line (231 GHz) with the Caltech interferometer in the Owens Valley, CA. They found a separation between the blue- and red-shifted peaks of about 80 AU, if MWC349 lies 1.2 kpc from the Sun. Furthermore, the two peaks lay along a line perpendicular to the axis of the bipolar radio continuum associated with the star. Figure 3.57 shows the approximate geometry.

Since the initial detection of RRLs from MWC349A in 1981, astronomers have observed recombination lines from H76α (15 GHz) to H6α

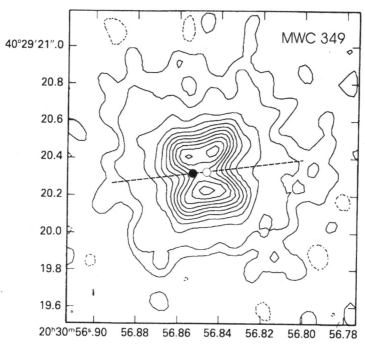

Fig. 3.57 The *circles* mark the relative locations of the *peaks* of the H30α line components observed by Planesas et al. (1992). The *left circle* corresponds to the red-shifted line component and the *right circle* corresponds to the blue-shifted component. The *broken line* marks a position angle of approximately 107°. The *background contours* mark the VLA observations of the MWC346 continuum emission at $\lambda = 2\,\mathrm{cm}$ (White and Becker, 1985). The *ordinate* is 1950.0 declination and the *abscissa* is 1950.0 RA. The absolute location of the circle structure with respect to the contours is a guess

($\approx 20{,}000\,\mathrm{GHz}$). Of these, the lines H40α ($99\,\mathrm{GHz}$) through H6α are masering to some extent. Furthermore, the average radial velocity of these profiles increases with frequency, as shown in Fig. 3.58. For double-peaked profiles, "average" means the average of the red- and blue-shifted velocity components when both are present.

Very generally, the model of the circumstellar, Keplerian disk viewed edge on accounts for the salient kinematic observations but does not model the masering itself nor the observed time variations. To investigate these, Hollenbach et al. (1994) considered photoevaporation of circumstellar disks around massive O stars to explain some of the details of ultracompact H II regions. Figure 3.59 illustrates their two models of accretion disks involving high-velocity winds. The weak wind model creates a thin, neutral accretion disk with a width that increases significantly with distance from the central star (the disk is not shown in the figure). This disk is sandwiched between an H II atmosphere maintained by Lyman continuum photons from the star. The critical radius r_g is where the sound speed equals the escape velocity from

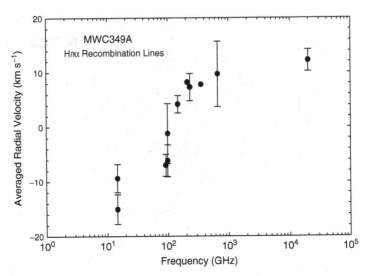

Fig. 3.58 The averaged radial velocity of H$n\alpha$ recombination lines from MWC349A plotted against their rest frequencies. Data listed by Gordon et al. (2001) and plotted in Gordon (2003)

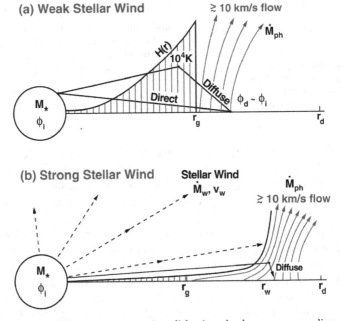

Fig. 3.59 Two models of an evaporating disk viewed edge on surrounding a massive O star, M$_*$, with a Lyman continuum flux density ϕ_i, stellar mass loss rate of $\dot{M}_*$, and disk mass loss by photoevaporation of $\dot{M}_{ph}$. (**a**) Star with a weak stellar wind, forming a 10^4 K H II disk with a radius r_g and a scale height $H(r)$. The UV photons evaporate the gas outside of r_g. (**b**) Star with strong stellar wind, forming a more flattened disk and pushing the H II material beyond r_g to where the ram pressure equals the thermal pressure from the ionized flow. Cartoon courtesy of Hollenbach (2002)

the disk, and marks the radius where evaporation begins. The bipolar outflow is probably responsible for the butterfly-shaped continuum emission shown in Fig. 3.57. The strong wind model entails a stellar wind sufficiently strong to blow the atmosphere of the accretion disk very far from the star; the ram pressure thereby narrowing the disk and accentuating the evaporative wind at its outer terminus.

Of these models, MWC349A seemed to fit the weak wind model best. The slow variation of the radio continuum emission with frequency – see Fig. 3.53 – suggests a significant disk component, i.e., one with adequate free–free absorption, must exist that in turn favors the weak wind model. Presumably, the variations in the radial velocities would result from corresponding variations in the locations where the densities are right for masing. Such an opacity variation in the disk could also be responsible for the variation in radial velocities seen in Fig. 3.58. In a differentially rotating disk, the larger optical depths would restrict the observed emission to the near parts of the circumstellar gas.

Figure 3.60 shows a model for an outflow derived (Gordon, 2003) from the RRL observations shown in Fig. 3.58. The lines are presumed to originate at

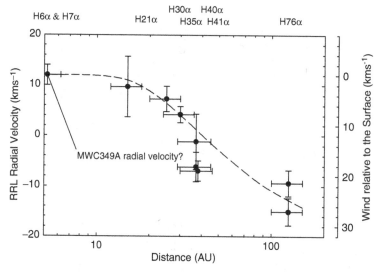

Fig. 3.60 Possible outflow from MWC349A deduced from observations of RRLs. *Points and error bars* mark the central velocities of RRLs from MWC349A with respect to the local standard of rest plotted against distance from MWC349A. Notations identify the specific RRLs observed. Distances obtain from a model that presumes each RRL originates from a location within a circumstellar disk where its optical depth is unity. The *right ordinate* indicates the velocity with respect to the H6α velocity. The *broken line* marks an approximate beta-law fit (Lamers and Cassinelli, 1999) to the observations. Note that the radial velocity of MWC349A with respect to the LSR appears to be 12 km s^{-1}. From Gordon (2003)

locations in the circumstellar disk where $\tau(\nu)_\ell = 1$ at their particular frequency. The analysis suggests a slow gas outflow from MWC349A that could be consistent with a wind driven by photoevaporation. The result also implies a radial velocity of $12\,\mathrm{km\,s^{-1}}$ with respect to the LSR for the MWC349A star itself.

The masering has been considered elsewhere. A series of papers by Ponomarev et al. (1994) and Strelnitski et al. (1996; 1996) describe how the masering lines could form in the circumstellar disk of MWC349A as sketched in Fig. 3.61 – a geometry first proposed by Elmegreen and Morris (1979) for disk systems in general. In this model, the maser amplification occurs within a ring. The path lengths within the ring on either side of the star are sufficient to generate intense, masering RRLs seen as red- and blue-shifted components in the line profiles. The path length along the sight line through the star is insufficient to generate lines of comparable intensity and, moreover, could have a broader line profile depending upon the velocity dispersion through the front/back ring segments.

Calculations showed that, with this ring model, an unsaturated maser could match the observed line profiles. Furthermore, the model limits the range of physical parameters for the ring that can match the observations.

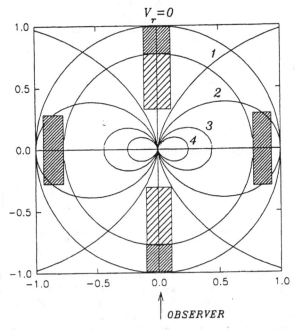

Fig. 3.61 Downward view of the differentially rotating, circumstellar disk of MWC349A. *Spider-leg curves* mark isochasms of the observed radial velocity in units of Doppler width. The maser amplification occurs within the two outer rings along sight lines marked by *crosshatching*. From Ponomarev et al. (1994)

What was not modeled was the asymmetrical conditions on either side of the ring that might account for the time variations of the red- and blue-shifted amplitudes and radial velocities. Presumably, these intensity variations are due to changes in local densities within the ring (Strelnitski, Ponomarev and Smith, 1996) resulting from anisotropies in the stellar wind and, probably, to concurrent but not necessarily simultaneous variations in the Lyman photons incident upon the local masering regions.

Finally, recent observations have detected a magnetic field in MWC349A (Thum and Morris, 1999). The H30α lines have circular polarization characteristic of a strong magnetic field (22 mG) causing a Zeeman effect. Equipartition of the energy density between this field and the thermal environment suggests that the line emitting gas has a temperature of about 5,000 K, which is somewhat lower than expected for H II gas but an entirely reasonable value.

In conclusion, RRLs have revealed an unusual stellar system in our Galaxy, one that not only may represent an evolutionary stage of ultracompact H II regions but also provides the physical environment adequate to engender high-gain hydrogen masers. With regard to masering RRLs, the only other object like MWC349A may be the star η Carina from which narrow, asymmetrical RRLs have also been found at millimeter wavelengths (Cox, Martín-Pintado, Bachiller, Bronfman, Cernicharo, Lyman and Roelfsema, 1995). Like the lines from MWC349A, these RRLs sit atop a broad line component and have peculiar ratios of β/α intensities. Also like MWC349A, the star appears to be undergoing substantial mass loss. Unlike MWC349A, the continuum emission may be time varying.

3.5 RRLs from Extragalactic Objects

In the second half of the 1970s, RRL studies expanded toward extragalactic objects. Except for the Magellanic Clouds, the first target was the edge-on galaxy M82, located at a distance of 3.2 Mpc from the Sun. Shaver et al. (1977) soon detected the H166α (1.4 GHz) line with the Westerbork Synthesis Telescope in the Netherlands. They proposed that stimulated emission due to the nonthermal continuum made the principal contribution to the line intensity.

Other observers quickly confirmed this detection. Bell and Seaquist (1977) detected the H102α (6.1 GHz) recombination line from M82 with the 46-m Algonquin radio telescope in Canada, and Chaisson and Rodríguez (1977) detected the H92α (8.3 GHz) line. Subsequent observations showed the radial velocity and line width of the H102α and H85α (10.5 GHz) lines from M82 $\approx$ 175 km s^{-1} and $\approx$ 150 km s^{-1}, respectively (Bell and Seaquist, 1978).

To investigate the circumstances of these RRLs further, Shaver et al. (1978) observed RRLs from M82 in three frequency ranges – 1.4, 4.9, and 14.7 GHz – with the 100-m Effelsberg telescope. The H166α (1.4 GHz) and

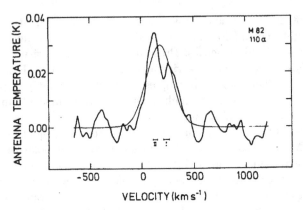

Fig. 3.62 The H110α spectrum of M82 with a resolution of 19.2 km s⁻¹, made with the 100-m radio telescope at Effelsberg, Germany. The velocity scale is heliocentric relative to the RRL rest frequency. The *numbers I and II* mark the radial velocities of the two H II regions in M82 studied optically by Recillas-Cruz and Peimbert (1970). From Shaver et al. (1978)

the H110α (4.9 GHz) lines were detected with good signal-to-noise ratios but the H76α (14.7 GHz) was not seen. Figure 3.62 shows the spectrum of the H110α line.

Comparison of all of these spectra from M82 confirmed that stimulated emission was a major factor in the line intensities. Its importance increased with wavelength – from 70% of the line intensity of the H102α line to 90% of the H166α line. At 14.7 GHz, the importance of stimulated emission had decreased so much that the H76α line could not even be detected.

Nearly simultaneously with the M82 detections, Seaquist and Bell (1977) detected the H102α line from the spiral galaxy NGC253, also located about 3 Mpc from the Sun. The radial velocity of the RRL was 132 ± 25 km s⁻¹ and the line width was 309 ± 65 km s⁻¹. Unlike the situation for M82, the NGC253 RRL seemed to be primarily spontaneous emission. Consequently, the authors proposed that the line originated from H II regions of $T_e \approx 5,000$ K embedded in the galaxy. In contrast, the M82 lines must have originated from gas in front of a background nonthermal source.

Based upon calculations, Shaver (1978) concluded that detections of RRLs with primarily spontaneous emission were possible only for galaxies with $D < 10$ Mpc. On the other hand, if the RRL emission were primarily stimulated emission, then more distant galaxies could be detected. These galaxies would need to contain intense sources of nonthermal emission, such as quasars and radio galaxies. Because nonthermal emission increases with wavelength, Shaver suggested that the best window for searching for RRLs from such galaxies would be 1 → 10 GHz. The line widths from such sources could be as large as 500 km s⁻¹.

Despite the large number of candidates, few galaxies were detected in RRL emission. New sources of extragalactic RRLs were detected only in the 1990s, more than 10 years after the detections of RRLs from M82 and NGC253. The first new detection was from NGC2146, a galaxy located about 13 Mpc from the Sun. This detection was the H53α (42.9 GHz), made with the 45-m radio telescope of the Nobeyama Radio Observatory in Japan (Puxley, Brand, Moore, Mountain and Nakai, 1991). Later, Anantharamaiah et al. (1993) used the VLA to detect the H92α (8.3 GHz) line in NGC3628, IC694, and NGC1365. The much greater angular resolution of the VLA at this frequency, about 3$''$, enabled localization of the emitting regions within these galaxies.

Figure 3.63 shows the continuum image of NGC3628 and the location of the H92α RRL emission. This galaxy lies 11 Mpc from the Sun and is seen nearly edge on with an inclination angle of 89°. Its rotating central disk, with dimensions of about 5$''$, emits the recombination lines. Figure 3.64 shows observations (Zhao, Anantharamaiah, Goss and Viallefond, 1997) made with the still higher resolution of 1.5$''$, which resolve three distinct region of H92α emission within the center: the nucleus, slightly SE, and slightly NW.

In the years following the initial detections, astronomers using the VLA detected H92α line emission from five more galaxies: Arp 220, M83, NGC2146 (Zhao, Anantharamaiah, Goss and Viallefond, 1996), NGC3690 (a second component of the interacting system IC694 + NGC3690) (Zhao et al., 1997), and NGC660 (Phookun, Anantharamaiah and Goss, 1998). In general, the observations of RRLs from extragalactic objects began at decimeter and centimeter wavelengths and expanded into millimeter wavelengths.

Table 3.9 lists the detections. The first three columns list the galaxy, its type, and its distance from the Sun. The remaining columns give the

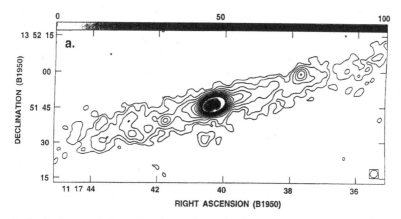

Fig. 3.63 An image of the continuum emission from NGC3628 obtained from VLA observations of the H92α RRL. The beam size is 3.5$'' \times$ 3.2$''$ and the contour levels are -0.1, 0.05, 0.1, 0.2, 0.35, 0.55, 0.8, 1.1, 1.5, 1.9, ... mJy beam^{-1}. *Gray-scale densities at the top* indicate the intensity of the RRL emission over the range $0 \rightarrow 100$ mJy beam^{-1}. From Anantharamaiah et al. (1993)

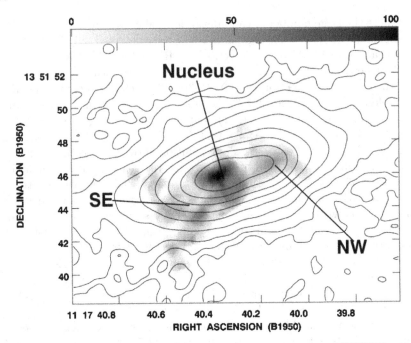

Fig. 3.64 The integrated H92α line image of the nuclear region of NGC3628 superimposed on the continuum emission. The *gray scale at top* shows the line emission $0 \rightarrow 100\,\text{mJy beam}^{-1}\text{km s}^{-1}$. The contour levels show the continuum intensity levels of 0.04, 0.08, 0.16, etc., to 10.3 mJy beam^{-1}. The beam size is $1.8'' \times 1.5''$. The three distinct RRL emission components are labeled: nucleus, SE, and NW. From Zhao et al. (1997)

transition, the line flux, the heliocentric velocity, the full line width at half-intensity, and the references. At present, RRLs have been detected from ten galaxies in addition to our own. The most distant is Arp 220. Its radial velocity $\approx 5,500\,\text{km s}^{-1}$, which corresponds to a distance of 73 Mpc for a Hubble constant of $75\,\text{km s}^{-1}\,\text{kpc}^{-1}$. Attempts to detect RRLs from the more distant galaxy NGC6240, lying at 100 Mpc, have not yet been successful (Zhao et al., 1996).

These observations led to a revision of the accepted nature of RRL formation in galaxies. As mentioned earlier, theoretical calculations (Shaver, 1978) suggested that stimulated emission would be a major factor in RRLs emitted from galaxies. Therefore, observers had believed that the lines would be more intense at lower frequencies. But, the observations indicated otherwise. In many case, the RRLs were strongest at higher frequencies where stimulated emission from external sources or from the Galactic nuclei would have to be negligibly small. For example, the detections listed in Table 3.9 for M82 and Arp 220 show the line fluxes to be greater at millimeter wavelengths than at decimeter or centimeter wavelengths. The situation is similar for NGC2146. The flux of the H53α line considerably exceeds that of the H92α line. In turn, the intensity of the H92α line in NGC253 can be explained if 75% of

Table 3.9 RRLs detected from extragalactic objects

Galaxy	Type	Dist. (Mpc)	RRL	$I_L\Delta\nu_L$ (10^{-19} Wm^{-2})	V_{Helio} (km s^{-1})	ΔV_L (km s^{-1})	References
M82	Ir	3.25	H27α	21.1±4.3			Seaquist et al. (1996)
			H30α	5.12±1.0			Seaquist et al. (1994)
			H41α	1.32±0.22			Seaquist et al. (1996)
			H53α	0.22±0.06			Puxley et al. (1989)
			H85α	$(4.6\pm0.7)\times10^{-3}$	180±10	149±25	Bell and Seaquist (1978)
			H102α	$(8.5\pm1.4)\times10^{-3}$	173±10	174±20	Bell and Seaquist (1978)
			H110α	$(1.05\pm0.1)\times10^{-2}$	163±10	293±20	Shaver et al. (1978)
			H166α	$>1.7\times10^{-3}$	150±40	250±40	Shaver et al. (1977; 1978)
			H166α	$(5.1\pm0.8)\times10^{-3}$	254±20	338±40	Shaver et al. (1978)
NGC253	Spiral	3.4	H92α	$(4.0\pm0.4)\times10^{-3}$	217±8	189±19	Anantharamaiah and Goss (1990)
			H102α	$(1.3\pm0.3)\times10^{-2}$	132±25	309±65	Seaquist and Bell (1977)
			H110α	$(2.9\pm0.6)\times10^{-3}$	209±13	185±32	Anantharamaiah and Goss (1990)
			H166α	$(3.8\pm1.0)\times10^{-3}$	195±36	220±87	Anantharamaiah and Goss (1990)
NGC2146	Spiral	13	H53α	$(9.5\pm1.7)\times10^{-2}$	≈ 880	≈ 300	Puxley et al. (1991)
			H92α	2.7×10^{-4}	960±7	200±95	Zhao et al. (1996)
NGC1365	Seyf. II	22.0	H92α	$(12\pm2)\times10^{-4}$	1,670±80	310±110	Anantharamaiah et al. (1993)
NGC3628	S3 pec	11.5	H92α	$(8.6\pm1.5)\times10^{-4}$	864±56	170±70	Anantharamaiah et al. (1993)
IC694[a]	Sc	40.3	H92α	$(3.9\pm1.0)\times10^{-4}$	3,020±90	350±110	Anantharamaiah et al. (1993)
NGC3690[a]	Sc		H92α	$(1.5\pm0.2)\times10^{-4}$	3,080±40	210±30	Zhao et al. (1997)
Arp 220	FIR	73	H31α	$>1.65\pm0.1$			Anantharamaiah et al. (2000)
			H40α	0.39±0.05	5,513	179	Anantharamaiah et al. (2000)
			H42α	0.22±0.02	5,424	210	Anantharamaiah et al. (2000)
			H92α	3.5×10^{-4}	5,560±70	320±120	Zhao et al. (1996)
			H92α	$(8\pm1.5)10^{-4}$	5,450±20	363±45	Anantharamaiah et al. (2000)
M83	SBc/b	5	H92α	2.8×10^{-4}	500±30	95±30	Zhao et al. (1996)
NGC660	SBa pec	11.3	H92α	5.6×10^{-4}	850	377	Phookun et al. (1998)

[a]IC694 and NGC3690 are an interacting system

the line intensity is contributed by spontaneous emission (Anantharamaiah and Goss, 1990). The situation is similar for observations of the H92α line in other galaxies.

To explain the observations of the H92α line from the galaxies NGC3628, NGC1365, and IC694, Anantharamaiah et al. (1993) suggested a model of a collection of discrete H II regions. According to this model, the principal part of the line emission is formed in 100–200 small ($1 \rightarrow 5\,\mathrm{pc}$), dense ($N_e = 5 \times 10^3 \rightarrow 5 \times 10^4\,\mathrm{cm}^{-3}$) H II regions with a temperature of 10^4 K. The volume filling factor of the H II regions is very small, $<10^{-5}$. In this model, only a small part of Galactic nuclear nonthermal radiation crosses the H II regions, and stimulated emission is not important. The RRL emission is primarily spontaneous, arising from the H II regions. Because the H II regions are optically thick even at centimeter wavelengths, the line emission arises from their outer layers. There, it can be amplified by stimulated emission due to the thermal continuum emission generated within the H II regions.

A similar model was developed for the NGC660 galaxy. Here, the best agreement with the H92α line observations was obtained from a model of several thousand H II regions with $N_e = 5{,}000\,\mathrm{cm}^{-3}$ and with diameters $\approx 1\,\mathrm{pc}$ (Phookun et al., 1998).

With this kind of model, it was possible to explain the observations of RRLs from M82 from centimeter wavelengths up to $\lambda \approx 1\,\mathrm{mm}$. The observed integrated line flux densities of the H53α (42.9 GHz), the H41α (92.03 GHz), and the H30α (231.9 GHz) lines are proportional to the squares of their line frequencies. This is what would be expected from pure spontaneous emission from optically thin H II regions in LTE (Seaquist, Kerton and Bell, 1994). At the same time, millimeter wave observations of RRLs from the other detected galaxies have shown that the multi-H II region model with the same density does not work. For these galaxies, the longer wavelength observations were inconsistent with the fits to the shorter wavelength data and vice versa. Fitting these galaxies will require more complicated models.

The RRL observations of M82 have been interpreted by a two-component model of ionized gas with an electron temperature of $5{,}000 \rightarrow 10{,}000$ K. One component consists of a low-density layer with a high filling factor; the other consists of a layer of compact ($d \leq 1\,\mathrm{pc}$), dense ($N_e > 10^{4.5}\,\mathrm{cm}^{-3}$) H II regions with a small filling factor. The idea is that RRL emission from the compact H II regions contains emission stimulated by the internal continuum while that from the densest regions – up to $10^7\,\mathrm{cm}^{-3}$ – may be undergoing maser amplification. This model rests on the observation that the integrated line flux density of the shortest wavelength RRLs (H27α at 316.4 GHz) from M82 was much larger than what would be expected from a square law fitted to the observed intensities of the low-frequency RRLs H30α, H41α, and H53α (Seaquist, Carlstrom, Bryant and Bell, 1996). This intensity excess implies that significant stimulated emission is present. Maser amplification of RRLs is known to exist in at least one Galactic source (see Sect. 3.4.3) and very well might exist in other galaxies.

A two-component model has also been proposed for the galaxy NGC2146. In this model, the H53α (43 GHz) lines are emitted by 100 compact H II regions with the very high electron densities of 10^5 cm^{-3} and individual sizes of about 0.2 pc. In contrast, the centimeter wavelength RRLs would come from a few hundred, less dense ($N_e = 5 \times 10^3 \rightarrow 10^4$ cm^{-3}) H II regions with sizes ≈ 1 pc. The temperature of these H II regions would be 5,000–7,000 K (Zhao et al., 1996).

The galaxy Arp 220 can be fitted by a similar model. The observed lines H92α (8.1 GHz), H42α (84 GHz), H40α (96 GHz), H31α (207 GHz), and the upper limit for the H167α (1.4 GHz) line fit a model containing two groups of H II regions with a temperature of 7,500 K. One group containing about 20,000 discrete H II regions with electron densities of about 10^3 cm^{-3} and a large filling factor (≈ 0.7) can explain the H92α line emission and the upper limit to the H167α line. The other group of about 1,000 H II regions with a density of about 2×10^4 cm^{-3}, sizes of about 0.1 pc, and a low filling factor ($\approx 10^{-5}$) can account for the shorter wavelength lines H42α, H40α, and H31α (Anantharamaiah, Viallefond, Mohan, Goss and Zhao, 2000).

In general, RRLs have been found in galaxies containing active star formation sites – the starburst galaxies – based upon an analysis of 15 galaxies where the H92α line was either observed or an upper limit determined. There seems to be a correlation of RRL intensities with the flux density of IR emission and the radio continuum at 8.4 GHz, which are the traditional indicators of the star formation rate (Phookun et al., 1998).

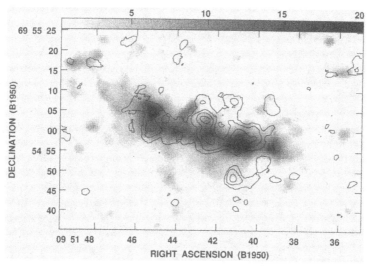

Fig. 3.65 Contours of the integrated H41α emission from M82 overlaid on a *gray-scale map* of its HCO$^+$(1 $\rightarrow$ 0) emission. The angular resolution is approximately 4″. From Seaquist et al. (1996)

In a number of galaxies, observations of the H92α line intensities correlate well with those of the HCN and HCO$^+$ lines (Zhao et al., 1996). Moreover, in M82, an excellent spatial correlation exists between the H41α line intensities and molecular line emission. Figure 3.65 shows an overlay of H41α emission upon HCO$^+$ emission. The emission centers are the same for both lines. There is similar agreement between the H41α and CO line emission (Seaquist et al., 1996). Molecular clouds are always associated with the formation of stars, and the correlation of the RRL emission with these clouds indicates that actual star formation is taking place within them.

To summarize, we note that, to date, detections of RRLs from extragalactic objects seem to come from starburst galaxies. These lines form in large complexes of H II regions located in the central parts of the galaxies, where stars are forming. The physical conditions of these regions are $T_e = 5{,}000 \to 10{,}000\,\mathrm{K}$, $N_e = 10^3 \to 10^7\,\mathrm{cm}^{-3}$, and the sizes of the H II regions are 0.2–5 pc – close to those of H II regions within our own Galaxy. The RRL emission is primarily spontaneous but also contains line emission stimulated by the continuum thermal emission of H II regions themselves. RRLs are most intense toward millimeter wavelengths. The densest H II regions may have masering RRLs. In contrast, RRL emission is weaker toward longer wavelengths.

The search for RRLs from quasars, carried out with the 100-m radio telescope at Effelsberg, was unsuccessful (Bell, Seaquist, Mebold, Reif and Shaver, 1984).

Appendixes

These appendixes give supplementary information. They include physical constants associated with radio recombination lines (RRLs); a table of line frequencies from 12 MHz to 29 THz for hydrogen, helium, carbon, and sulfur atoms; fine-structure components of lines associated with small principal quantum numbers; miscellaneous calculations relevant to line formation; a FORTRAN 77 computer code for calculating departure coefficients that index the relative population of atomic levels involved in RRLs; and a discussion of the relationship of the peculiar observational units used by astronomers to the more commonly used units of general physics.

Appendix A
Constants

A.1 Miscellaneous Constants

A.2 Rydberg Constants

A.2.1 Reduced Mass

The Bohr model used the center of the hydrogen mass as the centroid of the orbit. However, the reduced mass m_R of the nucleus M and the orbiting electron m gives better values. It results from

$$\frac{1}{m_R} = \frac{1}{m} + \frac{1}{M}, \quad \text{or} \tag{A.1}$$

$$m_R = \frac{mM}{m + M}. \tag{A.2}$$

Then, from (1.16) and (1.17), the correct value of R is

$$R = \frac{2\pi^2 m_R e^4}{ch^3} = R_\infty \frac{M}{M + m}. \tag{A.3}$$

Substituting the modern physical constants listed in Table A.1 into (A.3) gives a value of $R_\infty = 109737.3 \, \text{cm}^{-1}$.

Equation (A.3) enables a more general equation for the calculation of R for multielectron atoms and ions:

$$R = R_\infty \left[\frac{M_a - Zm}{M_a - (Z - 1)m} \right], \tag{A.4}$$

where M_a is the mass of neutral atom and Z is the nuclear charge. It is assumed that these atoms are hydrogenic, i.e., only one electron is in an excited level and $Z - 1$ other electrons are in atomic core. For neutral atoms,

Table A.1 Recommended values of physical units

Symbol	Identity	Value	CGS units	Basic units	Sources
α	Fine-structure constant	$7.297352533 \times 10^{-3}$			Mohr and Taylor (1999)
c	Vacuum speed of light	$2.99792458 \times 10^{10}$	$\mathrm{cm\,s}^{-1}$	$\mathrm{L\,T}^{-1}$	IAU resolution 6, 1973
e	Electronic charge	$-1.602176462 \times 10^{-19}$	C		
		$-4.8032041 \times 10^{-10}$	ESU		Mohr and Taylor (1999)
h	Planck's constant	6.626068×10^{-27}	$\mathrm{erg\,s}$	$\mathrm{M\,L\,L\,T}^{-1}$	Mohr and Taylor (1999)
k	Boltzmann's constant	$1.3806503 \times 10^{-16}$	$\mathrm{erg\,K}^{-1}$	$\mathrm{M\,L\,L\,T}^{-1}\,\mathrm{T}^{-1}\,\mathrm{K}^{-1}$	Mohr and Taylor (1999)
m	Electronic mass	$9.10938188 \times 10^{-28}$	g	M	Mohr and Taylor (1999)
M	Hydrogen mass	$1.67352499 \times 10^{-24}$	g	M	Mohr and Taylor (1999)
R_∞	Rydberg constant	$109{,}737.31568$	cm^{-1}	M^{-1}	Mohr and Taylor (1999)

Table A.2 Masses and Rydberg constants

Atom	Symbol	Mass (amu)	Rydberg constant (cm^{-1})
Hydrogen	H^1	1.007825035	109,677.576
Helium	He^4	4.00260324	109,722.27
Carbon	C^{12}	12.000000	109,732.30
Nitrogen	N^{14}	14.003074	109,733.01
Oxygen	O^{16}	15.994915	109,733.55
Sulfur	S^{32}	31.97207070	109,735.43

From Audi and Wapstra (1995) and (A.4)

the net negative charge of the inner electrons would screen the positive charge of the nucleus, so that the lone outer electron would see a single nuclear charge and $Z = 1$. For ions, $Z > 1$.

Therefore, from (1.17) and (A.4), the general expression for the frequency of an RRL becomes

$$\nu_{n_2 \to n_1} = R_\infty c Z^2 \left[\frac{M - (Z+1)m}{M - Zm} \right] \left(\frac{1}{n_1^2} - \frac{1}{n_2^2} \right) \tag{A.5}$$

or, in CGS units,

$$\frac{\nu_{n_2 \to n_1}}{[\text{Hz}]} = 3.28984196 \times 10^{15} Z^2$$

$$\times \left[\frac{M_u - 5.48579911 \times 10^{-4}(Z+1)}{M_u - 5.48579911 \times 10^{-4} Z} \right]$$

$$\times \left(\frac{1}{n_1{}^2} - \frac{1}{n_2{}^2} \right), \tag{A.6}$$

where M_u is the mass of the neutral atom in atomic mass units (m_u) and $R_\infty = 109,737.315685 \, \text{cm}^{-1}$.

A.2.2 Table of Rydberg Constants

Table A.2 lists Rydberg constants for the most abundant elements in the cosmos.

Appendix B
Tables of Line Frequencies

B.1 Frequencies Below 100 GHz

Table B.1 lists rest frequencies calculated for the $n\alpha$, $n\beta$, $n\gamma$, $n\delta$, and $n\epsilon$ transitions of atomic hydrogen for the range of principal quantum numbers $n = 40 \rightarrow 800$. The nomenclature is the usual. The term $n\alpha$ designates a transition from $n+1 \rightarrow n$; $n\beta$, from $n+2 \rightarrow n$; etc. The table also contains frequencies for the $n\alpha$ and $n\beta$ transitions of helium 4, and the $n\alpha$ transitions of carbon 12 and sulfur 32. The calculations use (A.5) and the constants listed in Tables A.1 and A.2. Below 100 GHz, frequency shifts due to the blending of fine-structure components are negligible. The shifts are less than 2 kHz for $n = 40$ and vary as n^{-5} (Lilley and Palmer, 1968).

Table B.1. H, ^{4}He, ^{12}C, and ^{32}S RRL rest frequencies for $n = 40 \rightarrow 800$

n	Line transition (frequencies in MHz)								
	Hnα	Hnβ	Hnγ	Hnδ	Hnϵ	Henα	Henβ	Cnα	Snα
40	99,022.96	191,057.4	276,745.8	356,658.4	431,303.0	99,063.31	191,135.3	99,072.36	99,075.19
41	92,034.44	177,722.8	257,635.5	332,280.1	402,109.6	92,071.94	177,795.3	92,080.35	92,082.98
42	85,688.39	165,601.1	240,245.6	310,075.2	375,495.1	85,723.31	165,668.5	85,731.14	85,733.59
43	79,912.66	154,557.2	224,386.8	289,806.7	351,180.7	79,945.22	154,620.2	79,952.52	79,954.81
44	74,644.57	144,474.1	209,894.1	271,268.0	328,922.8	74,674.98	144,533.0	74,681.81	74,683.94
45	69,829.56	135,249.5	196,623.4	254,278.3	308,508.5	69,858.01	135,304.6	69,864.39	69,866.39
46	65,419.94	126,793.9	184,448.7	238,679.0	289,750.6	65,446.60	126,845.5	65,452.58	65,454.45
47	61,373.94	119,028.8	173,259.0	224,330.6	272,484.2	61,398.95	119,077.3	61,404.56	61,406.31
48	57,654.83	111,885.1	162,956.7	211,110.3	256,564.0	57,678.32	111,930.7	57,683.59	57,685.24
49	54,230.25	105,301.9	153,455.5	198,909.2	241,861.2	54,252.35	105,344.8	54,257.30	54,258.85
50	51,071.61	99,225.21	144,678.9	187,630.9	228,261.4	51,092.43	99,265.65	51,097.09	51,098.55
51	48,153.60	93,607.32	136,559.3	177,189.8	215,663.2	48,173.22	93,645.47	48,177.62	48,179.00
52	45,453.72	88,405.69	129,036.2	167,509.6	203,975.8	45,472.24	88,441.72	45,476.40	45,477.70
53	42,951.97	83,582.47	122,055.8	158,522.1	193,118.5	42,969.47	83,616.53	42,973.40	42,974.63
54	40,630.50	79,103.86	115,570.1	150,166.5	183,018.7	40,647.06	79,136.09	40,650.77	40,651.93
55	38,473.36	74,939.62	109,536.0	142,388.2	173,611.5	38,489.04	74,970.16	38,492.55	38,493.65
56	36,466.26	71,062.65	103,914.8	135,138.2	164,838.5	36,481.12	71,091.60	36,484.45	36,485.50
57	34,596.38	67,448.58	98,671.90	128,372.3	156,647.1	34,610.48	67,476.07	34,613.64	34,614.63
58	32,852.20	64,075.51	93,775.88	122,050.7	148,989.9	32,865.58	64,101.62	32,868.59	32,869.53
59	31,223.32	60,923.68	89,198.55	116,137.7	141,824.0	31,236.04	60,948.50	31,238.89	31,239.78

(continued)

Table B.1. (continued)

n	Line transition (frequencies in MHz)								
	Hnα	Hnβ	Hnγ	Hnδ	Hnε	Henα	Henβ	Cnα	Snα
60	29,700.36	57,975.24	84,914.40	110,600.7	135,110.6	29,712.47	57,998.86	29,715.18	29,716.03
61	28,274.87	55,214.04	80,900.32	105,410.2	128,814.5	28,286.39	55,236.53	28,288.98	28,289.78
62	26,939.16	52,625.45	77,135.35	100,539.6	122,903.8	26,950.14	52,646.89	26,952.60	26,953.37
63	25,686.28	50,196.19	73,600.47	95,964.63	117,349.4	25,696.75	50,216.64	25,699.10	25,699.83
64	24,509.90	47,914.18	70,278.35	91,663.14	112,124.9	24,519.89	47,933.71	24,522.13	24,522.83
65	23,404.28	45,768.45	67,153.23	87,615.00	107,206.1	23,413.82	45,787.10	23,415.96	23,416.62
66	22,364.17	43,748.95	64,210.72	83,801.83	102,571.0	22,373.28	43,766.78	22,375.32	22,375.96
67	21,384.79	41,846.55	61,437.67	80,206.83	98,199.39	21,393.50	41,863.61	21,395.45	21,396.07
68	20,461.77	40,052.88	58,822.04	76,814.60	94,072.82	20,470.10	40,069.20	20,471.97	20,472.56
69	19,591.11	38,360.28	56,352.83	73,611.05	90,174.34	19,599.10	38,375.91	19,600.89	19,601.45
70	18,769.16	36,761.72	54,019.94	70,583.23	86,488.42	18,776.81	36,776.70	18,778.53	18,779.06
71	17,992.56	35,250.77	51,814.07	67,719.26	83,000.75	17,999.89	35,265.14	18,001.53	18,002.05
72	17,258.21	33,821.51	49,726.70	65,008.19	79,698.18	17,265.25	33,835.29	17,266.82	17,267.32
73	16,563.30	32,468.49	47,749.98	62,439.97	76,568.58	16,570.04	32,481.72	16,571.56	16,572.03
74	15,905.19	31,186.68	45,876.67	60,005.29	73,600.78	15,911.67	31,199.39	15,913.13	15,913.58
75	15,281.49	29,971.48	44,100.10	57,695.58	70,784.44	15,287.72	29,983.69	15,289.12	15,289.55
76	14,689.99	28,818.60	42,414.09	55,502.94	68,110.02	14,695.97	28,830.34	14,697.31	14,697.73
77	14,128.62	27,724.11	40,812.96	53,420.03	65,568.70	14,134.37	27,735.40	14,135.66	14,136.07
78	13,595.49	26,684.34	39,291.42	51,440.08	63,152.28	13,601.03	26,695.21	13,602.27	13,602.66
79	13,088.85	25,695.93	37,844.59	49,556.80	60,853.20	13,094.19	25,706.40	13,095.38	13,095.76
80	12,607.08	24,755.74	36,467.94	47,764.35	58,664.41	12,612.22	24,765.83	12,613.37	12,613.73
81	12,148.66	23,860.87	35,157.27	46,057.33	56,579.37	12,153.61	23,870.59	12,154.72	12,155.07
82	11,712.20	23,008.61	33,908.67	44,430.71	54,592.01	11,716.98	23,017.99	11,718.05	11,718.38
83	11,296.41	22,196.47	32,718.50	42,879.80	52,696.67	11,301.01	22,205.51	11,302.04	11,302.36
84	10,900.06	21,422.10	31,583.40	41,400.26	50,888.08	10,904.50	21,430.83	10,905.50	10,905.81
85	10,522.04	20,683.34	30,500.20	39,988.02	49,161.35	10,526.33	20,691.77	10,527.29	10,527.59
86	10,161.30	19,978.16	29,465.98	38,639.31	47,511.87	10,165.44	19,986.30	10,166.37	10,166.66
87	9,816.864	19,304.68	28,478.01	37,350.57	45,935.39	9,820.864	19,312.55	9,821.761	9,822.042
88	9,487.821	18,661.14	27,533.71	36,118.53	44,427.91	9,491.687	18,668.75	9,492.554	9,492.825
89	9,173.321	18,045.89	26,630.71	34,940.09	42,985.69	9,177.059	18,053.24	9,177.897	9,178.159
90	8,872.568	17,457.39	25,766.77	33,812.37	41,605.24	8,876.184	17,464.50	8,876.995	8,877.248
91	8,584.821	16,894.20	24,939.81	32,732.68	40,283.29	8,588.319	16,901.09	8,589.103	8,589.349
92	8,309.382	16,354.99	24,147.86	31,698.47	39,016.77	8,312.768	16,361.65	8,313.528	8,313.765
93	8,045.603	15,838.47	23,389.09	30,707.38	37,802.79	8,048.881	15,844.93	8,049.616	8,049.846
94	7,792.871	15,343.48	22,661.78	29,757.19	36,638.68	7,796.046	15,349.74	7,796.758	7,796.981
95	7,550.614	14,868.91	21,964.32	28,845.81	35,521.88	7,553.691	14,874.97	7,554.381	7,554.596
96	7,318.296	14,413.71	21,295.19	27,971.27	34,450.03	7,321.278	14,419.58	7,321.947	7,322.156
97	7,095.411	13,976.90	20,652.97	27,131.73	33,420.88	7,098.302	13,982.59	7,098.951	7,099.154
98	6,881.486	13,557.56	20,036.32	26,325.47	32,432.32	6,884.291	13,563.09	6,884.919	6,885.116
99	6,676.076	13,154.84	19,443.98	25,550.83	31,482.38	6,678.796	13,160.20	6,679.406	6,679.597
100	6,478.760	12,767.90	18,874.76	24,806.30	30,569.18	6,481.400	12,773.11	6,481.992	6,482.177
101	6,289.144	12,396.00	18,327.54	24,090.42	29,690.97	6,291.706	12,401.05	6,292.281	6,292.461
102	6,106.855	12,038.40	17,801.28	23,401.83	28,846.09	6,109.344	12,043.31	6,109.902	6,110.077
103	5,931.544	11,694.42	17,294.97	22,739.24	28,032.97	5,933.962	11,699.19	5,934.504	5,934.673
104	5,762.880	11,363.43	16,807.69	22,101.42	27,250.13	5,765.228	11,368.06	5,765.755	5,765.920
105	5,600.550	11,044.81	16,338.54	21,487.25	26,496.17	5,602.832	11,049.31	5,603.344	5,603.504
106	5,444.260	10,737.99	15,886.70	20,895.62	25,769.77	5,446.479	10,742.37	5,446.976	5,447.132
107	5,293.732	10,442.43	15,451.36	20,325.51	25,069.70	5,295.889	10,446.69	5,296.373	5,296.524
108	5,148.703	10,157.63	15,031.78	19,775.40	24,394.75	5,150.801	10,161.76	5,151.271	5,151.418
109	5,008.923	9,883.080	14,627.26	19,246.05	23,743.83	5,010.964	9,887.107	5,011.421	5,011.565
110	4,874.157	9,618.340	14,237.13	18,734.91	23,115.86	4,876.143	9,622.259	4,876.589	4,876.728
111	4,744.183	9,362.972	13,860.75	18,241.70	22,509.84	4,746.116	9,366.788	4,746.550	4,746.685
112	4,618.789	9,116.566	13,497.52	17,765.66	21,924.83	4,620.671	9,120.280	4,621.094	4,621.226
113	4,497.776	8,878.730	13,146.87	17,306.04	21,359.92	4,499.609	8,882.348	4,500.020	4,500.149
114	4,380.954	8,649.096	12,808.27	16,862.15	20,814.26	4,382.739	8,652.621	4,383.139	4,383.265
115	4,268.142	8,427.314	12,481.19	16,433.30	20,287.02	4,269.882	8,430.748	4,270.272	4,270.394
116	4,159.171	8,213.049	12,165.16	16,018.88	19,777.45	4,160.866	8,216.396	4,161.246	4,161.365
117	4,053.878	8,005.988	11,859.71	15,618.27	19,284.80	4,055.530	8,009.250	4,055.901	4,056.016
118	3,952.110	7,805.829	11,564.40	15,230.92	18,808.38	3,953.720	7,809.010	3,954.081	3,954.194

(continued)

Table B.1. (continued)

n	Line transition (frequencies in MHz)								
	H$n\alpha$	H$n\beta$	H$n\gamma$	H$n\delta$	H$n\epsilon$	He$n\alpha$	He$n\beta$	C$n\alpha$	S$n\alpha$
119	3,853.719	7,612.287	11,278.81	14,856.27	18,347.53	3,855.289	7,615.389	3,855.642	3,855.752
120	3,758.568	7,425.091	11,002.55	14,493.81	17,901.61	3,760.099	7,428.117	3,760.443	3,760.550
121	3,666.523	7,243.983	10,735.24	14,143.04	17,470.03	3,668.017	7,246.935	3,668.352	3,668.457
122	3,577.460	7,068.718	10,476.52	13,803.51	17,052.22	3,578.918	7,071.598	3,579.245	3,579.347
123	3,491.258	6,899.061	10,226.05	13,474.76	16,647.62	3,492.681	6,901.873	3,493.000	3,493.099
124	3,407.803	6,734.791	9,983.498	13,156.36	16,255.72	3,409.192	6,737.535	3,409.503	3,409.601
125	3,326.988	6,575.695	9,748.558	12,847.92	15,876.03	3,328.343	6,578.374	3,328.647	3,328.742
126	3,248.707	6,421.571	9,520.933	12,549.05	15,508.08	3,250.031	6,424.187	3,250.328	3,250.421
127	3,172.863	6,272.226	9,300.340	12,259.37	15,151.41	3,174.156	6,274.781	3,174.446	3,174.537
128	3,099.362	6,127.476	9,086.509	11,978.55	14,805.59	3,100.625	6,129.973	3,100.908	3,100.997
129	3,028.114	5,987.147	8,879.184	11,706.23	14,470.23	3,029.348	5,989.586	3,029.625	3,029.711
130	2,959.033	5,851.070	8,678.119	11,442.11	14,144.91	2,960.239	5,853.454	2,960.509	2,960.594
131	2,892.037	5,719.086	8,483.079	11,185.88	13,829.28	2,893.216	5,721.416	2,893.480	2,893.563
132	2,827.049	5,591.042	8,293.841	10,937.24	13,522.96	2,828.201	5,593.320	2,828.459	2,828.540
133	2,763.993	5,466.792	8,110.190	10,695.91	13,225.63	2,765.119	5,469.020	2,765.372	2,765.451
134	2,702.799	5,346.197	7,931.921	10,461.64	12,936.95	2,703.900	5,348.375	2,704.147	2,704.224
135	2,643.398	5,229.122	7,758.839	10,234.16	12,656.62	2,644.475	5,231.253	2,644.716	2,644.792
136	2,585.725	5,115.442	7,590.757	10,013.22	12,384.32	2,586.778	5,117.526	2,587.015	2,587.088
137	2,529.717	5,005.033	7,427.495	9,798.599	12,119.79	2,530.748	5,007.072	2,530.979	2,531.051
138	2,475.316	4,897.778	7,268.882	9,590.069	11,862.73	2,476.324	4,899.774	2,476.550	2,476.621
139	2,422.463	4,793.567	7,114.754	9,387.415	11,612.89	2,423.450	4,795.520	2,423.671	2,423.740
140	2,371.104	4,692.291	6,964.952	9,190.430	11,370.02	2,372.070	4,694.203	2,372.287	2,372.355
141	2,321.187	4,593.848	6,819.326	8,998.920	11,133.88	2,322.133	4,595.720	2,322.345	2,322.411
142	2,272.661	4,498.140	6,677.733	8,812.693	10,904.23	2,273.587	4,499.973	2,273.795	2,273.860
143	2,225.479	4,405.072	6,540.032	8,631.570	10,680.86	2,226.385	4,406.867	2,226.589	2,226.652
144	2,179.593	4,314.554	6,406.092	8,455.378	10,463.54	2,180.481	4,316.312	2,180.680	2,180.743
145	2,134.961	4,226.499	6,275.784	8,283.948	10,252.08	2,135.830	4,228.221	2,136.026	2,136.087
146	2,091.538	4,140.824	6,148.987	8,117.122	10,046.28	2,092.391	4,142.511	2,092.582	2,092.641
147	2,049.286	4,057.449	6,025.583	7,954.745	9,845.957	2,050.121	4,059.102	2,050.308	2,050.367
148	2,008.163	3,976.298	5,905.459	7,796.671	9,650.921	2,008.982	3,977.918	2,009.165	2,009.223
149	1,968.134	3,897.296	5,788.508	7,642.758	9,461.004	1,968.936	3,898.884	1,969.116	1,969.172
150	1,929.162	3,820.373	5,674.624	7,492.870	9,276.038	1,929.948	3,821.930	1,930.124	1,930.179
151	1,891.212	3,745.462	5,563.708	7,346.876	9,095.862	1,891.982	3,746.988	1,892.155	1,892.209
152	1,854.250	3,672.496	5,455.666	7,204.650	8,920.323	1,855.006	3,673.993	1,855.175	1,855.228
153	1,818.246	3,601.414	5,350.400	7,066.073	8,749.272	1,818.987	3,602.881	1,819.153	1,819.205
154	1,783.168	3,532.154	5,247.827	6,931.026	8,582.567	1,783.894	3,533.593	1,784.057	1,784.108
155	1,748.986	3,464.659	5,147.858	6,799.400	8,420.071	1,749.699	3,466.071	1,749.859	1,749.909
156	1,715.673	3,398.872	5,050.414	6,671.085	8,261.653	1,716.372	3,400.257	1,716.528	1,716.578
157	1,683.200	3,334.741	4,955.413	6,545.979	8,107.182	1,683.886	3,336.100	1,684.039	1,684.088
158	1,651.541	3,272.213	4,862.780	6,423.982	7,956.540	1,652.214	3,273.546	1,652.365	1,652.412
159	1,620.672	3,211.238	4,772.441	6,304.998	7,809.607	1,621.332	3,212.547	1,621.480	1,621.527
160	1,590.567	3,151.769	4,684.327	6,188.935	7,666.270	1,591.215	3,153.054	1,591.360	1,591.406
161	1,561.203	3,093.760	4,598.368	6,075.703	7,526.419	1,561.839	3,095.021	1,561.982	1,562.026
162	1,532.557	3,037.165	4,514.500	5,965.216	7,389.950	1,533.182	3,038.403	1,533.322	1,533.366
163	1,504.608	2,981.943	4,432.659	5,857.393	7,256.761	1,505.221	2,983.158	1,505.359	1,505.402
164	1,477.335	2,928.051	4,352.785	5,752.152	7,126.753	1,477.937	2,929.244	1,478.072	1,478.114
165	1,450.716	2,875.450	4,274.818	5,649.418	6,999.833	1,451.307	2,876.622	1,451.440	1,451.482
166	1,424.734	2,824.101	4,198.702	5,549.116	6,875.908	1,425.314	2,825.252	1,425.444	1,425.485
167	1,399.368	2,773.968	4,124.383	5,451.175	6,754.892	1,399.938	2,775.099	1,400.066	1,400.106
168	1,374.601	2,725.015	4,051.807	5,355.525	6,636.700	1,375.161	2,726.125	1,375.286	1,375.326
169	1,350.414	2,677.206	3,980.924	5,262.100	6,521.249	1,350.965	2,678.297	1,351.088	1,351.127
170	1,326.792	2,630.510	3,911.685	5,170.835	6,408.461	1,327.333	2,631.582	1,327.454	1,327.492
171	1,303.718	2,584.893	3,844.043	5,081.669	6,298.259	1,304.249	2,585.946	1,304.368	1,304.405
172	1,281.175	2,540.325	3,777.951	4,994.541	6,190.569	1,281.697	2,541.360	1,281.814	1,281.851
173	1,259.150	2,496.776	3,713.366	4,909.394	6,085.321	1,259.663	2,497.793	1,259.778	1,259.814
174	1,237.626	2,454.216	3,650.244	4,826.171	5,982.445	1,238.130	2,455.216	1,238.243	1,238.279
175	1,216.590	2,412.618	3,588.545	4,744.819	5,881.875	1,217.086	2,413.601	1,217.197	1,217.232
176	1,196.028	2,371.955	3,528.229	4,665.285	5,783.548	1,196.516	2,372.922	1,196.625	1,196.659
177	1,175.927	2,332.201	3,469.257	4,587.519	5,687.399	1,176.406	2,333.151	1,176.514	1,176.547

(continued)

Table B.1. (continued)

n	Line transition (frequencies in MHz)							
	Hnα	Hnβ	Hnγ	Hnδ	Hnϵ	Henα	Henβ	Cnα	Snα
178	1,156.274	2,293.330	3,411.592	4,511.472	5,593.371	1,156.745	2,294.265	1,156.851	1,156.884
179	1,137.056	2,255.318	3,355.198	4,437.097	5,501.403	1,137.520	2,256.237	1,137.624	1,137.656
180	1,118.262	2,218.142	3,300.040	4,364.347	5,411.442	1,118.718	2,219.046	1,118.820	1,118.852
181	1,099.880	2,181.778	3,246.085	4,293.179	5,323.431	1,100.328	2,182.667	1,100.429	1,100.460
182	1,081.898	2,146.205	3,193.300	4,223.551	5,237.318	1,082.339	2,147.080	1,082.438	1,082.469
183	1,064.307	2,111.401	3,141.652	4,155.420	5,153.053	1,064.740	2,112.262	1,064.838	1,064.868
184	1,047.094	2,077.346	3,091.113	4,088.746	5,070.586	1,047.521	2,078.192	1,047.617	1,047.647
185	1,030.251	2,044.019	3,041.652	4,023.492	4,989.870	1,030.671	2,044.851	1,030.765	1,030.795
186	1,013.767	2,011.401	2,993.240	3,959.618	4,910.857	1,014.180	2,012.220	1,014.273	1,014.302
187	997.6333	1,979.473	2,945.851	3,897.090	4,833.504	998.0398	1,980.280	998.1310	998.1595
188	981.8398	1,948.218	2,899.457	3,835.871	4,757.768	982.2399	1,949.012	982.3296	982.3576
189	966.3779	1,917.617	2,854.031	3,775.928	4,683.605	966.7717	1,918.398	966.8600	966.8876
190	951.2389	1,887.653	2,809.550	3,717.227	4,610.976	951.6265	1,888.423	951.7135	951.7407
191	936.4146	1,858.311	2,765.988	3,659.737	4,539.841	936.7961	1,859.068	936.8817	936.9085
192	921.8966	1,829.574	2,723.323	3,603.427	4,470.162	922.2723	1,830.319	922.3565	922.3829
193	907.6773	1,801.426	2,681.530	3,548.265	4,401.902	908.0471	1,802.160	908.1301	908.1560
194	893.7488	1,773.853	2,640.588	3,494.224	4,335.024	894.1130	1,774.576	894.1947	894.2203
195	880.1039	1,746.839	2,600.476	3,441.275	4,269.495	880.4626	1,747.551	880.5430	880.5682
196	866.7354	1,720.372	2,561.171	3,389.391	4,205.279	867.0886	1,721.073	867.1678	867.1925
197	853.6362	1,694.436	2,522.655	3,338.544	4,142.345	853.9841	1,695.126	854.0621	854.0865
198	840.7997	1,669.019	2,484.908	3,288.709	4,080.661	841.1423	1,669.699	841.2192	841.2432
199	828.2193	1,644.108	2,447.909	3,239.861	4,020.195	828.5567	1,644.778	828.6324	828.6561
200	815.8886	1,619.690	2,411.642	3,191.976	3,960.918	816.2210	1,620.350	816.2956	816.3189
201	803.8014	1,595.753	2,376.087	3,145.030	3,902.801	804.1290	1,596.404	804.2024	804.2254
202	791.9519	1,572.286	2,341.228	3,099.000	3,845.816	792.2746	1,572.927	792.3469	792.3696
203	780.3341	1,549.277	2,307.048	3,053.864	3,789.934	780.6521	1,549.908	780.7234	780.7457
204	768.9424	1,526.714	2,273.530	3,009.600	3,735.130	769.2558	1,527.336	769.3260	769.3480
205	757.7714	1,504.587	2,240.658	2,966.188	3,681.378	758.0802	1,505.200	758.1495	758.1711
206	746.8158	1,482.886	2,208.416	2,923.606	3,628.652	747.1201	1,483.490	747.1884	747.2097
207	736.0703	1,461.600	2,176.790	2,881.836	3,576.928	736.3703	1,462.196	736.4375	736.4586
208	725.5300	1,440.720	2,145.766	2,840.858	3,526.183	725.8257	1,441.307	725.8920	725.9127
209	715.1900	1,420.236	2,115.328	2,800.653	3,476.393	715.4814	1,420.814	715.5468	715.5672
210	705.0455	1,400.138	2,085.463	2,761.203	3,427.536	705.3328	1,400.708	705.3973	705.4174
211	695.0920	1,380.417	2,056.157	2,722.490	3,379.590	695.3753	1,380.980	695.4388	695.4587
212	685.3250	1,361.065	2,027.398	2,684.498	3,332.535	685.6043	1,361.620	685.6669	685.6865
213	675.7401	1,342.073	1,999.173	2,647.210	3,286.349	676.0155	1,342.620	676.0772	676.0965
214	666.3331	1,323.433	1,971.470	2,610.609	3,241.012	666.6046	1,323.972	666.6655	666.6846
215	657.0999	1,305.136	1,944.276	2,574.679	3,196.506	657.3677	1,305.668	657.4277	657.4465
216	648.0365	1,287.176	1,917.579	2,539.406	3,152.811	648.3006	1,287.700	648.3598	648.3784
217	639.1391	1,269.543	1,891.370	2,504.775	3,109.909	639.3995	1,270.060	639.4579	639.4762
218	630.4038	1,252.231	1,865.636	2,470.770	3,067.782	630.6606	1,252.741	630.7183	630.7363
219	621.8269	1,235.232	1,840.366	2,437.378	3,026.412	622.0803	1,235.735	622.1371	622.1549
220	613.4049	1,218.539	1,815.551	2,404.585	2,985.783	613.6549	1,219.036	613.7109	613.7285
221	605.1343	1,202.146	1,791.180	2,372.378	2,945.877	605.3809	1,202.636	605.4362	605.4535
222	597.0118	1,186.046	1,767.243	2,340.743	2,906.680	597.2551	1,186.529	597.3096	597.3267
223	589.0340	1,170.232	1,743.731	2,309.668	2,868.175	589.2740	1,170.708	589.3278	589.3446
224	581.1976	1,154.697	1,720.634	2,279.141	2,830.348	581.4345	1,155.168	581.4876	581.5042
225	573.4997	1,139.437	1,697.944	2,249.150	2,793.182	573.7334	1,139.901	573.7858	573.8022
226	565.9371	1,124.444	1,675.650	2,219.682	2,756.665	566.1677	1,124.902	566.2195	566.2356
227	558.5069	1,109.713	1,653.745	2,190.728	2,720.781	558.7345	1,110.165	558.7856	558.8015
228	551.2062	1,095.238	1,632.221	2,162.274	2,685.517	551.4308	1,095.685	551.4812	551.4970
229	544.0322	1,081.014	1,611.068	2,134.311	2,650.861	544.2539	1,081.455	544.3036	544.3192
230	536.9822	1,067.036	1,590.279	2,106.828	2,616.798	537.2010	1,067.470	537.2501	537.2654
231	530.0534	1,053.297	1,569.846	2,079.815	2,583.316	530.2694	1,053.726	530.3178	530.3330
232	523.2433	1,039.793	1,549.762	2,053.262	2,550.403	523.4566	1,040.216	523.5044	523.5193
233	516.5494	1,026.519	1,530.019	2,027.159	2,518.046	516.7599	1,026.937	516.8071	516.8219
234	509.9692	1,013.470	1,510.610	2,001.497	2,486.235	510.1770	1,013.883	510.2236	510.2382
235	503.5003	1,000.641	1,491.528	1,976.266	2,454.958	503.7055	1,001.048	503.7515	503.7659
236	497.1404	988.0274	1,472.766	1,951.457	2,424.202	497.3429	988.4300	497.3884	497.4026

(continued)

Table B.1. (continued)

n	Line transition (frequencies in MHz)								
	$Hn\alpha$	$Hn\beta$	$Hn\gamma$	$Hn\delta$	$Hn\epsilon$	$Hen\alpha$	$Hen\beta$	$Cn\alpha$	$Sn\alpha$
237	490.8871	975.6252	1,454.317	1,927.062	2,393.959	491.0871	976.0228	491.1319	491.1460
238	484.7382	963.4298	1,436.175	1,903.072	2,364.216	484.9357	963.8224	484.9800	484.9939
239	478.6916	951.4367	1,418.334	1,879.478	2,334.964	478.8867	951.8244	478.9304	478.9441
240	472.7452	939.6419	1,400.786	1,856.273	2,306.193	472.9378	940.0248	472.9810	472.9945
241	466.8968	928.0413	1,383.528	1,833.448	2,277.893	467.0870	928.4195	467.1297	467.1431
242	461.1445	916.6308	1,366.551	1,810.996	2,250.053	461.3324	917.0043	461.3746	461.3877
243	455.4863	905.4066	1,349.851	1,788.909	2,222.666	455.6719	905.7756	455.7135	455.7266
244	449.9203	894.3650	1,333.422	1,767.179	2,195.721	450.1037	894.7294	450.1448	450.1576
245	444.4447	883.5021	1,317.259	1,745.801	2,169.210	444.6258	883.8621	444.6664	444.6791
246	439.0575	872.8145	1,301.356	1,724.765	2,143.124	439.2364	873.1701	439.2765	439.2891
247	433.7570	862.2985	1,285.708	1,704.067	2,117.455	433.9338	862.6499	433.9734	433.9858
248	428.5415	851.9508	1,270.310	1,683.698	2,092.194	428.7161	852.2980	428.7553	428.7676
249	423.4093	841.7680	1,255.156	1,663.652	2,067.333	423.5818	842.1111	423.6205	423.6326
250	418.3587	831.7469	1,240.243	1,643.924	2,042.865	418.5292	832.0858	418.5674	418.5794
251	413.3882	821.8842	1,225.565	1,624.506	2,018.781	413.5566	822.2191	413.5944	413.6062
252	408.4960	812.1768	1,211.118	1,605.393	1,995.074	408.6625	812.5077	408.6998	408.7115
253	403.6808	802.6217	1,196.897	1,586.578	1,971.737	403.8453	802.9487	403.8822	403.8937
254	398.9409	793.2158	1,182.897	1,568.056	1,948.763	399.1035	793.5391	399.1399	399.1513
255	394.2749	783.9564	1,169.116	1,549.822	1,926.144	394.4356	784.2759	394.4716	394.4829
256	389.6815	774.8406	1,155.547	1,531.869	1,903.874	389.8403	775.1563	389.8759	389.8870
257	385.1591	765.8655	1,142.188	1,514.192	1,881.946	385.3160	766.1776	385.3512	385.3622
258	380.7064	757.0285	1,129.033	1,496.787	1,860.353	380.8615	757.3370	380.8963	380.9072
259	376.3221	748.3269	1,116.080	1,479.647	1,839.090	376.4754	748.6319	376.5098	376.5206
260	372.0048	739.7582	1,103.325	1,462.768	1,818.149	372.1564	740.0597	372.1904	372.2011
261	367.7534	731.3198	1,090.763	1,446.144	1,797.525	367.9032	731.6178	367.9368	367.9473
262	363.5664	723.0093	1,078.391	1,429.771	1,777.212	363.7146	723.3039	363.7478	363.7582
263	359.4428	714.8242	1,066.205	1,413.645	1,757.203	359.5893	715.1155	359.6222	359.6324
264	355.3814	706.7622	1,054.202	1,397.760	1,737.494	355.5262	707.0502	355.5587	355.5688
265	351.3808	698.8210	1,042.379	1,382.113	1,718.079	351.5240	699.1057	351.5561	351.5662
266	347.4401	690.9983	1,030.732	1,366.698	1,698.952	347.5817	691.2799	347.6135	347.6234
267	343.5581	683.2919	1,019.258	1,351.511	1,680.107	343.6981	683.5703	343.7295	343.7394
268	339.7338	675.6997	1,007.953	1,336.549	1,661.541	339.8722	675.9750	339.9033	339.9130
269	335.9659	668.2196	996.8154	1,321.807	1,643.247	336.1028	668.4919	336.1335	336.1431
270	332.2536	660.8494	985.8409	1,307.281	1,625.220	332.3890	661.1187	332.4194	332.4289
271	328.5958	653.5873	975.0270	1,292.966	1,607.456	328.7297	653.8536	328.7597	328.7691
272	324.9915	646.4312	964.3706	1,278.860	1,589.950	325.1239	646.6946	325.1536	325.1629
273	321.4397	639.3791	953.8690	1,264.959	1,572.698	321.5707	639.6397	321.6000	321.6092
274	317.9394	632.4293	943.5192	1,251.258	1,555.694	318.0690	632.6870	318.0981	318.1071
275	314.4898	625.5798	933.3187	1,237.755	1,538.934	314.6180	625.8347	314.6467	314.6557
276	311.0900	618.8289	923.2647	1,224.445	1,522.415	311.2167	619.0811	311.2452	311.2541
277	307.7389	612.1748	913.3546	1,211.325	1,506.131	307.8643	612.4242	307.8924	307.9012
278	304.4358	605.6157	903.5858	1,198.392	1,490.078	304.5599	605.8625	304.5877	304.5964
279	301.1799	599.1500	893.9558	1,185.642	1,474.253	301.3026	599.3941	301.3301	301.3387
280	297.9701	592.7760	884.4622	1,173.073	1,458.650	298.0916	593.0175	298.1188	298.1273
281	294.8059	586.4921	875.1026	1,160.680	1,443.268	294.9260	586.7311	294.9529	294.9614
282	291.6862	580.2967	865.8745	1,148.462	1,428.101	291.8051	580.5332	291.8318	291.8401
283	288.6105	574.1883	856.7757	1,136.414	1,413.145	288.7281	574.4222	288.7545	288.7627
284	285.5778	568.1653	847.8040	1,124.535	1,398.398	285.6942	568.3968	285.7203	285.7284
285	282.5875	562.2262	838.9571	1,112.820	1,383.855	282.7026	562.4553	282.7284	282.7365
286	279.6387	556.3696	830.2329	1,101.268	1,369.514	279.7527	556.5963	279.7782	279.7862
287	276.7309	550.5941	821.6292	1,089.875	1,355.369	276.8437	550.8185	276.8690	276.8769
288	273.8632	544.8983	813.1440	1,078.638	1,341.419	273.9748	545.1203	273.9999	274.0077
289	271.0350	539.2807	804.7752	1,067.556	1,327.660	271.1455	539.5005	271.1703	271.1780
290	268.2457	533.7401	796.5208	1,056.625	1,314.088	268.3550	533.9576	268.3795	268.3872
291	265.4945	528.2752	788.3790	1,045.842	1,300.700	265.6026	528.4904	265.6269	265.6345
292	262.7807	522.8846	780.3478	1,035.206	1,287.494	262.8878	523.0976	262.9118	262.9193
293	260.1038	517.5671	772.4252	1,024.713	1,274.466	260.2098	517.7780	260.2336	260.2410
294	257.4632	512.3214	764.6096	1,014.362	1,261.613	257.5681	512.5302	257.5917	257.5990
295	254.8582	507.1464	756.8991	1,004.150	1,248.933	254.9620	507.3531	254.9853	254.9926

(continued)

Table B.1. (continued)

n	Line transition (frequencies in MHz)								
	H$n\alpha$	H$n\beta$	H$n\gamma$	H$n\delta$	H$n\epsilon$	He$n\alpha$	He$n\beta$	C$n\alpha$	S$n\alpha$
296	252.2882	502.0409	749.2918	994.0744	1,236.421	252.3910	502.2454	252.4141	252.4213
297	249.7527	497.0036	741.7862	984.1331	1,224.077	249.8544	497.2061	249.8773	249.8844
298	247.2510	492.0335	734.3805	974.3240	1,211.896	247.3517	492.2340	247.3743	247.3814
299	244.7826	487.1295	727.0730	964.6448	1,199.876	244.8823	487.3280	244.9047	244.9117
300	242.3469	482.2905	719.8622	955.0934	1,188.015	242.4457	482.4870	242.4678	242.4748
301	239.9435	477.5153	712.7464	945.6676	1,176.309	240.0413	477.7098	240.0632	240.0701
302	237.5718	472.8029	705.7241	936.3655	1,164.757	237.6686	472.9956	237.6903	237.6971
303	235.2312	468.1524	698.7937	927.1850	1,153.355	235.3270	468.3431	235.3485	235.3552
304	232.9212	463.5626	691.9538	918.1241	1,142.102	233.0161	463.7515	233.0374	233.0441
305	230.6414	459.0326	685.2029	909.1809	1,130.995	230.7354	459.2197	230.7565	230.7630
306	228.3912	454.5615	678.5395	900.3535	1,120.031	228.4843	454.7467	228.5052	228.5117
307	226.1703	450.1483	671.9622	891.6399	1,109.209	226.2624	450.3317	226.2831	226.2896
308	223.9780	445.7920	665.4697	883.0385	1,098.525	224.0693	445.9736	224.0897	224.0961
309	221.8140	441.4917	659.0605	874.5473	1,087.979	221.9043	441.6716	221.9246	221.9310
310	219.6777	437.2465	652.7334	866.1647	1,077.567	219.7672	437.4247	219.7873	219.7936
311	217.5688	433.0556	646.4870	857.8888	1,067.287	217.6575	433.2321	217.6774	217.6836
312	215.4868	428.9181	640.3200	849.7181	1,057.138	215.5746	429.0929	215.5943	215.6005
313	213.4313	424.8332	634.2312	841.6508	1,047.117	213.5183	425.0063	213.5378	213.5439
314	211.4019	420.7999	628.2194	833.6852	1,037.222	211.4880	420.9714	211.5073	211.5134
315	209.3981	416.8176	622.2834	825.8199	1,027.451	209.4834	416.9874	209.5025	209.5085
316	207.4195	412.8853	616.4218	818.0532	1,017.803	207.5040	413.0535	207.5230	207.5289
317	205.4658	409.0023	610.6337	810.3836	1,008.275	205.5495	409.1690	205.5683	205.5742
318	203.5365	405.1679	604.9178	802.8096	998.8662	203.6195	405.3330	203.6381	203.6439
319	201.6314	401.3813	599.2731	795.3297	989.5737	201.7135	401.5448	201.7320	201.7377
320	199.7499	397.6417	593.6983	787.9424	980.3962	199.8313	397.8037	199.8496	199.8553
321	197.8918	393.9484	588.1925	780.6463	971.3318	197.9724	394.1089	197.9905	197.9962
322	196.0566	390.3007	582.7545	773.4400	962.3788	196.1365	390.4597	196.1544	196.1600
323	194.2441	386.6979	577.3834	766.3221	953.5354	194.3232	386.8555	194.3410	194.3465
324	192.4538	383.1393	572.0781	759.2914	944.8001	192.5322	383.2954	192.5498	192.5553
325	190.6855	379.6242	566.8375	752.3463	936.1712	190.7632	379.7789	190.7806	190.7861
326	188.9388	376.1520	561.6608	745.4857	927.6470	189.0157	376.3053	189.0330	189.0384
327	187.2133	372.7221	556.5470	738.7083	919.2260	187.2896	372.8739	187.3067	187.3120
328	185.5088	369.3337	551.4950	732.0128	910.9067	185.5844	369.4842	185.6013	185.6066
329	183.8249	365.9862	546.5040	725.3979	902.6874	183.8998	366.1353	183.9166	183.9218
330	182.1613	362.6791	541.5730	718.8625	894.5667	182.2355	362.8269	182.2522	182.2574
331	180.5178	359.4117	536.7012	712.4054	886.5431	180.5913	359.5581	180.6078	180.6130
332	178.8939	356.1834	531.8876	706.0254	878.6152	178.9668	356.3286	178.9832	178.9883
333	177.2895	352.9937	527.1314	699.7213	870.7816	177.3617	353.1375	177.3779	177.3830
334	175.7042	349.8419	522.4318	693.4921	863.0408	175.7758	349.9845	175.7919	175.7969
335	174.1377	346.7276	517.7879	687.3366	855.3914	174.2087	346.8689	174.2246	174.2296
336	172.5899	343.6501	513.1988	681.2537	847.8323	172.6602	343.7902	172.6760	172.6809
337	171.0603	340.6090	508.6638	675.2424	840.3619	171.1300	340.7478	171.1456	171.1505
338	169.5487	337.6036	504.1821	669.3016	832.9790	169.6178	337.7411	169.6333	169.6381
339	168.0549	334.6334	499.7529	663.4303	825.6823	168.1234	334.7698	168.1387	168.1435
340	166.5786	331.6980	495.3754	657.6275	818.4707	166.6464	331.8332	166.6617	166.6664
341	165.1195	328.7969	491.0489	651.8921	811.3428	165.1868	328.9308	165.2019	165.2066
342	163.6774	325.9294	486.7726	646.2233	804.2974	163.7441	326.0623	163.7590	163.7637
343	162.2520	323.0953	482.5459	640.6200	797.3333	162.3182	323.2269	162.3330	162.3376
344	160.8432	320.2938	478.3679	635.0812	790.4494	160.9087	320.4243	160.9234	160.9280
345	159.4506	317.5247	474.2380	629.6062	783.6445	159.5156	317.6541	159.5302	159.5347
346	158.0741	314.7874	470.1556	624.1939	776.9175	158.1385	314.9157	158.1529	158.1575
347	156.7133	312.0815	466.1198	618.8435	770.2673	156.7772	312.2087	156.7915	156.7960
348	155.3682	309.4065	462.1301	613.5540	763.6928	155.4315	309.5326	155.4457	155.4501
349	154.0383	306.7620	458.1858	608.3246	757.1929	154.1011	306.8870	154.1152	154.1196
350	152.7236	304.1475	454.2863	603.1546	750.7665	152.7859	304.2714	152.7998	152.8042
351	151.4239	301.5627	450.4309	598.0429	744.4127	151.4856	301.6856	151.4994	151.5037
352	150.1388	299.0070	446.6190	592.9888	738.1303	150.2000	299.1289	150.2137	150.2180
353	148.8682	296.4802	442.8500	587.9915	731.9185	148.9289	296.6010	148.9425	148.9468
354	147.6120	293.9818	439.1233	583.0502	725.7761	147.6721	294.1016	147.6856	147.6898

(continued)

Table B.1. (continued)

n	Line transition (frequencies in MHz)								
	Hnα	Hnβ	Hnγ	Hnδ	Hnε	Henα	Henβ	Cnα	Snα
355	146.3698	291.5113	435.4383	578.1642	719.7023	146.4294	291.6301	146.4428	146.4470
356	145.1415	289.0685	431.7944	573.3325	713.6961	145.2007	289.1863	145.2139	145.2181
357	143.9270	286.6528	428.1910	568.5546	707.7565	143.9856	286.7697	143.9988	144.0029
358	142.7259	284.2641	424.6276	563.8296	701.8827	142.7841	284.3799	142.7971	142.8012
359	141.5382	281.9017	421.1037	559.1568	696.0737	141.5958	282.0166	141.6088	141.6128
360	140.3636	279.5655	417.6186	554.5355	690.3286	140.4208	279.6795	140.4336	140.4376
361	139.2020	277.2551	414.1719	549.9650	684.6465	139.2587	277.3680	139.2714	139.2754
362	138.0531	274.9700	410.7631	545.4446	679.0267	138.1094	275.0820	138.1220	138.1259
363	136.9169	272.7099	407.3915	540.9736	673.4682	136.9727	272.8211	136.9852	136.9891
364	135.7931	270.4746	404.0567	536.5513	667.9701	135.8484	270.5848	135.8608	135.8647
365	134.6815	268.2636	400.7582	532.1771	662.5318	134.7364	268.3729	134.7487	134.7526
366	133.5821	266.0767	397.4955	527.8503	657.1524	133.6365	266.1851	133.6487	133.6526
367	132.4946	263.9134	394.2682	523.5703	651.8310	132.5486	264.0210	132.5607	132.5645
368	131.4189	261.7736	391.0757	519.3364	646.5669	131.4724	261.8803	131.4844	131.4882
369	130.3547	259.6568	387.9175	515.1481	641.3594	130.4079	259.7626	130.4198	130.4235
370	129.3021	257.5628	384.7933	511.0046	636.2076	129.3548	257.6678	129.3666	129.3703
371	128.2607	255.4912	381.7026	506.9055	631.1109	128.3130	255.5954	128.3247	128.3284
372	127.2305	253.4418	378.6448	502.8502	626.0684	127.2824	253.5451	127.2940	127.2976
373	126.2113	251.4143	375.6196	498.8379	621.0796	126.2627	251.5167	126.2743	126.2779
374	125.2030	249.4083	372.6266	494.8683	616.1436	125.2540	249.5100	125.2654	125.2690
375	124.2053	247.4236	369.6653	490.9406	611.2598	124.2560	247.5245	124.2673	124.2709
376	123.2183	245.4599	366.7353	487.0544	606.4274	123.2685	245.5600	123.2798	123.2833
377	122.2417	243.5170	363.8361	483.2091	601.6459	122.2915	243.6162	122.3026	122.3061
378	121.2753	241.5945	360.9675	479.4042	596.9144	121.3247	241.6929	121.3358	121.3393
379	120.3191	239.6921	358.1289	475.6391	592.2325	120.3682	239.7898	120.3792	120.3826
380	119.3730	237.8097	355.3200	471.9134	587.5994	119.4216	237.9066	119.4325	119.4360
381	118.4367	235.9470	352.5404	468.2264	583.0145	118.4850	236.0431	118.4958	118.4992
382	117.5102	234.1036	349.7897	464.5778	578.4772	117.5581	234.1990	117.5689	117.5722
383	116.5934	232.2794	347.0676	460.9670	573.9869	116.6409	232.3741	116.6516	116.6549
384	115.6861	230.4742	344.3736	457.3935	569.5429	115.7332	230.5681	115.7438	115.7471
385	114.7881	228.6875	341.7074	453.8568	565.1446	114.8349	228.7807	114.8454	114.8486
386	113.8994	226.9193	339.0687	450.3565	560.7916	113.9458	227.0118	113.9562	113.9595
387	113.0199	225.1693	336.4571	446.8921	556.4831	113.0659	225.2610	113.0763	113.0795
388	112.1494	223.4372	333.8722	443.4632	552.2186	112.1951	223.5283	112.2053	112.2085
389	111.2878	221.7229	331.3138	440.0692	547.9976	111.3332	221.8132	111.3433	111.3465
390	110.4350	220.0260	328.7814	436.7098	543.8196	110.4800	220.1156	110.4901	110.4933
391	109.5909	218.3464	326.2748	433.3845	539.6839	109.6356	218.4354	109.6456	109.6488
392	108.7554	216.6838	323.7936	430.0929	535.5900	108.7998	216.7721	108.8097	108.8128
393	107.9284	215.0381	321.3375	426.8345	531.5374	107.9724	215.1258	107.9822	107.9853
394	107.1097	213.4091	318.9061	423.6090	527.5256	107.1534	213.4960	107.1632	107.1662
395	106.2993	211.7964	316.4992	420.4159	523.5541	106.3426	211.8827	106.3524	106.3554
396	105.4971	210.1999	314.1165	417.2547	519.6223	105.5401	210.2856	105.5497	105.5527
397	104.7029	208.6195	311.7577	414.1252	515.7298	104.7455	208.7045	104.7551	104.7581
398	103.9166	207.0548	309.4224	411.0270	511.8761	103.9590	207.1392	103.9684	103.9714
399	103.1382	205.5058	307.1103	407.9595	508.0607	103.1802	205.5895	103.1897	103.1926
400	102.3676	203.9721	304.8213	404.9225	504.2831	102.4093	204.0553	102.4186	102.4216
401	101.6046	202.4537	302.5549	401.9156	500.5429	101.6460	202.5362	101.6553	101.6582
402	100.8492	200.9504	300.3110	398.9383	496.8396	100.8902	201.0322	100.8995	100.9023
403	100.1012	199.4618	298.0892	395.9904	493.1727	100.1420	199.5431	100.1511	100.1540
404	99.36063	197.9880	295.8892	393.0715	489.5418	99.40111	198.0686	99.41019	99.41303
405	98.62734	196.5286	293.7109	390.1812	485.9465	98.66753	196.6087	98.67654	98.67936
406	97.90125	195.0835	291.5538	387.3191	482.3863	97.94114	195.1630	97.95009	97.95289
407	97.18227	193.6526	289.4179	384.4850	478.8607	97.22187	193.7315	97.23075	97.23353
408	96.47031	192.2356	287.3027	381.6785	475.3695	96.50962	192.3139	96.51844	96.52120
409	95.76529	190.8324	285.2082	378.8992	471.9121	95.80432	190.9102	95.81307	95.81581
410	95.06713	189.4429	283.1339	376.1468	468.4882	95.10587	189.5201	95.11455	95.11727
411	94.37573	188.0668	281.0797	373.4210	465.0973	94.41419	188.1434	94.42281	94.42551
412	93.69102	186.7039	279.0453	370.7215	461.7390	93.72920	186.7800	93.73776	93.74044
413	93.01292	185.3543	277.0305	368.0480	458.4130	93.05083	185.4298	93.05933	93.06198

(continued)

Table B.1. (continued)

n	Line transition (frequencies in MHz)							
	H$n\alpha$	H$n\beta$	H$n\gamma$	H$n\delta$	H$n\epsilon$	He$n\alpha$	He$n\beta$	C$n\alpha$	S$n\alpha$
414	92.34135	184.0176	275.0351	365.4001	455.1189	92.37898	184.0926	92.38742	92.39006
415	91.67623	182.6937	273.0587	362.7775	451.8562	91.71359	182.7682	91.72196	91.72458
416	91.01748	181.3825	271.1013	360.1800	448.6247	91.05457	181.4564	91.06289	91.06549
417	90.36503	180.0838	269.1625	357.6072	445.4239	90.40185	180.1572	90.41011	90.41269
418	89.71879	178.7975	267.2422	355.0589	442.2535	89.75535	178.8704	89.76355	89.76612
419	89.07871	177.5234	265.3401	352.5347	439.1131	89.11501	177.5957	89.12315	89.12569
420	88.44470	176.2614	263.4560	350.0344	436.0024	88.48074	176.3332	88.48882	88.49135
421	87.81669	175.0113	261.5897	347.5577	432.9209	87.85247	175.0826	87.86050	87.86301
422	87.19461	173.7730	259.7410	345.1043	429.8685	87.23014	173.8438	87.23811	87.24060
423	86.57840	172.5464	257.9096	342.6739	426.8447	86.61368	172.6167	86.62159	86.62406
424	85.96797	171.3313	256.0955	340.2663	423.8491	86.00300	171.4011	86.01086	86.01332
425	85.36328	170.1275	254.2983	337.8812	420.8816	85.39806	170.1968	85.40586	85.40830
426	84.76424	168.9350	252.5179	335.5183	417.9417	84.79878	169.0039	84.80652	84.80895
427	84.17079	167.7537	250.7541	333.1774	415.0291	84.20509	167.8220	84.21278	84.21519
428	83.58287	166.5833	249.0066	330.8583	412.1434	83.61693	166.6512	83.62457	83.62695
429	83.00041	165.4238	247.2754	328.5606	409.2845	83.03423	165.4912	83.04182	83.04419
430	82.42335	164.2750	245.5602	326.2841	406.4520	82.45694	164.3419	82.46447	82.46683
431	81.85163	163.1368	243.8608	324.0286	403.6455	81.88498	163.2033	81.89246	81.89480
432	81.28518	162.0091	242.1770	321.7939	400.8649	81.31831	162.0752	81.32574	81.32806
433	80.72395	160.8918	240.5087	319.5797	398.1097	80.75685	160.9574	80.76422	80.76653
434	80.16788	159.7848	238.8557	317.3857	395.3797	80.20054	159.8499	80.20787	80.21016
435	79.61690	158.6878	237.2178	315.2118	392.6746	79.64934	158.7525	79.65661	79.65889
436	79.07095	157.6009	235.5949	313.0577	389.9941	79.10317	157.6652	79.11040	79.11266
437	78.52999	156.5239	233.9867	310.9231	387.3379	78.56199	156.5877	78.56917	78.57141
438	77.99395	155.4567	232.3931	308.8080	384.7059	78.02573	155.5201	78.03286	78.03509
439	77.46278	154.3992	230.8140	306.7119	382.0976	77.49434	154.4621	77.50142	77.50363
440	76.93642	153.3512	229.2491	304.6348	379.5128	76.96777	153.4137	76.97480	76.97700
441	76.41481	152.3127	227.6984	302.5764	376.9513	76.44595	152.3748	76.45294	76.45512
442	75.89792	151.2836	226.1616	300.5365	374.4129	75.92884	151.3452	75.93578	75.93795
443	75.38567	150.2637	224.6386	298.5149	371.8971	75.41639	150.3249	75.42328	75.42543
444	74.87802	149.2529	223.1293	296.5114	369.4038	74.90854	149.3138	74.91538	74.91752
445	74.37492	148.2512	221.6334	294.5258	366.9328	74.40523	148.3117	74.41203	74.41415
446	73.87632	147.2585	220.1509	292.5579	364.4838	73.90643	147.3185	73.91318	73.91529
447	73.38217	146.2746	218.6816	290.6075	362.0565	73.41207	146.3342	73.41878	73.42087
448	72.89241	145.2994	217.2253	288.6743	359.6507	72.92211	145.3586	72.92877	72.93086
449	72.40700	144.3329	215.7819	286.7583	357.2662	72.43651	144.3917	72.44312	72.44519
450	71.92589	143.3749	214.3513	284.8592	354.9027	71.95520	143.4334	71.96177	71.96383
451	71.44904	142.4254	212.9333	282.9768	352.5600	71.47815	142.4835	71.48468	71.48672
452	70.97639	141.4843	211.5278	281.1110	350.2379	71.00531	141.5419	71.01179	71.01382
453	70.50790	140.5514	210.1346	279.2616	347.9362	70.53663	140.6087	70.54307	70.54508
454	70.04352	139.6267	208.7537	277.4283	345.6546	70.07206	139.6836	70.07846	70.08046
455	69.58321	138.7101	207.3848	275.6110	343.3928	69.61157	138.7667	69.61793	69.61991
456	69.12693	137.8016	206.0278	273.8096	341.1508	69.15510	137.8577	69.16142	69.16339
457	68.67463	136.9009	204.6827	272.0239	338.9282	68.70261	136.9567	68.70889	68.71085
458	68.22626	136.0081	203.3492	270.2536	336.7249	68.25406	136.0635	68.26030	68.26225
459	67.78179	135.1230	202.0273	268.4987	334.5407	67.80941	135.1780	67.81561	67.81754
460	67.34118	134.2455	200.7169	266.7589	332.3753	67.36862	134.3002	67.37477	67.37669
461	66.90437	133.3757	199.4177	265.0341	330.2286	66.93163	133.4300	66.93775	66.93966
462	66.47133	132.5134	198.1298	263.3242	328.1003	66.49842	132.5674	66.50449	66.50639
463	66.04202	131.6584	196.8529	261.6289	325.9902	66.06893	131.7121	66.07497	66.07686
464	65.61640	130.8108	195.5869	259.9482	323.8982	65.64314	130.8641	65.64914	65.65101
465	65.19443	129.9705	194.3318	258.2818	321.8241	65.22100	130.0235	65.22696	65.22882
466	64.77607	129.1374	193.0874	256.6297	319.7676	64.80247	129.1900	64.80839	64.81024
467	64.36129	128.3113	191.8536	254.9916	317.7287	64.38751	128.3636	64.39340	64.39523
468	63.95003	127.4923	190.6303	253.3674	315.7070	63.97609	127.5443	63.98194	63.98376
469	63.54228	126.6803	189.4174	251.7570	313.7024	63.56817	126.7319	63.57398	63.57579
470	63.13798	125.8751	188.2147	250.1602	311.7148	63.16371	125.9264	63.16948	63.17128
471	62.73710	125.0767	187.0222	248.5768	309.7440	62.76267	125.1277	62.76840	62.77019
472	62.33961	124.2851	185.8397	247.0069	307.7897	62.36502	124.3357	62.37071	62.37249

(continued)

Table B.1. (continued)

n	Line transition (frequencies in MHz)								
	Hnα	Hnβ	Hnγ	Hnδ	Hnϵ	Henα	Henβ	Cnα	Snα
473	61.94548	123.5001	184.6672	245.4501	305.8518	61.97072	123.5505	61.97638	61.97815
474	61.55465	122.7218	183.5046	243.9063	303.9301	61.57974	122.7718	61.58536	61.58712
475	61.16711	121.9499	182.3517	242.3755	302.0246	61.19204	121.9996	61.19763	61.19937
476	60.78282	121.1845	181.2084	240.8574	300.1349	60.80758	121.2339	60.81314	60.81488
477	60.40173	120.4256	180.0746	239.3521	298.2609	60.42635	120.4746	60.43187	60.43359
478	60.02383	119.6729	178.9503	237.8592	296.4025	60.04829	119.7217	60.05377	60.05549
479	59.64907	118.9265	177.8354	236.3787	294.5596	59.67338	118.9750	59.67883	59.68053
480	59.27743	118.1863	176.7296	234.9105	292.7318	59.30158	118.2344	59.30700	59.30869
481	58.90886	117.4522	175.6331	233.4544	290.9192	58.93287	117.5001	58.93825	58.93993
482	58.54335	116.7242	174.5455	232.0103	289.1215	58.56720	116.7718	58.57255	58.57423
483	58.18085	116.0022	173.4670	230.5781	287.3386	58.20456	116.0495	58.20988	58.21154
484	57.82134	115.2861	172.3973	229.1577	285.5703	57.84490	115.3331	57.85019	57.85184
485	57.46479	114.5759	171.3364	227.7489	283.8164	57.48820	114.6226	57.49345	57.49510
486	57.11116	113.8716	170.2841	226.3517	282.0770	57.13443	113.9180	57.13965	57.14128
487	56.76042	113.1730	169.2405	224.9658	280.3517	56.78355	113.2191	56.78874	56.79036
488	56.41256	112.4801	168.2054	223.5912	278.6404	56.43554	112.5259	56.44070	56.44231
489	56.06753	111.7928	167.1787	222.2279	276.9431	56.09037	111.8384	56.09550	56.09710
490	55.72530	111.1112	166.1603	220.8755	275.2595	55.74801	111.1564	55.75310	55.75470
491	55.38586	110.4350	165.1502	219.5342	273.5895	55.40843	110.4800	55.41349	55.41507
492	55.04917	109.7644	164.1483	218.2036	271.9330	55.07160	109.8091	55.07663	55.07820
493	54.71520	109.0991	163.1545	216.8838	270.2898	54.73750	109.1436	54.74250	54.74406
494	54.38393	108.4393	162.1686	215.5746	268.6599	54.40609	108.4834	54.41106	54.41261
495	54.05532	107.7847	161.1907	214.2760	267.0430	54.07735	107.8286	54.08229	54.08384
496	53.72936	107.1354	160.2206	212.9877	265.4391	53.75126	107.1790	53.75617	53.75770
497	53.40602	106.4913	159.2583	211.7098	263.8480	53.42778	106.5347	53.43266	53.43419
498	53.08526	105.8523	158.3037	210.4420	262.2696	53.10689	105.8955	53.11174	53.11326
499	52.76707	105.2185	157.3568	209.1844	260.7038	52.78857	105.2614	52.79339	52.79490
500	52.45141	104.5897	156.4173	207.9367	259.1504	52.47279	104.6323	52.47758	52.47908
501	52.13827	103.9659	155.4853	206.6990	257.6093	52.15952	104.0083	52.16428	52.16577
502	51.82762	103.3470	154.5607	205.4710	256.0804	51.84873	103.3892	51.85347	51.85495
503	51.51942	102.7331	153.6434	204.2528	254.5636	51.54042	102.7750	51.54513	51.54660
504	51.21367	102.1240	152.7334	203.0442	253.0588	51.23454	102.1656	51.23922	51.24068
505	50.91033	101.5197	151.8305	201.8451	251.5657	50.93108	101.5611	50.93573	50.93719
506	50.60939	100.9202	150.9348	200.6554	250.0845	50.63001	100.9613	50.63463	50.63608
507	50.31081	100.3254	150.0460	199.4751	248.6148	50.33131	100.3663	50.33590	50.33734
508	50.01457	99.73522	149.1643	198.3040	247.1566	50.03495	99.77587	50.03952	50.04095
509	49.72066	99.14969	148.2894	197.1420	245.7098	49.74092	99.19010	49.74546	49.74688
510	49.42904	98.56874	147.4213	195.9891	244.2742	49.44918	98.60890	49.45370	49.45511
511	49.13970	99.99231	146.5601	194.8452	242.8498	49.15972	98.03224	49.16421	49.16562
512	48.85261	97.42037	145.7055	193.7101	241.4365	48.87252	97.46007	48.87698	48.87838
513	48.56776	96.85287	144.8575	192.5839	240.0341	48.58755	96.89234	48.59199	48.59337
514	48.28511	96.28977	144.0161	191.4664	238.6426	48.30479	96.32901	48.30920	48.31058
515	48.00466	95.73103	143.1813	190.3575	237.2618	48.02422	95.77004	48.02861	48.02998
516	47.72637	95.17660	142.3528	189.2571	235.8916	47.74582	95.21538	47.75018	47.75154
517	47.45023	94.62644	141.5308	188.1652	234.5320	47.46956	94.66500	47.47390	47.47526
518	47.17622	94.08052	140.7150	187.0817	233.1827	47.19544	94.11886	47.19975	47.20110
519	46.90431	93.53879	139.9055	186.0065	231.8439	46.92342	93.57691	46.92771	46.92905
520	46.63448	93.00121	139.1022	184.9396	230.5152	46.65349	93.03911	46.65775	46.65908
521	46.36673	92.46774	138.3051	183.8807	229.1967	46.38562	92.50542	46.38986	46.39118
522	46.10101	91.93834	137.5140	182.8300	227.8882	46.11980	91.97581	46.12401	46.12533
523	45.83733	91.41298	136.7289	181.7872	226.5897	45.85601	91.45023	45.86020	45.86151
524	45.57565	90.89161	135.9499	180.7523	225.3010	45.59422	90.92865	45.59839	45.59969
525	45.31596	90.37420	135.1767	179.7253	224.0220	45.33443	90.41103	45.33857	45.33986
526	45.05824	89.86071	134.4093	178.7061	222.7527	45.07660	89.89733	45.08072	45.08201
527	44.80247	89.35110	133.6478	177.6945	221.4930	44.82073	89.38751	44.82482	44.82610
528	44.54863	88.84534	132.8920	176.6905	220.2428	44.56678	88.88154	44.57086	44.57213
529	44.29671	88.34339	132.1419	175.6942	219.0020	44.31476	88.37939	44.31881	44.32007
530	44.04668	87.84521	131.3974	174.7052	217.7704	44.06463	87.88101	44.06865	44.06991
531	43.79853	87.35077	130.6586	173.7237	216.5481	43.81638	87.38636	43.82038	43.82163

(continued)

Table B.1. (continued)

n	Line transition (frequencies in MHz)								
	Hnα	Hnβ	Hnγ	Hnδ	Hnε	Henα	Henβ	Cnα	Snα
532	43.55224	86.86003	129.9252	172.7496	215.3349	43.56999	86.89543	43.57397	43.57521
533	43.30779	86.37297	129.1973	171.7827	214.1308	43.32544	86.40816	43.32940	43.33064
534	43.06517	85.88954	128.4749	170.8230	212.9356	43.08272	85.92453	43.08666	43.08789
535	42.82436	85.40970	127.7578	169.8704	211.7493	42.84181	85.44451	42.84573	42.84695
536	42.58534	84.93344	127.0461	168.9249	210.5718	42.60270	84.96805	42.60659	42.60780
537	42.34810	84.46071	126.3396	167.9864	209.4030	42.36535	84.49513	42.36923	42.37044
538	42.11261	83.99149	125.6383	167.0549	208.2428	42.12977	84.02571	42.13362	42.13483
539	41.87887	83.52573	124.9423	166.1302	207.0912	41.89594	83.55976	41.89976	41.90096
540	41.64686	83.06341	124.2513	165.2124	205.9481	41.66383	83.09725	41.66763	41.66882
541	41.41655	82.60449	123.5655	164.3012	204.8134	41.43343	82.63815	41.43721	41.43840
542	41.18794	82.14895	122.8847	163.3968	203.6870	41.20472	82.18243	41.20849	41.20967
543	40.96101	81.69675	122.2089	162.4990	202.5688	40.97770	81.73004	40.98144	40.98262
544	40.73574	81.24787	121.5380	161.6078	201.4588	40.75234	81.28098	40.75607	40.75723
545	40.51213	80.80227	120.8720	160.7231	200.3569	40.52863	80.83519	40.53234	40.53349
546	40.29014	80.35992	120.2109	159.8448	199.2630	40.30656	80.39266	40.31024	40.31139
547	40.06978	79.92079	119.5546	158.9729	198.1771	40.08610	79.95336	40.08977	40.09091
548	39.85102	79.48486	118.9031	158.1073	197.0990	39.86725	79.51725	39.87090	39.87204
549	39.63384	79.05209	118.2563	157.2480	196.0288	39.64999	79.08430	39.65362	39.65475
550	39.41825	78.62246	117.6142	156.3950	194.9663	39.43431	78.65450	39.43791	39.43904
551	39.20421	78.19594	116.9767	155.5480	193.9114	39.22019	78.22780	39.22377	39.22489
552	38.99172	77.77249	116.3438	154.7072	192.8642	39.00761	77.80418	39.01118	39.01229
553	38.78077	77.35210	115.7155	153.8725	191.8245	38.79657	77.38362	38.80011	38.80122
554	38.57133	76.93473	115.0917	153.0437	190.7922	38.58705	76.96608	38.59057	38.59168
555	38.36340	76.52036	114.4724	152.2209	189.7673	38.37903	76.55154	38.38254	38.38363
556	38.15696	76.10896	113.8575	151.4039	188.7498	38.17251	76.13997	38.17600	38.17709
557	37.95200	75.70050	113.2470	150.5928	187.7395	37.96746	75.73135	37.97093	37.97202
558	37.74850	75.29496	112.6408	149.7875	186.7364	37.76389	75.32564	37.76733	37.76841
559	37.54646	74.89231	112.0390	148.9879	185.7405	37.56176	74.92283	37.56519	37.56626
560	37.34586	74.49253	111.4414	148.1940	184.7516	37.36107	74.52289	37.36449	37.36555
561	37.14668	74.09559	110.8481	147.4057	183.7697	37.16181	74.12578	37.16521	37.16627
562	36.94891	73.70147	110.2590	146.6230	182.7948	36.96397	73.73150	36.96735	36.96840
563	36.75255	73.31013	109.6741	145.8459	181.8267	36.76753	73.34000	36.77089	36.77194
564	36.55758	72.92156	109.0933	145.0742	180.8655	36.57249	72.95127	36.57582	36.57686
565	36.36398	72.53573	108.5166	144.3079	179.9110	36.37880	72.56529	36.38212	36.38316
566	36.17175	72.15262	107.9439	143.5471	178.9633	36.18649	72.18202	36.18979	36.19083
567	35.98087	71.77220	107.3753	142.7915	178.0222	35.99553	71.80144	35.99882	35.99985
568	35.79133	71.39445	106.8107	142.0413	177.0877	35.80591	71.42354	35.80919	35.81021
569	35.60312	71.01935	106.2500	141.2963	176.1597	35.61763	71.04829	35.62088	35.62190
570	35.41623	70.64687	105.6932	140.5566	175.2382	35.43066	70.67565	35.43390	35.43491
571	35.23064	70.27699	105.1403	139.8219	174.3230	35.24500	70.30562	35.24822	35.24922
572	35.04635	69.90969	104.5913	139.0924	173.4143	35.06063	69.93817	35.06383	35.06483
573	34.86334	69.54494	104.0461	138.3680	172.5119	34.87754	69.57328	34.88073	34.88173
574	34.68160	69.18272	103.5046	137.6485	171.6157	34.69573	69.21092	34.69890	34.69989
575	34.50112	68.82302	102.9669	136.9341	170.7257	34.51518	68.85107	34.51834	34.51932
576	34.32190	68.46581	102.4330	136.2246	169.8418	34.33588	68.49371	34.33902	34.34000
577	34.14391	68.11106	101.9027	135.5200	168.9641	34.15782	68.13882	34.16094	34.16192
578	33.96715	67.75876	101.3760	134.8202	168.0924	33.98099	67.78637	33.98410	33.98507
579	33.79161	67.40889	100.8530	134.1252	167.2266	33.80538	67.43636	33.80847	33.80943
580	33.61728	67.06142	100.3336	133.4350	166.3668	33.63098	67.08875	33.63405	33.63501
581	33.44414	66.71633	99.81775	132.7496	165.5129	33.45777	66.74352	33.46083	33.46178
582	33.27219	66.37361	99.30543	132.0688	164.6649	33.28575	66.40066	33.28879	33.28974
583	33.10142	66.03323	98.79660	131.3927	163.8226	33.11491	66.06014	33.11793	33.11888
584	32.93181	65.69518	98.29124	130.7211	162.9860	32.94523	65.72195	32.94824	32.94918
585	32.76337	65.35943	97.78933	130.0542	162.1551	32.77672	65.38606	32.77971	32.78065
586	32.59606	65.02596	97.29082	129.3918	161.3299	32.60935	65.05246	32.61232	32.61326
587	32.42990	64.69476	96.79570	128.7338	160.5102	32.44311	64.72112	32.44608	32.44700
588	32.26486	64.36580	96.30393	128.0804	159.6961	32.27801	64.39203	32.28096	32.28188
589	32.10094	64.03907	95.81549	127.4313	158.8875	32.11402	64.06517	32.11696	32.11787
590	31.93813	63.71455	95.33034	126.7866	158.0844	31.95114	63.74051	31.95406	31.95498

(continued)

Table B.1. (continued)

n	Line transition (frequencies in MHz)								
	Hnα	Hnβ	Hnγ	Hnδ	Hnϵ	Henα	Henβ	Cnα	Snα
591	31.77642	63.39221	94.84847	126.1463	157.2866	31.78937	63.41805	31.79227	31.79318
592	31.61580	63.07205	94.36984	125.5102	156.4942	31.62868	63.09775	31.63157	31.63247
593	31.45625	62.75404	93.89442	124.8784	155.7072	31.46907	62.77961	31.47195	31.47285
594	31.29779	62.43816	93.42219	124.2509	154.9254	31.31054	62.46361	31.31340	31.31429
595	31.14038	62.12440	92.95312	123.6276	154.1488	31.15307	62.14972	31.15591	31.15680
596	30.98403	61.81274	92.48719	123.0084	153.3774	30.99665	61.83793	30.99948	31.00037
597	30.82872	61.50316	92.02436	122.3933	152.6111	30.84128	61.52822	30.84410	30.84498
598	30.67444	61.19565	91.56462	121.7824	151.8499	30.68694	61.22058	30.68975	30.69062
599	30.52120	60.89018	91.10794	121.1755	151.0938	30.53364	60.91499	30.53643	30.53730
600	30.36898	60.58674	90.65428	120.5726	150.3427	30.38135	60.61142	30.38413	30.38499
601	30.21776	60.28531	90.20364	119.9737	149.5966	30.23007	60.30987	30.23284	30.23370
602	30.06755	59.98588	89.75597	119.3788	148.8554	30.07980	60.01032	30.08255	30.08341
603	29.91833	59.68843	89.31127	118.7878	148.1191	29.93052	59.71275	29.93326	29.93411
604	29.77010	59.39294	88.86950	118.2007	147.3876	29.78223	59.41714	29.78495	29.78580
605	29.62284	59.09940	88.43063	117.6175	146.6609	29.63491	59.12348	29.63762	29.63847
606	29.47656	58.80779	87.99465	117.0381	145.9390	29.48857	58.83175	29.49126	29.49210
607	29.33123	58.51809	87.56153	116.4625	145.2219	29.34318	58.54194	29.34587	29.34670
608	29.18686	58.23030	87.13125	115.8906	144.5094	29.19876	58.25403	29.20142	29.20226
609	29.04344	57.94439	86.70379	115.3226	143.8016	29.05527	57.96800	29.05793	29.05876
610	28.90095	57.66035	86.27911	114.7582	143.0984	28.91273	57.68384	28.91537	28.91620
611	28.75940	57.37816	85.85721	114.1974	142.3998	28.77112	57.40154	28.77374	28.77457
612	28.61876	57.09781	85.43805	113.6404	141.7057	28.63043	57.12108	28.63304	28.63386
613	28.47905	56.81928	85.02161	113.0869	141.0161	28.49065	56.84244	28.49325	28.49407
614	28.34024	56.54257	84.60788	112.5371	140.3310	28.35179	56.56561	28.35438	28.35519
615	28.20233	56.26764	84.19683	111.9908	139.6503	28.21382	56.29057	28.21640	28.21720
616	28.06531	55.99450	83.78843	111.4480	138.9740	28.07675	56.01731	28.07931	28.08012
617	27.92918	55.72312	83.38267	110.9087	138.3021	27.94057	55.74583	27.94312	27.94392
618	27.79393	55.45349	82.97953	110.3729	137.6345	27.80526	55.47609	27.80780	27.80859
619	27.65956	55.18560	82.57899	109.8406	136.9712	27.67083	55.20809	27.67335	27.67414
620	27.52604	54.91943	82.18101	109.3116	136.3121	27.53726	54.94181	27.53977	27.54056
621	27.39339	54.65497	81.78559	108.7861	135.6573	27.40455	54.67724	27.40705	27.40784
622	27.26158	54.39220	81.39270	108.2639	135.0067	27.27269	54.41437	27.27518	27.27596
623	27.13062	54.13112	81.00233	107.7451	134.3602	27.14168	54.15318	27.14416	27.14493
624	27.00050	53.87171	80.61445	107.2295	133.7178	27.01150	53.89366	27.01397	27.01474
625	26.87121	53.61395	80.22904	106.7173	133.0795	26.88216	53.63579	26.88461	26.88538
626	26.74274	53.35783	79.84608	106.2083	132.4453	26.75364	53.37957	26.75608	26.75685
627	26.61509	53.10334	79.46556	105.7025	131.8151	26.62593	53.12498	26.62837	26.62913
628	26.48825	52.85047	79.08745	105.2000	131.1889	26.49904	52.87200	26.50146	26.50222
629	26.36222	52.59920	78.71173	104.7006	130.5666	26.37296	52.62063	26.37537	26.37612
630	26.23698	52.34952	78.33839	104.2044	129.9483	26.24767	52.37085	26.25007	26.25082
631	26.11254	52.10141	77.96741	103.7113	129.3339	26.12318	52.12264	26.12556	26.12631
632	25.98888	51.85488	77.59877	103.2213	128.7233	25.99947	51.87601	26.00184	26.00259
633	25.86600	51.60989	77.23245	102.7344	128.1166	25.87654	51.63092	25.87890	25.87964
634	25.74389	51.36645	76.86843	102.2506	127.5137	25.75438	51.38738	25.75674	25.75747
635	25.62256	51.12453	76.50669	101.7698	126.9145	25.63300	51.14537	25.63534	25.63607
636	25.50198	50.88414	76.14722	101.2920	126.3191	25.51237	50.90487	25.51470	25.51543
637	25.38216	50.64524	75.79000	100.8172	125.7275	25.39250	50.66588	25.39482	25.39555
638	25.26309	50.40784	75.43501	100.3453	125.1395	25.27338	50.42838	25.27569	25.27641
639	25.14476	50.17192	75.08223	99.87641	124.5552	25.15500	50.19237	25.15730	25.15802
640	25.02717	49.93748	74.73165	99.41042	123.9745	25.03737	49.95783	25.03965	25.04037
641	24.91031	49.70449	74.38325	98.94732	123.3974	24.92046	49.72474	24.92274	24.92345
642	24.79418	49.47294	74.03701	98.48710	122.8239	24.80428	49.49310	24.80655	24.80725
643	24.67877	49.24284	73.69292	98.02972	122.2539	24.68882	49.26290	24.69108	24.69178
644	24.56407	49.01415	73.35096	97.57517	121.6875	24.57408	49.03413	24.57632	24.57703
645	24.45008	48.78689	73.01110	97.12343	121.1246	24.46005	48.80677	24.46228	24.46298
646	24.33680	48.56102	72.67335	96.67447	120.5651	24.34672	48.58081	24.34894	24.34964
647	24.22422	48.33655	72.33767	96.22828	120.0090	24.23409	48.35624	24.23630	24.23700
648	24.11233	48.11345	72.00406	95.78483	119.4564	24.12215	48.13306	24.12436	24.12505
649	24.00113	47.89173	71.67250	95.34409	118.9072	24.01091	47.91125	24.01310	24.01379

(continued)

Table B.1. (continued)

n	Line transition (frequencies in MHz)							
	H$n\alpha$	H$n\beta$	H$n\gamma$	H$n\delta$	H$n\epsilon$	He$n\alpha$	He$n\beta$	C$n\alpha$	S$n\alpha$
650	23.89061	47.67137	71.34297	94.90606	118.3613	23.90034	47.69080	23.90253	23.90321
651	23.78077	47.45236	71.01545	94.47071	117.8188	23.79046	47.47170	23.79263	23.79331
652	23.67160	47.23469	70.68994	94.03801	117.2796	23.68124	47.25394	23.68340	23.68408
653	23.56309	47.01835	70.36642	93.60796	116.7436	23.57269	47.03751	23.57485	23.57552
654	23.45525	46.80332	70.04486	93.18052	116.2109	23.46481	46.82239	23.46695	23.46762
655	23.34807	46.58961	69.72526	92.75568	115.6815	23.35758	46.60859	23.35972	23.36038
656	23.24154	46.37719	69.40761	92.33342	115.1553	23.25101	46.39609	23.25313	23.25380
657	23.13566	46.16607	69.09188	91.91371	114.6322	23.14508	46.18488	23.14720	23.14786
658	23.03041	45.95622	68.77806	91.49655	114.1123	23.03980	45.97495	23.04190	23.04256
659	22.92581	45.74765	68.46614	91.08191	113.5956	22.93515	45.76629	22.93725	22.93790
660	22.82184	45.54033	68.15610	90.66977	113.0820	22.83114	45.55889	22.83322	22.83387
661	22.71849	45.33427	67.84793	90.26011	112.5714	22.72775	45.35274	22.72983	22.73048
662	22.61577	45.12944	67.54162	89.85292	112.0639	22.62499	45.14783	22.62705	22.62770
663	22.51367	44.92585	67.23715	89.44817	111.5595	22.52284	44.94416	22.52490	22.52554
664	22.41218	44.72348	66.93451	89.04585	111.0581	22.42131	44.74170	22.42336	22.42400
665	22.31130	44.52233	66.63367	88.64594	110.5597	22.32039	44.54047	22.32243	22.32307
666	22.21102	44.32237	66.33464	88.24842	110.0643	22.22008	44.34044	22.22211	22.22274
667	22.11135	44.12362	66.03740	87.85328	109.5718	22.12036	44.14160	22.12238	22.12301
668	22.01227	43.92605	65.74193	87.46049	109.0823	22.02124	43.94395	22.02325	22.02388
669	21.91378	43.72966	65.44822	87.07003	108.5957	21.92271	43.74748	21.92471	21.92534
670	21.81588	43.53444	65.15625	86.68190	108.1120	21.82477	43.55218	21.82676	21.82739
671	21.71856	43.34038	64.86602	86.29607	107.6311	21.72741	43.35804	21.72939	21.73001
672	21.62182	43.14747	64.57752	85.91253	107.1531	21.63063	43.16505	21.63260	21.63322
673	21.52565	42.95570	64.29072	85.53126	106.6779	21.53442	42.97320	21.53639	21.53700
674	21.43005	42.76507	64.00561	85.15224	106.2055	21.43878	42.78249	21.44074	21.44135
675	21.33502	42.57556	63.72219	84.77546	105.7359	21.34371	42.59291	21.34566	21.34627
676	21.24054	42.38717	63.44044	84.40089	105.2691	21.24920	42.40445	21.25114	21.25175
677	21.14663	42.19990	63.16035	84.02853	104.8050	21.15525	42.21709	21.15718	21.15778
678	21.05327	42.01372	62.88190	83.65836	104.3436	21.06185	42.03084	21.06377	21.06437
679	20.96045	41.82864	62.60509	83.29036	103.8850	20.96899	41.84568	20.97091	20.97151
680	20.86818	41.64464	62.32990	82.92451	103.4290	20.87669	41.66161	20.87859	20.87919
681	20.77646	41.46172	62.05633	82.56080	102.9757	20.78492	41.47861	20.78682	20.78741
682	20.68526	41.27987	61.78435	82.19922	102.5250	20.69369	41.29669	20.69558	20.69617
683	20.59461	41.09908	61.51395	81.83974	102.0770	20.60300	41.11583	20.60488	20.60547
684	20.50448	40.91935	61.24514	81.48236	101.6315	20.51283	40.93602	20.51471	20.51529
685	20.41487	40.74066	60.97788	81.12706	101.1887	20.42319	40.75726	20.42506	20.42564
686	20.32579	40.56301	60.71218	80.77381	100.7484	20.33407	40.57954	20.33593	20.33651
687	20.23722	40.38639	60.44802	80.42262	100.3107	20.24547	40.40285	20.24732	20.24790
688	20.14917	40.21080	60.18540	80.07346	99.87549	20.15738	40.22719	20.15922	20.15980
689	20.06163	40.03622	59.92429	79.72632	99.44281	20.06980	40.05254	20.07164	20.07221
690	19.97459	39.86266	59.66469	79.38118	99.01262	19.98273	39.87890	19.98456	19.98513
691	19.88806	39.69009	59.40658	79.03803	98.58491	19.89617	39.70626	19.89798	19.89855
692	19.80203	39.51852	59.14997	78.69685	98.15967	19.81010	39.53462	19.81191	19.81247
693	19.71649	39.34794	58.89482	78.35764	97.73686	19.72453	39.36397	19.72633	19.72689
694	19.63145	39.17833	58.64115	78.02037	97.31648	19.63944	39.19430	19.64124	19.64180
695	19.54689	39.00970	58.38893	77.68504	96.89851	19.55485	39.02560	19.55664	19.55720
696	19.46282	38.84204	58.13815	77.35162	96.48293	19.47075	38.85787	19.47252	19.47308
697	19.37922	38.67533	57.88881	77.02011	96.06972	19.38712	38.69109	19.38889	19.38945
698	19.29611	38.50958	57.64089	76.69049	95.65886	19.30397	38.52528	19.30574	19.30629
699	19.21347	38.34478	57.39438	76.36275	95.25035	19.22130	38.36040	19.22306	19.22361
700	19.13130	38.18091	57.14928	76.03688	94.84416	19.13910	38.19647	19.14085	19.14140
701	19.04961	38.01798	56.90557	75.71285	94.44027	19.05737	38.03347	19.05911	19.05965
702	18.96837	37.85597	56.66325	75.39067	94.03868	18.97610	37.87139	18.97783	18.97838
703	18.88760	37.69488	56.42230	75.07031	93.63936	18.89529	37.71024	18.89702	18.89756
704	18.80728	37.53470	56.18271	74.75176	93.24229	18.81494	37.55000	18.81666	18.81720
705	18.72742	37.37543	55.94448	74.43501	92.84747	18.73505	37.39066	18.73676	18.73730
706	18.64801	37.21706	55.70759	74.12005	92.45488	18.65561	37.23223	18.65731	18.65785
707	18.56905	37.05958	55.47204	73.80687	92.06449	18.57662	37.07468	18.57831	18.57884
708	18.49053	36.90299	55.23782	73.49544	91.67630	18.49807	36.91803	18.49976	18.50029

(continued)

Table B.1. (continued)

n	Line transition (frequencies in MHz)								
	Hnα	Hnβ	Hnγ	Hnδ	Hnϵ	Henα	Henβ	Cnα	Snα
709	18.41246	36.74728	55.00491	73.18577	91.29029	18.41996	36.76226	18.42164	18.42217
710	18.33482	36.59245	54.77331	72.87783	90.90644	18.34230	36.60736	18.34397	18.34450
711	18.25763	36.43848	54.54301	72.57162	90.52474	18.26506	36.45333	18.26673	18.26726
712	18.18086	36.28538	54.31399	72.26712	90.14518	18.18827	36.30017	18.18993	18.19045
713	18.10452	36.13313	54.08626	71.96432	89.76773	18.11190	36.14786	18.11355	18.11407
714	18.02861	35.98174	53.85980	71.66321	89.39239	18.03596	35.99640	18.03761	18.03812
715	17.95313	35.83119	53.63460	71.36378	89.01914	17.96044	35.84579	17.96208	17.96260
716	17.87806	35.68147	53.41066	71.06602	88.64797	17.88535	35.69601	17.88698	17.88749
717	17.80341	35.53259	53.18796	70.76990	88.27885	17.81067	35.54707	17.81230	17.81280
718	17.72918	35.38454	52.96649	70.47544	87.91178	17.73641	35.39896	17.73803	17.73853
719	17.65536	35.23731	52.74626	70.18260	87.54674	17.66256	35.25167	17.66417	17.66467
720	17.58195	35.09089	52.52724	69.89138	87.18372	17.58911	35.10519	17.59072	17.59122
721	17.50895	34.94529	52.30943	69.60178	86.82271	17.51608	34.95953	17.51768	17.51818
722	17.43634	34.80049	52.09283	69.31377	86.46369	17.44345	34.81467	17.44504	17.44554
723	17.36414	34.65649	51.87742	69.02734	86.10664	17.37122	34.67061	17.37281	17.37330
724	17.29234	34.51328	51.66320	68.74250	85.75155	17.29939	34.52734	17.30097	17.30146
725	17.22094	34.37086	51.45015	68.45921	85.39842	17.22795	34.38486	17.22953	17.23002
726	17.14992	34.22922	51.23828	68.17749	85.04722	17.15691	34.24317	17.15848	17.15897
727	17.07930	34.08836	51.02756	67.89730	84.69795	17.08626	34.10225	17.08782	17.08831
728	17.00906	33.94827	50.81800	67.61865	84.35058	17.01599	33.96210	17.01755	17.01803
729	16.93921	33.80894	50.60959	67.34152	84.00511	16.94611	33.82272	16.94766	16.94814
730	16.86974	33.67038	50.40232	67.06591	83.66153	16.87661	33.68410	16.87815	16.87864
731	16.80065	33.53258	50.19617	66.79179	83.31982	16.80749	33.54624	16.80903	16.80951
732	16.73193	33.39552	49.99115	66.51917	82.97996	16.73875	33.40913	16.74028	16.74076
733	16.66359	33.25922	49.78724	66.24803	82.64195	16.67038	33.27277	16.67190	16.67238
734	16.59562	33.12365	49.58444	65.97836	82.30578	16.60239	33.13715	16.60390	16.60438
735	16.52802	32.98882	49.38274	65.71016	81.97142	16.53476	33.00226	16.53627	16.53674
736	16.46079	32.85471	49.18213	65.44340	81.63888	16.46750	32.86810	16.46900	16.46947
737	16.39392	32.72134	48.98261	65.17809	81.30813	16.40060	32.73467	16.40210	16.40257
738	16.32742	32.58869	48.78416	64.91421	80.97916	16.33407	32.60196	16.33556	16.33603
739	16.26127	32.45675	48.58679	64.65175	80.65197	16.26790	32.46997	16.26938	16.26985
740	16.19548	32.32552	48.39048	64.39070	80.32654	16.20208	32.33869	16.20356	16.20402
741	16.13004	32.19500	48.19524	64.13106	80.00285	16.13661	32.20812	16.13809	16.13855
742	16.06496	32.06518	48.00102	63.87281	79.68091	16.07150	32.07825	16.07297	16.07343
743	16.00022	31.93606	47.80785	63.61595	79.36068	16.00674	31.94907	16.00821	16.00866
744	15.93584	31.80763	47.61572	63.36046	79.04217	15.94233	31.82059	15.94379	15.94424
745	15.87179	31.67989	47.42462	63.10634	78.72537	15.87826	31.69280	15.87971	15.88017
746	15.80809	31.55283	47.23454	62.85357	78.41025	15.81454	31.56569	15.81598	15.81643
747	15.74474	31.42645	47.04548	62.60216	78.09682	15.75115	31.43926	15.75259	15.75304
748	15.68171	31.30074	46.85742	62.35208	77.78505	15.68810	31.31350	15.68954	15.68999
749	15.61903	31.17571	46.67037	62.10333	77.47494	15.62539	31.18841	15.62682	15.62727
750	15.55668	31.05134	46.48430	61.85591	77.16647	15.56302	31.06399	15.56444	15.56488
751	15.49466	30.92763	46.29923	61.60979	76.85964	15.50097	30.94023	15.50239	15.50283
752	15.43297	30.80457	46.11514	61.36499	76.55444	15.43926	30.81712	15.44067	15.44111
753	15.37160	30.68217	45.93202	61.12147	76.25085	15.37787	30.69467	15.37927	15.37971
754	15.31057	30.56041	45.74987	60.87924	75.94886	15.31680	30.57287	15.31820	15.31864
755	15.24985	30.43930	45.56868	60.63830	75.64847	15.25606	30.45171	15.25746	15.25789
756	15.18945	30.31883	45.38845	60.39862	75.34966	15.19564	30.33118	15.19703	15.19747
757	15.12938	30.19899	45.20916	60.16020	75.05241	15.13554	30.21130	15.13692	15.13736
758	15.06962	30.07979	45.03082	59.92303	74.75673	15.07576	30.09204	15.07713	15.07756
759	15.01017	29.96121	44.85342	59.68712	74.46260	15.01629	29.97341	15.01766	15.01809
760	14.95104	29.84325	44.67695	59.45243	74.17002	14.95713	29.85541	14.95849	14.95892
761	14.89221	29.72591	44.50140	59.21898	73.87896	14.89828	29.73802	14.89964	14.90007
762	14.83370	29.60918	44.32677	58.98675	73.58942	14.83974	29.62125	14.84110	14.84152
763	14.77549	29.49307	44.15305	58.75573	73.30140	14.78151	29.50509	14.78286	14.78328
764	14.71758	29.37756	43.98024	58.52591	73.01487	14.72358	29.38953	14.72492	14.72535
765	14.65998	29.26266	43.80833	58.29729	72.72984	14.66595	29.27458	14.66729	14.66771
766	14.60268	29.14835	43.63731	58.06986	72.44629	14.60863	29.16023	14.60996	14.61038
767	14.54567	29.03464	43.46719	57.84361	72.16421	14.55160	29.04647	14.55293	14.55334

(continued)

Table B.1. (continued)

n	Line transition (frequencies in MHz)								
	Hnα	Hnβ	Hnγ	Hnδ	Hnϵ	Henα	Henβ	Cnα	Snα
768	14.48896	28.92151	43.29794	57.61854	71.88360	14.49487	28.93330	14.49619	14.49661
769	14.43255	28.80898	43.12958	57.39463	71.60443	14.43843	28.82072	14.43975	14.44016
770	14.37643	28.69703	42.96208	57.17188	71.32671	14.38229	28.70872	14.38360	14.38401
771	14.32060	28.58565	42.79545	56.95028	71.05043	14.32643	28.59730	14.32774	14.32815
772	14.26506	28.47486	42.62969	56.72983	70.77557	14.27087	28.48646	14.27217	14.27258
773	14.20980	28.36463	42.46477	56.51051	70.50212	14.21559	28.37619	14.21689	14.21730
774	14.15483	28.25497	42.30071	56.29232	70.23009	14.16060	28.26649	14.16189	14.16230
775	14.10014	28.14588	42.13749	56.07526	69.95945	14.10589	28.15735	14.10718	14.10759
776	14.04574	28.03735	41.97511	55.85930	69.69020	14.05146	28.04877	14.05274	14.05315
777	13.99161	27.92937	41.81357	55.64446	69.42232	13.99731	27.94076	13.99859	13.99899
778	13.93776	27.82195	41.65285	55.43071	69.15582	13.94344	27.83329	13.94472	13.94511
779	13.88419	27.71508	41.49295	55.21806	68.89069	13.88985	27.72638	13.89112	13.89151
780	13.83089	27.60876	41.33387	55.00650	68.62690	13.83653	27.62001	13.83779	13.83819
781	13.77787	27.50298	41.17560	54.79601	68.36446	13.78348	27.51419	13.78474	13.78513
782	13.72511	27.39774	41.01814	54.58660	68.10336	13.73070	27.40890	13.73196	13.73235
783	13.67263	27.29303	40.86149	54.37825	67.84359	13.67820	27.30415	13.67945	13.67984
784	13.62041	27.18886	40.70562	54.17096	67.58513	13.62596	27.19994	13.62720	13.62759
785	13.56845	27.08522	40.55056	53.96473	67.32799	13.57398	27.09625	13.57522	13.57561
786	13.51676	26.98210	40.39627	53.75954	67.07215	13.52227	26.99310	13.52351	13.52389
787	13.46534	26.87951	40.24277	53.55538	66.81760	13.47082	26.89046	13.47205	13.47244
788	13.41417	26.77743	40.09005	53.35226	66.56434	13.41964	26.78835	13.42086	13.42125
789	13.36326	26.67588	39.93809	53.15017	66.31236	13.36871	26.68675	13.36993	13.37031
790	13.31261	26.57483	39.78691	52.94910	66.06165	13.31804	26.58566	13.31925	13.31963
791	13.26222	26.47430	39.63648	52.74904	65.81220	13.26762	26.48508	13.26883	13.26921
792	13.21208	26.37427	39.48682	52.54998	65.56400	13.21746	26.38501	13.21867	13.21905
793	13.16219	26.27474	39.33790	52.35193	65.31706	13.16755	26.28545	13.16876	13.16913
794	13.11255	26.17572	39.18974	52.15487	65.07135	13.11789	26.18638	13.11909	13.11947
795	13.06316	26.07719	39.04232	51.95879	64.82687	13.06849	26.08781	13.06968	13.07005
796	13.01402	25.97915	38.89563	51.76370	64.58361	13.01933	25.98974	13.02052	13.02089
797	12.96513	25.88161	38.74968	51.56959	64.34157	12.97041	25.89215	12.97160	12.97197
798	12.91648	25.78455	38.60446	51.37644	64.10074	12.92174	25.79506	12.92292	12.92329
799	12.86807	25.68798	38.45996	51.18426	63.86111	12.87332	25.69845	12.87449	12.87486
800	12.81991	25.59189	38.31619	50.99303	63.62267	12.82513	25.60232	12.82630	12.82667

B.2 Frequencies Above 100 GHz

Although negligible at 100 GHz, the shifts in rest frequencies due to blending of the fine-structure lines become increasingly significant at higher frequencies. To explore these differences, Towle et al. (1996) have calculated rest frequencies of recombination lines from both the classical Rydberg equation (A.5) and the Dirac equation that describes the fine-structure lines. Their calculations include frequencies from 100 GHz to 29 THz (about $\lambda = 10\,\mu m$) and estimate relative strengths for all of these lines.

Unlike the Rydberg equation, Dirac theory gives the energy E_{nj} of a discrete state of a hydrogenic atom as

$$E_{nj} = \frac{m_R c^2}{\left(1 + X_{nj}^2\right)} - \frac{m Z^4 \alpha^2 h c R_\infty}{4 M n^4}, \qquad (B.1)$$

where

$$X_{nj} = \frac{Z\alpha}{n - k + (k^2 - Z^2\alpha^2)^{1/2}}. \tag{B.2}$$

The rightmost term of (B.1) is a correction for the energy shift (not splitting) of a principal quantum level n due to nuclear motion, described by (42.2), (42.5), and (42.7) of Bethe and Salpeter (1957). Here, n is the principal quantum number, the angular momentum quantum number $j = 1/2, 3/2, \ldots, n - 1/2$, $k = j + 1/2$, and α is the fine-structure constant given in Table A.1. The parameter m_R is the reduced mass given by (A.2). The energy of each line component results from $E_{n'j'} - E_{nj} = h\nu_{nj \to n'j'}$. The line strengths are given as $g'A$ in units of $10^4 \, s^{-1}$, where g' is the statistical weight of the upper quantum level and A is the spontaneous emission rate for that transition.

These calculations do not include relativistic effects arising from the high velocity of the electron around the nucleus, because they lead to frequency shifts that are much smaller than the fine-structure shifts and, hence, too small to be measured in the spectra of astronomical objects.

Table B.2 reproduces the entries of Table 1 of Towle et al. (1996). The column headings are frequency, intensity in the units described above, lower principal quantum number n, and change in principal quantum number Δn. The frequencies are calculated from the Rydberg equation (A.5) for H, ^{4}He, ^{12}C, and ^{32}S. For the carbon and sulfur, the quoted line intensities assume hydrogenic atoms.

Table B.2. Rydberg H, ^{4}He, ^{12}C, and ^{32}S RRL rest frequencies for 100 GHz to 29 THz

ν/MHz	Int.	n	Δn		ν/MHz	Int.	n	Δn	
100,527.40	0.2	82	10	H	122,050.75	1.7	58	4	H
100,539.63	1.4	62	4	H	122,055.83	2.9	53	3	H
100,580.60	1.4	62	4	He	122,100.48	1.7	58	4	He
101,769.92	0.6	70	6	H	122,105.57	2.9	53	3	He
102,253.46	0.3	79	9	H	122,903.80	1.1	62	5	H
102,571.00	0.9	66	5	H	122,953.88	1.1	62	5	He
102,612.79	0.9	66	5	He	123,157.57	0.3	74	9	H
103,252.92	0.5	73	7	H	124,689.39	0.3	76	10	H
103,267.29	0.4	76	8	H	124,746.74	23.2	37	1	H
103,914.85	2.5	56	3	H	124,797.57	23.2	37	1	He
103,957.19	2.5	56	3	He	124,808.97	23.2	37	1	C
104,091.24	0.2	81	10	H	124,812.54	23.2	37	1	S
105,301.86	5.4	49	2	H	125,414.84	0.4	71	8	H
105,344.77	5.4	49	2	He	125,975.28	0.8	65	6	H
105,410.22	1.4	61	4	H	126,541.30	0.6	68	7	H
105,453.18	1.4	61	4	He	126,793.88	6.5	46	2	H
106,032.09	0.3	78	9	H	126,845.54	6.5	46	2	He
106,079.54	0.7	69	6	H	128,008.66	0.4	73	9	H
106,737.36	19.9	39	1	H	128,372.26	1.7	57	4	H
106,780.86	19.9	39	1	He	128,424.57	1.7	57	4	He
106,790.61	19.9	39	1	C	128,814.50	1.1	61	5	H
106,793.66	19.9	39	1	S	128,866.99	1.1	61	5	He

(continued)

Table B.2. (continued)

ν/MHz	Int.	n	Δn		ν/MHz	Int.	n	Δn	
107,206.11	0.9	65	5	H	129,036.19	3.1	52	3	H
107,249.80	0.9	65	5	He	129,088.77	3.1	52	3	He
107,252.38	0.4	75	8	H	129,448.85	0.3	75	10	H
107,422.29	0.5	72	7	H	130,588.52	0.5	70	8	H
107,825.75	0.2	80	10	H	131,716.02	0.8	64	6	H
109,536.01	2.6	55	3	H	132,020.90	0.6	67	7	H
109,580.64	2.6	55	3	He	133,118.21	0.4	72	9	H
109,999.40	0.3	77	9	H	134,453.98	0.3	74	10	H
110,600.68	1.5	60	4	H	135,110.59	1.2	60	5	H
110,636.11	0.7	68	6	H	135,138.16	1.8	56	4	H
110,645.75	1.5	60	4	He	135,165.64	1.2	60	5	He
111,445.37	0.4	74	8	H	135,193.23	1.8	56	4	He
111,741.29	0.3	79	10	H	135,249.50	7.0	45	2	H
111,819.35	0.5	71	7	H	135,286.04	25.2	36	1	H
111,885.08	5.8	48	2	H	135,304.61	7.0	45	2	He
111,930.67	5.8	48	2	He	135,341.17	25.2	36	1	He
112,124.91	1.0	64	5	H	135,353.53	25.2	36	1	C
112,170.60	1.0	64	5	He	135,357.40	25.2	36	1	S
114,167.35	0.3	76	9	H	136,051.02	0.5	69	8	H
115,274.41	21.5	38	1	H	136,559.29	3.2	51	3	H
115,321.38	21.5	38	1	He	136,614.94	3.2	51	3	He
115,331.91	21.5	38	1	C	137,811.19	0.9	63	6	H
115,335.21	21.5	38	1	S	137,821.77	0.6	66	7	H
115,457.60	0.7	67	6	H	138,503.70	0.4	71	9	H
115,570.12	2.7	54	3	H	139,720.87	0.3	73	10	H
115,617.22	2.7	54	3	He	141,822.80	0.5	68	8	H
115,848.95	0.3	78	10	H	141,824.00	1.3	59	5	H
115,860.00	0.4	73	8	H	141,881.79	1.3	59	5	He
116,137.71	1.6	59	4	H	142,388.20	1.9	55	4	H
116,185.04	1.6	59	4	He	142,446.23	1.9	55	4	He
116,459.90	0.5	70	7	H	143,967.84	0.7	65	7	H
117,349.42	1.0	63	5	H	144,184.01	0.4	70	9	H
117,397.24	1.0	63	5	He	144,288.58	0.9	62	6	H
118,548.79	0.3	75	9	H	144,474.12	7.4	44	2	H
119,028.76	6.1	47	2	H	144,532.99	7.4	44	2	He
119,077.27	6.1	47	2	He	144,678.94	3.4	50	3	H
120,160.71	0.3	77	10	H	144,737.89	3.4	50	3	He
120,511.14	0.4	72	8	H	145,266.88	0.3	72	10	H
120,563.56	0.7	66	6	H	147,046.89	27.4	35	1	H
121,361.03	0.6	69	7	H	147,106.81	27.4	35	1	He
147,120.24	27.4	35	1	C	183,018.71	1.6	54	5	H
147,124.45	27.4	35	1	S	183,093.28	1.6	54	5	He
147,926.09	0.5	67	8	H	183,586.30	1.1	57	6	H
148,989.91	1.3	58	5	H	184,341.46	0.6	62	8	H
149,050.62	1.3	58	5	He	184,448.70	4.4	46	3	H
150,166.51	2.0	54	4	H	184,523.87	4.4	46	3	He
150,179.63	0.4	69	9	H	185,571.75	0.4	66	10	H
150,227.70	2.0	54	4	He	185,735.95	0.5	64	9	H
150,485.18	0.7	64	7	H	187,630.91	2.6	50	4	H

(continued)

Table B.2. (continued)

ν/MHz	Int.	n	Δn		ν/MHz	Int.	n	Δn	
151,110.77	0.3	71	10	H	187,707.36	2.6	50	4	He
151,178.67	0.9	61	6	H	189,738.18	0.9	59	7	H
153,455.46	3.6	49	3	H	191,057.40	9.8	40	2	H
153,517.99	3.6	49	3	He	191,135.25	9.8	40	2	He
154,385.07	0.5	66	8	H	191,656.74	35.7	32	1	H
154,557.22	7.9	43	2	H	191,734.84	35.7	32	1	He
154,620.20	7.9	43	2	He	191,752.35	35.7	32	1	C
156,512.78	0.4	68	9	H	191,757.83	35.7	32	1	S
156,647.13	1.4	57	5	H	193,025.22	0.7	61	8	H
156,710.96	1.4	57	5	He	193,113.39	1.2	56	6	H
157,272.86	0.4	70	10	H	193,118.48	1.7	53	5	H
157,402.30	0.7	63	7	H	193,197.17	1.7	53	5	He
158,514.87	1.0	60	6	H	193,694.54	0.4	65	10	H
158,522.09	2.2	53	4	H	194,164.02	0.5	63	9	H
158,586.69	2.2	53	4	He	196,623.43	4.7	45	3	H
160,211.52	29.9	34	1	H	196,703.55	4.7	45	3	He
160,276.80	29.9	34	1	He	198,909.18	2.7	49	4	H
160,291.45	29.9	34	1	C	198,990.24	2.7	49	4	He
160,296.02	29.9	34	1	S	199,186.10	0.9	58	7	H
161,226.05	0.6	65	8	H	202,263.82	0.7	60	8	H
162,956.69	3.9	48	3	H	202,299.25	0.5	64	10	H
163,023.09	3.9	48	3	He	203,110.63	0.6	62	9	H
163,207.58	0.5	67	9	H	203,311.88	1.3	55	6	H
163,775.12	0.4	69	10	H	203,975.82	1.8	52	5	H
164,750.35	0.8	62	7	H	204,058.93	1.8	52	5	He
164,838.52	1.5	56	5	H	205,760.32	10.6	39	2	H
164,905.69	1.5	56	5	He	205,844.17	10.6	39	2	He
165,601.05	8.5	42	2	H	209,272.58	1.0	57	7	H
165,668.53	8.5	42	2	He	209,894.06	5.0	44	3	H
166,333.90	1.0	59	6	H	209,979.59	5.0	44	3	He
167,509.55	2.3	52	4	H	210,501.78	39.2	31	1	H
167,577.81	2.3	52	4	He	210,587.56	39.2	31	1	He
168,477.74	0.6	64	8	H	210,606.80	39.2	31	1	C
170,290.26	0.5	66	9	H	210,612.81	39.2	31	1	S
170,641.40	0.4	68	10	H	211,110.29	2.9	48	4	H
172,563.46	0.8	61	7	H	211,196.32	2.9	48	4	He
173,259.01	4.1	47	3	H	211,422.24	0.5	63	10	H
173,329.61	4.1	47	3	He	212,102.35	0.7	59	8	H
173,611.52	1.5	55	5	H	212,616.34	0.6	61	9	H
173,682.26	1.5	55	5	He	214,242.02	1.3	54	6	H
174,676.19	1.1	58	6	H	215,663.15	1.9	51	5	H
174,995.82	32.6	33	1	H	215,751.03	1.9	51	5	He
175,067.12	32.6	33	1	He	220,052.56	1.0	56	7	H
175,083.12	32.6	33	1	C	221,103.19	0.5	62	10	H
175,088.12	32.6	33	1	S	222,011.77	11.4	38	2	H
176,171.46	0.6	63	8	H	222,102.24	11.4	38	2	He
177,189.79	2.4	51	4	H	222,590.38	0.8	58	8	H
177,261.99	2.4	51	4	He	222,725.59	0.6	60	9	H
177,722.83	9.1	41	2	H	224,330.63	3.1	47	4	H
177,789.35	0.5	65	9	H	224,386.78	5.3	43	3	H

(continued)

Table B.2. (continued)

ν/MHz	Int.	n	Δn		ν/MHz	Int.	n	Δn	
177,795.25	9.1	41	2	He	224,422.04	3.1	47	4	He
177,897.57	0.4	67	10	H	224,478.21	5.3	43	3	He
180,879.03	0.8	60	7	H	225,970.68	1.4	53	6	H
228,261.41	2.0	50	5	H	289,750.57	2.6	46	5	H
228,354.42	2.0	50	5	He	289,806.72	3.9	43	4	H
231,385.50	0.5	61	10	H	289,868.64	2.6	46	5	He
231,586.76	1.1	55	7	H	289,924.81	3.9	43	4	He
231,900.94	43.2	30	1	H	293,653.03	0.7	56	10	H
231,995.44	43.2	30	1	He	297,794.76	7.1	39	3	H
232,016.63	43.2	30	1	C	297,916.11	7.1	39	3	He
232,023.26	43.2	30	1	S	299,156.42	0.8	54	9	H
233,487.14	0.7	59	9	H	299,515.98	1.9	48	6	H
233,782.48	0.8	57	8	H	302,647.71	1.1	52	8	H
238,572.20	1.5	52	6	H	303,201.03	1.4	50	7	H
238,678.95	3.3	46	4	H	307,258.41	15.8	34	2	H
238,776.21	3.3	46	4	He	307,383.61	15.8	34	2	He
240,021.14	12.3	37	2	H	308,508.51	2.7	45	5	H
240,118.95	12.3	37	2	He	308,634.22	2.7	45	5	He
240,245.62	5.7	42	3	H	308,722.11	0.7	55	10	H
240,343.51	5.7	42	3	He	310,075.17	4.2	42	4	H
241,861.16	2.1	49	5	H	310,201.52	4.2	42	4	He
241,959.71	2.1	49	5	He	315,169.23	0.9	53	9	H
242,316.70	0.5	60	10	H	316,415.44	59.0	27	1	H
243,942.39	1.1	54	7	H	316,544.38	59.0	27	1	He
244,954.55	0.7	58	9	H	316,573.30	59.0	27	1	C
245,738.84	0.9	56	8	H	316,582.34	59.0	27	1	S
252,129.42	1.6	51	6	H	317,937.95	2.0	47	6	H
253,948.90	0.6	59	10	H	319,578.00	1.1	51	8	H
254,278.26	3.5	45	4	H	320,965.02	1.5	49	7	H
254,381.87	3.5	45	4	He	321,034.73	7.6	38	3	H
256,302.05	47.8	29	1	H	321,165.54	7.6	38	3	He
256,406.49	47.8	29	1	He	324,842.70	0.7	54	10	H
256,429.91	47.8	29	1	C	328,922.83	2.9	44	5	H
256,437.24	47.8	29	1	S	329,056.86	2.9	44	5	He
256,564.01	2.3	48	5	H	332,280.06	4.5	41	4	H
256,668.56	2.3	48	5	He	332,348.08	0.9	52	9	H
257,186.76	0.7	57	9	H	332,415.46	4.5	41	4	He
257,193.99	1.2	53	7	H	335,207.34	17.2	33	2	H
257,635.49	6.1	41	3	H	335,343.93	17.2	33	2	He
257,740.47	6.1	41	3	He	337,797.41	1.2	50	8	H
258,525.92	0.9	55	8	H	337,904.17	2.1	46	6	H
260,032.78	13.4	36	2	H	340,146.48	1.6	48	7	H
260,138.74	13.4	36	2	He	342,108.39	0.8	53	10	H
266,339.33	0.6	58	10	H	346,758.51	8.2	37	3	H
266,734.77	1.6	50	6	H	346,899.81	8.2	37	3	He
270,248.75	0.8	56	9	H	350,801.31	1.0	51	9	H
271,268.00	3.7	44	4	H	351,180.66	3.1	43	5	H
271,378.54	3.7	44	4	He	351,323.76	3.1	43	5	He
271,424.40	1.2	52	7	H	353,622.77	65.9	26	1	H

(continued)

Table B.2. (continued)

ν/MHz	Int.	n	Δn		ν/MHz	Int.	n	Δn	
272,217.26	1.0	54	8	H	353,766.86	65.9	26	1	He
272,484.23	2.4	47	5	H	353,799.18	65.9	26	1	C
272,595.26	2.4	47	5	He	353,809.29	65.9	26	1	S
276,745.79	6.6	40	3	H	356,658.45	4.9	40	4	H
276,858.56	6.6	40	3	He	356,803.78	4.9	40	4	He
279,550.93	0.6	57	10	H	357,431.28	1.3	49	8	H
282,332.93	14.5	35	2	H	359,580.12	2.2	45	6	H
282,447.97	14.5	35	2	He	360,622.95	0.8	52	10	H
282,491.66	1.7	49	6	H	360,889.92	1.6	47	7	H
284,212.20	0.8	55	9	H	366,652.55	18.8	32	2	H
284,250.59	53.0	28	1	H	366,801.96	18.8	32	2	He
284,366.42	53.0	28	1	He	370,649.61	1.0	50	9	H
284,392.39	53.0	28	1	C	375,307.18	8.9	36	3	H
284,400.52	53.0	28	1	S	375,460.12	8.9	36	3	He
286,725.80	1.3	51	7	H	375,495.11	3.3	42	5	H
286,894.35	1.0	53	8	H	375,648.12	3.3	42	5	He
378,619.84	1.3	48	8	H	507,175.51	94.5	23	1	H
380,501.68	0.9	51	10	H	507,382.18	94.5	23	1	He
383,153.08	2.4	44	6	H	507,428.53	94.5	23	1	C
383,357.89	1.8	46	7	H	507,443.02	94.5	23	1	S
383,483.16	5.2	39	4	H	514,137.35	1.8	43	8	H
383,639.42	5.2	39	4	He	522,354.07	7.1	35	4	H
392,027.66	1.1	49	9	H	522,566.92	7.1	35	4	He
396,900.86	73.9	25	1	H	526,864.07	12.5	32	3	H
397,062.59	73.9	25	1	He	527,078.77	12.5	32	3	He
397,098.86	73.9	25	1	C	527,832.01	1.5	44	9	H
397,110.20	73.9	25	1	S	528,903.49	2.4	41	7	H
401,520.42	1.4	47	8	H	536,769.92	1.2	45	10	H
401,872.93	0.9	50	10	H	537,815.90	4.7	37	5	H
402,109.61	3.5	41	5	H	538,035.06	4.7	37	5	He
402,158.52	20.7	31	2	H	538,040.38	3.3	39	6	H
402,273.47	3.5	41	5	He	540,552.64	27.8	28	2	H
402,322.40	20.7	31	2	He	540,772.91	27.8	28	2	He
407,079.66	9.7	35	3	H	548,754.13	1.9	42	8	H
407,245.54	9.7	35	3	He	562,290.95	1.6	43	9	H
407,733.72	1.9	45	7	H	566,552.51	2.6	40	7	H
408,835.48	2.5	43	6	H	567,291.18	7.7	34	4	H
413,069.17	5.6	38	4	H	567,522.35	7.7	34	4	He
413,237.49	5.6	38	4	He	570,783.98	1.3	44	10	H
415,086.11	1.2	48	9	H	577,154.34	13.7	31	3	H
424,879.86	1.0	49	10	H	577,389.52	13.7	31	3	He
426,309.86	1.5	46	8	H	577,896.49	107.6	22	1	H
431,303.02	3.8	40	5	H	578,131.97	107.6	22	1	He
431,478.77	3.8	40	5	He	578,184.79	107.6	22	1	C
434,224.69	2.0	44	7	H	578,201.30	107.6	22	1	S
436,869.05	2.7	42	6	H	578,670.22	3.6	38	6	H
439,993.78	1.2	47	9	H	581,067.51	5.1	36	5	H
442,402.72	22.7	30	2	H	581,304.28	5.1	36	5	He
442,544.45	10.5	34	3	H	586,558.32	2.1	41	8	H

(continued)

Table B.2. (continued)

ν/MHz	Int.	n	Δn		ν/MHz	Int.	n	Δn	
442,583.00	22.7	30	2	He	599,825.74	1.7	42	9	H
442,724.78	10.5	34	3	He	600,666.03	30.9	27	2	H
445,781.47	6.1	37	4	H	600,910.80	30.9	27	2	He
445,963.11	6.1	37	4	He	607,744.67	1.4	43	10	H
447,540.30	83.4	24	1	H	607,869.93	2.8	39	7	H
447,722.67	83.4	24	1	He	617,540.26	8.4	33	4	H
447,763.57	83.4	24	1	C	617,791.90	8.4	33	4	He
447,776.36	83.4	24	1	S	623,504.30	3.8	37	6	H
449,682.49	1.0	48	10	H	627,926.45	2.2	40	8	H
453,187.45	1.6	45	8	H	629,091.43	5.5	35	5	H
463,065.73	2.1	43	7	H	629,347.78	5.5	35	5	He
463,395.81	4.1	39	5	H	634,059.46	15.0	30	3	H
463,584.64	4.1	39	5	He	634,317.83	15.0	30	3	He
466,940.36	1.3	46	9	H	640,788.57	1.8	41	9	H
467,529.55	2.9	41	6	H	647,979.34	1.5	42	10	H
476,460.04	1.1	47	10	H	653,314.78	3.0	38	7	H
482,044.55	6.6	36	4	H	662,404.20	123.4	21	1	H
482,240.97	6.6	36	4	He	662,674.12	123.4	21	1	He
482,254.22	11.4	33	3	H	662,734.66	123.4	21	1	C
482,378.29	1.7	44	8	H	662,753.59	123.4	21	1	S
482,450.73	11.4	33	3	He	670,038.21	34.4	26	2	H
488,202.99	25.1	29	2	H	670,311.24	34.4	26	2	He
488,401.93	25.1	29	2	He	673,101.94	4.1	36	6	H
494,523.88	2.3	42	7	H	673,289.87	2.4	39	8	H
496,139.42	1.4	45	9	H	673,910.96	9.2	32	4	H
498,757.56	4.4	38	5	H	674,185.57	9.2	32	4	He
498,960.80	4.4	38	5	He	682,565.59	6.0	34	5	H
501,132.57	3.1	40	6	H	682,843.73	6.0	34	5	He
505,413.72	1.1	46	10	H	685,581.28	1.9	40	9	H
691,860.18	1.6	41	10	H	985,749.88	2.7	35	9	H
698,704.78	16.6	29	3	H	1,019,698.79	6.3	31	6	H
698,989.49	16.6	29	3	He	1,040,131.11	193.7	18	1	H
703,416.95	3.2	37	7	H	1,040,554.95	193.7	18	1	He
723,144.34	2.5	38	8	H	1,040,650.01	193.7	18	1	C
728,114.39	4.5	35	6	H	1,040,679.73	193.7	18	1	S
734,663.81	2.0	39	9	H	1,049,218.15	4.8	32	7	H
737,365.86	10.0	31	4	H	1,060,394.45	2.4	35	10	H
737,666.32	10.0	31	4	He	1,063,321.73	3.7	33	8	H
739,811.53	1.7	40	10	H	1,065,357.33	9.4	29	5	H
742,287.00	6.5	33	5	H	1,065,791.45	9.4	29	5	He
742,589.47	6.5	33	5	He	1,066,048.75	2.9	34	9	H
750,523.63	38.5	25	2	H	1,066,939.07	25.3	25	3	H
750,829.45	38.5	25	2	He	1,067,373.83	25.3	25	3	He
758,790.34	3.5	36	7	H	1,085,072.00	55.7	22	2	H
764,229.59	142.3	20	1	H	1,085,514.15	55.7	22	2	He
764,541.01	142.3	20	1	He	1,088,869.02	14.8	27	4	H
764,610.85	142.3	20	1	C	1,089,312.72	14.8	27	4	He
764,632.69	142.3	20	1	S	1,116,313.69	6.9	30	6	H
772,453.58	18.3	28	3	H	1,144,445.52	5.2	31	7	H

(continued)

Table B.2. (continued)

ν/MHz	Int.	n	Δn		ν/MHz	Int.	n	Δn	
772,768.34	18.3	28	3	He	1,145,961.40	2.6	34	10	H
778,061.52	2.7	37	8	H	1,155,356.17	3.2	33	9	H
788,564.28	2.2	38	9	H	1,155,955.51	4.0	32	8	H
789,302.95	4.9	34	6	H	1,174,612.10	10.3	28	5	H
792,318.64	1.8	39	10	H	1,175,090.74	10.3	28	5	He
809,055.28	11.0	30	4	H	1,198,063.93	28.3	24	3	H
809,197.00	7.1	32	5	H	1,198,552.13	28.3	24	3	He
809,384.96	11.0	30	4	He	1,210,590.85	16.4	26	4	H
809,526.74	7.1	32	5	He	1,211,084.15	16.4	26	4	He
820,148.83	3.7	35	7	H	1,225,568.85	7.5	29	6	H
838,702.99	2.9	36	8	H	1,229,033.63	228.8	17	1	H
844,441.16	43.4	24	2	H	1,229,534.44	228.8	17	1	He
844,785.26	43.4	24	2	He	1,229,646.76	228.8	17	1	C
847,891.08	2.3	37	9	H	1,229,681.88	228.8	17	1	S
849,938.22	1.9	38	10	H	1,240,300.69	63.6	21	2	H
856,968.08	20.3	27	3	H	1,240,806.09	63.6	21	2	He
857,317.29	20.3	27	3	He	1,241,044.56	2.8	33	10	H
857,561.41	5.3	33	6	H	1,251,599.73	5.7	30	7	H
884,412.74	7.8	31	5	H	1,254,978.47	3.5	32	9	H
884,773.13	7.8	31	5	He	1,259,719.93	4.4	31	8	H
888,047.07	165.4	19	1	H	1,299,370.81	11.4	27	5	H
888,325.91	4.0	34	7	H	1,299,900.28	11.4	27	5	He
888,408.94	165.4	19	1	He	1,347,012.91	3.0	32	10	H
888,490.10	165.4	19	1	C	1,349,607.92	8.3	28	6	H
888,515.47	165.4	19	1	S	1,351,189.66	18.3	25	4	H
890,361.52	12.1	29	4	H	1,351,616.68	32.0	23	3	H
890,724.32	12.1	29	4	He	1,351,740.25	18.3	25	4	He
905,837.23	3.2	35	8	H	1,352,167.44	32.0	23	3	He
913,311.02	2.1	37	10	H	1,366,457.29	3.8	31	9	H
913,347.56	2.5	36	9	H	1,372,615.74	6.2	29	7	H
933,943.74	5.7	32	6	H	1,376,346.46	4.8	30	8	H
954,288.80	22.6	26	3	H	1,426,633.79	73.2	20	2	H
954,677.66	22.6	26	3	He	1,427,215.13	73.2	20	2	He
954,715.82	49.0	23	2	H	1,442,491.79	12.7	26	5	H
955,104.85	49.0	23	2	He	1,443,079.59	12.7	26	5	He
964,298.77	4.4	33	7	H	1,465,480.25	3.3	31	10	H
969,266.80	8.5	30	5	H	1,466,610.22	273.0	16	1	H
969,661.76	8.5	30	5	He	1,467,207.84	273.0	16	1	He
980,360.35	3.4	34	8	H	1,467,341.88	273.0	16	1	C
982,955.36	13.3	28	4	H	1,467,383.78	273.0	16	1	S
983,177.12	2.2	36	10	H	1,491,027.55	9.1	27	6	H
983,355.91	13.3	28	4	He	1,491,620.87	4.1	30	9	H
1,507,901.78	5.2	29	8	H	2,497,434.74	8.6	24	8	H
1,509,819.44	6.8	28	7	H	2,511,705.79	33.9	20	4	H
1,514,479.38	20.5	24	4	H	2,512,729.28	33.9	20	4	He
1,515,096.50	20.5	24	4	He	2,562,207.53	11.5	23	7	H
1,532,612.30	36.2	22	3	H	2,576,758.51	5.7	25	10	H
1,533,236.82	36.2	22	3	He	2,591,917.36	22.6	21	5	H
1,598,358.23	3.6	30	10	H	2,592,973.54	22.6	21	5	He

(continued)

Table B.2. (continued)

ν/MHz	Int.	n	Δn		ν/MHz	Int.	n	Δn	
1,607,491.71	14.1	25	5	H	2,599,551.38	15.9	22	6	H
1,608,146.74	14.1	25	5	He	2,680,152.85	498.5	13	1	H
1,632,648.51	4.5	29	9	H	2,681,244.98	498.5	13	1	He
1,652,276.67	84.7	19	2	H	2,681,489.92	498.5	13	1	C
1,652,949.95	84.7	19	2	He	2,681,566.50	498.5	13	1	S
1,652,993.58	10.1	26	6	H	2,689,091.48	7.3	24	9	H
1,656,866.32	5.8	28	8	H	2,692,407.78	63.5	18	3	H
1,666,023.36	7.5	27	7	H	2,693,504.90	63.5	18	3	He
1,705,239.44	23.1	23	4	H	2,695,643.85	138.1	16	2	H
1,705,934.31	23.1	23	4	He	2,696,742.28	138.1	16	2	He
1,747,476.20	41.3	21	3	H	2,794,108.47	9.6	23	8	H
1,747,922.92	3.9	29	10	H	2,864,087.29	6.3	24	10	H
1,748,188.27	41.3	21	3	He	2,883,801.96	13.0	22	7	H
1,769,610.90	329.3	15	1	H	2,892,577.36	39.0	19	4	H
1,770,331.99	329.3	15	1	He	2,893,756.04	39.0	19	4	He
1,770,493.72	329.3	15	1	C	2,945,540.13	18.0	21	6	H
1,770,544.28	329.3	15	1	S	2,959,246.10	25.8	20	5	H
1,792,152.36	4.9	28	9	H	2,960,451.95	25.8	20	5	He
1,798,729.96	15.8	24	5	H	3,004,610.25	8.2	23	9	H
1,799,462.92	15.8	24	5	He	3,140,104.01	10.8	22	8	H
1,826,234.88	6.3	27	8	H	3,157,211.81	74.4	17	3	H
1,839,392.65	11.3	25	6	H	3,158,498.33	74.4	17	3	He
1,844,650.31	8.3	26	7	H	3,196,266.99	7.0	23	10	H
1,916,899.10	4.3	28	10	H	3,236,221.12	165.7	15	2	H
1,928,178.19	98.9	18	2	H	3,237,539.83	165.7	15	2	He
1,928,963.89	98.9	18	2	He	3,261,955.58	14.6	21	7	H
1,929,513.16	26.1	22	4	H	3,354,811.98	45.2	18	4	H
1,930,299.41	26.1	22	4	He	3,356,146.96	20.4	20	6	H
1,973,281.77	5.4	27	9	H	3,356,179.02	45.2	18	4	He
2,004,530.28	47.3	20	3	H	3,372,004.96	9.2	22	9	H
2,005,347.10	47.3	20	3	He	3,377,764.65	628.0	12	1	H
2,019,646.13	7.0	26	8	H	3,379,141.04	628.0	12	1	He
2,021,654.89	17.7	23	5	H	3,379,449.74	628.0	12	1	C
2,022,478.69	17.7	23	5	He	3,379,546.25	628.0	12	1	S
2,049,894.43	9.3	25	7	H	3,399,752.87	29.6	19	5	H
2,055,032.01	12.6	24	6	H	3,401,138.22	29.6	19	5	He
2,108,567.81	4.7	27	10	H	3,546,206.17	12.2	21	8	H
2,162,210.55	402.3	14	1	H	3,582,506.74	7.9	22	10	H
2,163,091.63	402.3	14	1	He	3,709,769.73	16.6	20	7	H
2,163,289.23	402.3	14	1	C	3,735,774.96	87.9	16	3	H
2,163,351.01	402.3	14	1	S	3,737,297.23	87.9	16	3	He
2,179,857.65	6.0	26	9	H	3,802,508.21	10.3	21	9	H
2,195,016.51	29.7	21	4	H	3,847,293.17	23.4	19	6	H
2,195,910.95	29.7	21	4	He	3,921,441.40	52.8	17	4	H
2,241,551.17	7.8	25	8	H	3,923,039.34	52.8	17	4	He
2,269,164.74	116.3	17	2	H	3,931,821.45	201.1	14	2	H
2,270,089.39	116.3	17	2	He	3,932,708.47	34.2	18	5	H
2,283,135.93	20.0	22	5	H	3,933,423.62	201.1	14	2	He
2,284,066.28	20.0	22	5	He	3,934,310.99	34.2	18	5	He
2,286,932.96	10.3	24	7	H	4,026,185.17	13.8	20	8	H

(continued)

Table B.2. (continued)

ν/MHz	Int.	n	Δn		ν/MHz	Int.	n	Δn	
2,305,905.48	14.1	23	6	H	4,034,409.16	8.8	21	10	H
2,314,680.87	54.6	19	3	H	4,244,194.03	19.0	19	7	H
2,315,624.07	54.6	19	3	He	4,310,435.76	11.6	20	9	H
2,326,904.54	5.2	26	10	H	4,340,288.00	806.5	11	1	H
2,416,546.99	6.6	25	9	H	4,342,056.61	806.5	11	1	He
4,342,453.27	806.5	11	1	C	8,555,643.48	51.2	14	6	H
4,342,577.28	806.5	11	1	S	8,905,138.72	23.5	15	9	H
4,439,883.98	26.9	18	6	H	9,307,618.15	79.9	13	5	H
4,465,254.75	105.0	15	3	H	9,311,410.88	79.9	13	5	He
4,467,074.28	105.0	15	3	He	9,319,873.08	40.9	14	7	H
4,566,737.81	10.0	20	10	H	9,352,679.03	19.9	15	10	H
4,583,845.61	39.8	17	5	H	9,982,277.28	33.3	14	8	H
4,585,713.46	39.8	17	5	He	9,989,738.95	133.0	12	4	H
4,597,816.80	15.7	19	8	H	9,993,809.64	133.0	12	4	He
4,623,822.03	62.2	16	4	H	10,046,823.18	511.2	10	2	H
4,625,706.18	62.2	16	4	He	10,050,917.12	511.2	10	2	He
4,842,363.40	247.5	13	2	H	10,347,749.26	61.6	13	6	H
4,844,336.60	247.5	13	2	He	10,398,205.50	242.6	11	3	H
4,887,424.29	21.8	18	7	H	10,402,442.63	242.6	11	3	He
4,914,232.24	13.2	19	9	H	10,560,173.77	27.7	14	9	H
5,161,742.09	31.2	17	6	H	10,782,575.38	1998.3	8	1	H
5,198,482.83	11.3	19	10	H	10,786,969.13	1998.3	8	1	He
5,284,325.14	18.0	18	8	H	10,787,954.56	1998.3	8	1	C
5,388,051.62	46.7	16	5	H	10,788,262.64	1998.3	8	1	S
5,390,247.18	46.7	16	5	He	11,067,349.28	23.3	14	10	H
5,398,431.67	126.8	14	3	H	11,235,796.33	49.0	13	7	H
5,400,631.46	126.8	14	3	He	11,456,349.17	97.9	12	5	H
5,505,385.86	73.9	15	4	H	11,461,017.48	97.9	12	5	He
5,507,629.23	73.9	15	4	He	12,000,025.93	39.8	13	8	H
5,637,947.91	15.1	18	9	H	12,560,416.05	166.6	11	4	H
5,668,917.61	25.2	17	7	H	12,565,534.25	166.6	11	4	He
5,706,535.18	1059.7	10	1	H	12,662,430.13	32.9	13	9	H
5,708,860.51	1059.7	10	1	He	12,685,382.80	75.1	12	6	H
5,709,382.04	1059.7	10	1	C	13,240,326.62	27.6	13	10	H
5,709,545.09	1059.7	10	1	S	13,419,247.92	681.1	9	2	H
5,954,363.35	12.9	18	10	H	13,424,587.83	312.2	10	3	H
6,050,455.82	36.5	16	6	H	13,424,716.08	681.1	9	2	He
6,057,917.50	309.2	12	2	H	13,430,058.17	312.2	10	3	He
6,060,386.02	309.2	12	2	He	13,725,513.91	59.4	12	7	H
6,116,457.91	20.7	17	8	H	14,330,026.95	121.8	11	5	H
6,393,432.93	55.3	15	5	H	14,335,866.25	121.8	11	5	He
6,396,038.17	55.3	15	5	He	14,613,560.98	48.1	12	8	H
6,513,358.77	17.4	17	9	H	15,377,790.58	39.6	12	9	H
6,611,974.30	155.0	13	3	H	15,727,285.82	2909.8	7	1	H
6,614,668.59	155.0	13	3	He	15,733,694.47	2909.8	7	1	He
6,627,465.30	88.8	14	4	H	15,735,131.80	2909.8	7	1	C
6,628,352.31	29.3	16	7	H	15,735,581.17	2909.8	7	1	S
6,630,165.90	88.8	14	4	He	15,796,637.17	92.9	11	6	H
6,866,981.54	14.8	17	10	H	16,040,194.78	33.1	12	10	H

(continued)

Table B.2. (continued)

ν/MHz	Int.	n	Δn		ν/MHz	Int.	n	Δn	
7,135,527.82	24.1	16	8	H	16,104,740.68	212.5	10	4	H
7,157,662.52	43.0	15	6	H	16,111,303.14	212.5	10	4	He
7,583,068.13	0.1	16	9	H	17,025,670.80	73.1	11	7	H
7,667,596.41	6.1	14	5	H	17,759,535.92	411.2	9	3	H
7,670,720.85	66.1	14	5	He	17,766,772.69	411.2	9	3	He
7,712,712.74	1431.1	9	1	H	18,065,801.91	58.8	11	8	H
7,715,855.57	1431.1	9	1	He	18,266,951.23	154.2	10	5	H
7,716,560.44	1431.1	9	1	C	18,274,394.77	154.2	10	5	He
7,716,780.81	1431.1	9	1	S	18,495,288.12	935.7	8	2	H
7,718,052.65	393.4	11	2	H	18,502,824.70	935.7	8	2	He
7,721,197.65	393.4	11	2	He	18,953,848.98	48.2	11	9	H
7,820,066.72	34.5	15	7	H	19,718,078.57	40.2	11	10	H
7,979,968.99	17.1	16	10	H	20,036,562.13	116.8	10	6	H
8,078,584.52	107.9	13	4	H	21,137,300.57	277.0	9	4	H
8,081,876.43	107.9	13	4	He	21,145,913.73	277.0	9	4	He
8,220,128.05	192.3	12	3	H	21,503,172.35	91.3	10	7	H
8,223,477.64	192.3	12	3	He	22,732,205.98	73.1	10	8	H
8,397,963.21	28.2	15	8	H	23,772,337.09	59.6	10	9	H
23,817,453.42	199.2	9	5	H					
23,827,158.71	199.2	9	5	He					
24,201,823.30	557.1	8	3	H					
24,211,685.22	557.1	8	3	He					
24,231,670.00	4472.0	6	1	H					
24,241,544.08	4472.0	6	1	He					
24,243,758.62	4472.0	6	1	C					
24,244,450.98	4472.0	6	1	S					
24,660,384.16	49.4	10	10	H					
25,979,663.97	149.7	9	6	H					
26,509,861.20	1,335.1	7	2	H					
26,520,663.61	1,335.1	7	2	He					
27,749,274.87	116.2	9	7	H					
28,542,111.30	370.9	8	4	H					
28,553,741.82	370.9	8	4	He					
29,215,885.09	92.4	9	8	H					

Table B.3 compares the rest frequencies of hydrogen RRLs calculated with the Dirac equation and with the Rydberg equation. It shows the content of Table 2 of Towle et al. (1996). From left to right, the columns are the lower principal quantum number n, the change in principal quantum number Δn, the angular momentum quantum number j_{Imax} of the most intense fine-structure line, the frequency ν_{Imax} of that component, the line frequency ν_{Ryd} in MHz calculated by the Rydberg equation, the difference between ν_{Imax} and ν_{Ryd} in MHz, the intensity I_{max} of ν_{Imax} as $g'A$ in units of $10^4\,\mathrm{s}^{-1}$, the frequency $\bar{\nu}$ in MHz of the intensity-weighted mean of all fine-structure components, the minimum frequency ν_{min} in MHz, and the minimum frequency ν_{min} in MHz of the $j' - j = +1$ fine-structure components – a measure of half-width of the highly symmetric RRLs due to structure lines that appear at small principal quantum numbers.

Table B.3. Dirac hydrogen RRL frequencies for 100 GHz to 29 THz

n	Δn	j_{Imax}	ν_{Imax}	ν_{Ryd}	ν_{diff}	I_{max}	$\overline{\nu}$	ν_{min}
62	4	39	100,539.63	100,539.63	0.00	0.04	100,539.63	100,539.63
66	5	39	102,571.00	102,571.00	0.00	0.03	102,571.00	102,571.00
56	3	39	103,914.85	103,914.85	0.00	0.08	103,914.85	103,914.85
49	2	39	105,301.87	105,301.86	0.00	0.22	105,301.87	105,301.86
61	4	39	105,410.22	105,410.22	0.00	0.05	105,410.23	105,410.22
39	1	38	106,737.37	106,737.36	0.00	1.63	106,737.37	106,737.37
65	5	39	107,206.12	107,206.11	0.00	0.03	107,206.12	107,206.11
55	3	38	109,536.01	109,536.01	0.00	0.09	109,536.01	109,536.01
60	4	38	110,600.68	110,600.68	0.00	0.05	110,600.68	110,600.68
48	2	38	111,885.08	111,885.08	0.00	0.24	111,885.08	111,885.08
64	5	38	112,124.91	112,124.91	0.00	0.03	112,124.91	112,124.91
38	1	37	115,274.41	115,274.41	0.00	1.80	115,274.41	115,274.41
54	3	38	115,570.12	115,570.12	0.00	0.10	115,570.13	115,570.12
59	4	38	116,137.72	116,137.71	0.00	0.05	116,137.72	116,137.71
63	5	37	117,349.42	117,349.42	0.00	0.03	117,349.43	117,349.42
47	2	37	119,028.77	119,028.76	0.00	0.26	119,028.77	119,028.77
58	4	37	122,050.75	122,050.75	0.00	0.06	122,050.75	122,050.75
53	3	37	122,055.83	122,055.83	0.00	0.10	122,055.84	122,055.83
62	5	37	122,903.80	122,903.80	0.00	0.03	122,903.80	122,903.80
37	1	36	124,746.74	124,746.74	0.00	2.00	124,746.75	124,746.74
46	2	37	126,793.88	126,793.88	0.00	0.28	126,793.88	126,793.88
57	4	36	128,372.26	128,372.26	0.00	0.06	128,372.27	128,372.26
61	5	36	128,814.50	128,814.50	0.00	0.04	128,814.51	128,814.50
52	3	36	129,036.20	129,036.19	0.00	0.11	129,036.20	129,036.19
60	5	36	135,110.59	135,110.59	0.00	0.04	135,110.59	135,110.59
56	4	36	135,138.16	135,138.16	0.00	0.06	135,138.17	135,138.16
45	2	36	135,249.50	135,249.50	0.00	0.30	135,249.50	135,249.50
36	1	35	135,286.04	135,286.04	0.00	2.22	135,286.05	135,286.04
51	3	36	136,559.30	136,559.29	0.00	0.12	136,559.30	136,559.29
59	5	35	141,824.00	141,824.00	0.00	0.04	141,824.00	141,824.00
55	4	35	142,388.21	142,388.20	0.00	0.07	142,388.21	142,388.20
44	2	35	144,474.12	144,474.12	0.00	0.33	144,474.13	144,474.12
50	3	35	144,678.94	144,678.94	0.00	0.13	144,678.94	144,678.94
35	1	34	147,046.89	147,046.89	0.00	2.48	147,046.90	147,046.89
58	5	35	148,989.92	148,989.91	0.00	0.04	148,989.92	148,989.91
54	4	34	150,166.51	150,166.51	0.00	0.07	150,166.52	150,166.51
49	3	34	153,455.47	153,455.46	0.00	0.14	153,455.47	153,455.46
43	2	34	154,557.23	154,557.22	0.00	0.36	154,557.23	154,557.22
57	5	34	156,647.14	156,647.13	0.00	0.05	156,647.14	156,647.13
53	4	34	158,522.10	158,522.09	0.00	0.08	158,522.10	158,522.09
34	1	33	160,211.52	160,211.52	0.00	2.77	160,211.54	160,211.52
48	3	34	162,956.69	162,956.69	0.00	0.15	162,956.70	162,956.69
56	5	33	164,838.53	164,838.52	0.00	0.05	164,838.53	164,838.52
42	2	34	165,601.05	165,601.05	0.00	0.40	165,601.06	165,601.05
52	4	33	167,509.56	167,509.55	0.00	0.08	167,509.56	167,509.55
47	3	33	173,259.02	173,259.01	0.00	0.17	173,259.02	173,259.02
55	5	33	173,611.52	173,611.52	0.00	0.05	173,611.53	173,611.52
33	1	32	174,995.82	174,995.82	0.01	3.11	174,995.83	174,995.82
51	4	33	177,189.80	177,189.79	0.00	0.09	177,189.80	177,189.79
41	2	33	177,722.84	177,722.83	0.00	0.44	177,722.85	177,722.84

(continued)

Table B.3. (continued)

n	Δn	j_{Imax}	ν_{Imax}	ν_{Ryd}	ν_{diff}	I_{max}	$\overline{\nu}$	ν_{min}
54	5	32	183,018.71	183,018.71	0.00	0.06	183,018.71	183,018.71
46	3	32	184,448.71	184,448.70	0.01	0.18	184,448.72	184,448.71
50	4	32	187,630.91	187,630.91	0.01	0.10	187,630.92	187,630.91
40	2	32	191,057.40	191,057.40	0.01	0.48	191,057.41	191,057.40
32	1	31	191,656.74	191,656.74	0.01	3.50	191,656.76	191,656.74
53	5	32	193,118.48	193,118.48	0.01	0.06	193,118.49	193,118.48
45	3	32	196,623.44	196,623.43	0.01	0.20	196,623.45	196,623.43
49	4	31	198,909.19	198,909.18	0.01	0.11	198,909.20	198,909.19
52	5	31	203,975.82	203,975.82	0.01	0.07	203,975.83	203,975.82
39	2	31	205,760.33	205,760.32	0.01	0.53	205,760.34	205,760.32
44	3	31	209,894.07	209,894.06	0.01	0.21	209,894.08	209,894.06
31	1	30	210,501.79	210,501.78	0.01	3.95	210,501.81	210,501.79
48	4	31	211,110.30	211,110.29	0.01	0.11	211,110.30	211,110.29
51	5	30	215,663.16	215,663.15	0.01	0.07	215,663.16	215,663.15
38	2	30	222,011.77	222,011.77	0.01	0.59	222,011.79	222,011.77
47	4	30	224,330.64	224,330.63	0.01	0.12	224,330.64	224,330.63
43	3	30	224,386.78	224,386.78	0.01	0.23	224,386.79	224,386.78
50	5	30	228,261.41	228,261.41	0.01	0.08	228,261.42	228,261.41
30	1	29	231,900.95	231,900.94	0.01	4.48	231,900.97	231,900.95
46	4	29	238,678.96	238,678.95	0.01	0.14	238,678.97	238,678.95
37	2	30	240,021.15	240,021.14	0.01	0.65	240,021.16	240,021.15
42	3	29	240,245.62	240,245.62	0.01	0.26	240,245.63	240,245.62
49	5	29	241,861.16	241,861.16	0.01	0.09	241,861.17	241,861.16
45	4	29	254,278.27	254,278.26	0.01	0.15	254,278.28	254,278.26
29	1	28	256,302.06	256,302.05	0.01	5.10	256,302.08	256,302.06
48	5	29	256,564.02	256,564.01	0.01	0.09	256,564.03	256,564.01
41	3	29	257,635.50	257,635.49	0.01	0.28	257,635.51	257,635.49
36	2	29	260,032.79	260,032.78	0.01	0.72	260,032.80	260,032.78
44	4	28	271,268.01	271,268.00	0.01	0.16	271,268.02	271,268.00
47	5	28	272,484.24	272,484.23	0.01	0.10	272,484.24	272,484.23
40	3	28	276,745.80	276,745.79	0.01	0.31	276,745.81	276,745.79
35	2	28	282,332.94	282,332.93	0.01	0.81	282,332.95	282,332.93
28	1	27	284,250.60	284,250.59	0.01	5.84	284,250.63	284,250.60
46	5	28	289,750.58	289,750.57	0.01	0.11	289,750.59	289,750.57
43	4	28	289,806.73	289,806.72	0.01	0.17	289,806.74	289,806.72
39	3	27	297,794.77	297,794.76	0.01	0.34	297,794.78	297,794.76
34	2	27	307,258.42	307,258.41	0.01	0.90	307,258.44	307,258.41
45	5	27	308,508.52	308,508.51	0.01	0.12	308,508.53	308,508.51
42	4	27	310,075.18	310,075.17	0.01	0.19	310,075.19	310,075.17
27	1	26	316,415.46	316,415.44	0.01	6.71	316,415.49	316,415.46
38	3	27	321,034.74	321,034.73	0.01	0.38	321,034.75	321,034.73
44	5	26	328,922.84	328,922.83	0.01	0.13	328,922.85	328,922.83
41	4	26	332,280.07	332,280.06	0.01	0.21	332,280.08	332,280.06
33	2	26	335,207.35	335,207.34	0.01	1.01	335,207.37	335,207.34
37	3	26	346,758.52	346,758.51	0.01	0.42	346,758.54	346,758.51
43	5	26	351,180.67	351,180.66	0.01	0.14	351,180.68	351,180.66
26	1	25	353,622.78	353,622.77	0.02	7.75	353,622.82	353,622.78
40	4	26	356,658.46	356,658.45	0.01	0.23	356,658.48	356,658.45
32	2	26	366,652.57	366,652.55	0.01	1.14	366,652.59	366,652.56
36	3	25	375,307.20	375,307.18	0.01	0.46	375,307.22	375,307.19

(continued)

Table B.3. (continued)

n	Δn	j_{Imax}	ν_{Imax}	ν_{Ryd}	ν_{diff}	I_{max}	$\overline{\nu}$	ν_{min}
42	5	25	375, 495.13	375, 495.11	0.01	0.15	375, 495.14	375, 495.11
39	4	25	383, 483.17	383, 483.16	0.01	0.25	383, 483.19	383, 483.16
25	1	24	396, 900.88	396, 900.86	0.02	9.00	396, 900.92	396, 900.87
41	5	25	402, 109.63	402, 109.61	0.01	0.17	402, 109.64	402, 109.61
31	2	25	402, 158.54	402, 158.52	0.02	1.29	402, 158.57	402, 158.53
35	3	25	407, 079.68	407, 079.66	0.02	0.52	407, 079.70	407, 079.67
38	4	24	413, 069.18	413, 069.17	0.02	0.28	413, 069.20	413, 069.17
40	5	24	431, 303.03	431, 303.02	0.02	0.18	431, 303.05	431, 303.02
30	2	24	442, 402.74	442, 402.72	0.02	1.47	442, 402.78	442, 402.73
34	3	24	442, 544.47	442, 544.45	0.02	0.58	442, 544.49	442, 544.45
37	4	24	445, 781.48	445, 781.47	0.02	0.31	445, 781.51	445, 781.47
24	1	23	447, 540.33	447, 540.30	0.02	10.51	447, 540.39	447, 540.33
39	5	23	463, 395.83	463, 395.81	0.02	0.20	463, 395.85	463, 395.81
36	4	23	482, 044.57	482, 044.55	0.02	0.34	482, 044.59	482, 044.55
33	3	23	482, 254.24	482, 254.22	0.02	0.65	482, 254.27	482, 254.23
29	2	23	488, 203.02	488, 202.99	0.02	1.67	488, 203.05	488, 203.00
38	5	23	498, 757.58	498, 757.56	0.02	0.22	498, 757.61	498, 757.57
23	1	22	507, 175.54	507, 175.51	0.03	12.35	507, 175.61	507, 175.54
35	4	23	522, 354.09	522, 354.07	0.02	0.38	522, 354.12	522, 354.07
32	3	23	526, 864.10	526, 864.07	0.02	0.73	526, 864.13	526, 864.08
37	5	22	537, 815.93	537, 815.90	0.03	0.25	537, 815.96	537, 815.91
28	2	22	540, 552.67	540, 552.64	0.03	1.91	540, 552.71	540, 552.65
34	4	22	567, 291.21	567, 291.18	0.03	0.43	567, 291.24	567, 291.19
31	3	22	577, 154.37	577, 154.34	0.03	0.82	577, 154.41	577, 154.34
22	1	21	577, 896.53	577, 896.49	0.04	14.61	577, 896.61	577, 896.52
36	5	22	581, 067.53	581, 067.51	0.03	0.27	581, 067.56	581, 067.51
27	2	22	600, 666.06	600, 666.03	0.03	2.20	600, 666.12	600, 666.04
33	4	21	617, 540.30	617, 540.26	0.03	0.48	617, 540.33	617, 540.27
35	5	21	629, 091.47	629, 091.43	0.03	0.30	629, 091.50	629, 091.44
30	3	21	634, 059.50	634, 059.46	0.04	0.93	634, 059.55	634, 059.47
21	1	20	662, 404.25	662, 404.20	0.05	7.42	662, 404.36	662, 404.24
26	2	21	670, 038.25	670, 038.21	0.04	2.55	670, 038.32	670, 038.23
32	4	21	673, 911.00	673, 910.96	0.04	0.54	673, 911.04	673, 910.97
34	5	21	682, 565.63	682, 565.59	0.04	0.34	682, 565.67	682, 565.59
29	3	20	698, 704.82	698, 704.78	0.04	1.06	698, 704.87	698, 704.79
31	4	20	737, 365.90	737, 365.86	0.04	0.61	737, 365.95	737, 365.87
33	5	20	742, 287.04	742, 287.00	0.04	0.38	742, 287.09	742, 287.01
25	2	20	750, 523.67	750, 523.63	0.05	2.97	750, 523.75	750, 523.64
20	1	19	764, 229.65	764, 229.59	0.06	20.93	764, 229.79	764, 229.64
28	3	20	772, 453.63	772, 453.58	0.05	1.21	772, 453.69	772, 453.59
30	4	19	809, 055.33	809, 055.28	0.06	0.69	809, 055.39	809, 055.29
32	5	19	809, 197.05	809, 197.00	0.05	0.42	809, 197.10	809, 197.01
24	2	19	844, 441.22	844, 441.16	0.06	3.47	844, 441.32	844, 441.18
27	3	19	856, 968.14	856, 968.08	0.06	1.40	856, 968.22	856, 968.09
31	5	19	884, 412.80	884, 412.74	0.06	0.48	884, 412.86	884, 412.75
19	1	18	888, 047.15	888, 047.07	0.08	5.38	888, 047.32	888, 047.14
29	4	19	890, 361.57	890, 361.52	0.06	0.78	890, 361.64	890, 361.53
26	3	18	954, 288.87	954, 288.80	0.07	1.61	954, 288.96	954, 288.82
23	2	18	954, 715.89	954, 715.82	0.08	4.07	954, 716.01	954, 715.84
30	5	18	969, 266.87	969, 266.80	0.07	0.54	969, 266.94	969, 266.81

(continued)

Table B.3. (continued)

n	Δn	j_{Imax}	ν_{Imax}	ν_{Ryd}	ν_{diff}	I_{max}	$\overline{\nu}$	ν_{min}
28	4	18	982,955.44	982,955.36	0.07	0.89	982,955.51	982,955.38
15	1	17	1,040,131.21	1,040,131.11	0.10	31.06	1,040,131.44	1,040,131.19
29	5	18	1,065,357.41	1,065,357.33	0.08	0.61	1,065,357.49	1,065,357.34
25	3	18	1,066,939.15	1,066,939.07	0.08	1.87	1,066,939.26	1,066,939.09
22	2	18	1,085,072.08	1,085,072.00	0.08	4.84	1,085,072.23	1,085,072.03
27	4	18	1,088,869.10	1,088,869.02	0.08	1.02	1,088,869.20	1,088,869.04
28	5	17	1,174,612.20	1,174,612.10	0.10	0.70	1,174,612.29	1,174,612.12
24	3	17	1,198,064.03	1,198,063.93	0.10	2.18	1,198,064.16	1,198,063.95
26	4	17	1,210,590.95	1,210,590.85	0.10	1.17	1,210,591.06	1,210,590.87
17	1	16	1,229,033.77	1,229,033.63	0.14	38.44	1,229,034.05	1,229,033.74
21	2	17	1,240,300.80	1,240,300.69	0.11	5.79	1,240,300.98	1,240,300.73
27	5	17	1,299,370.91	1,299,370.81	0.10	0.79	1,299,371.03	1,299,370.82
25	4	16	1,351,189.78	1,351,189.66	0.13	1.36	1,351,189.91	1,351,189.68
23	3	16	1,351,616.81	1,351,616.68	0.13	2.55	1,351,616.96	1,351,616.70
20	2	16	1,426,633.94	1,426,633.79	0.14	6.98	1,426,634.16	1,426,633.85
26	5	16	1,442,491.92	1,442,491.79	0.13	0.92	1,442,492.06	1,442,491.81
16	1	15	1,466,610.41	1,466,610.22	0.19	48.14	1,466,610.79	1,466,610.36
24	4	16	1,514,479.51	1,514,479.38	0.14	1.58	1,514,479.68	1,514,479.40
22	3	16	1,532,612.45	1,532,612.30	0.14	3.02	1,532,612.65	1,532,612.34
25	5	15	1,607,491.88	1,607,491.71	0.17	1.06	1,607,492.02	1,607,491.73
19	2	15	1,652,276.85	1,652,276.67	0.19	8.46	1,652,277.14	1,652,276.73
23	4	15	1,705,239.63	1,705,239.44	0.18	1.85	1,705,239.82	1,705,239.48
21	3	15	1,747,476.39	1,747,476.20	0.19	3.60	1,747,476.64	1,747,476.25
15	1	14	1,769,611.17	1,769,610.90	0.27	61.11	1,769,611.68	1,769,611.10
24	5	15	1,798,730.15	1,798,729.96	0.19	1.23	1,798,730.34	1,798,729.99
18	2	14	1,928,178.44	1,928,178.19	0.26	10.34	1,928,178.79	1,928,178.27
22	4	14	1,929,513.40	1,929,513.16	0.24	2.17	1,929,513.62	1,929,513.20
20	3	14	2,004,530.53	2,004,530.28	0.25	4.31	2,004,530.83	2,004,530.34
23	5	14	2,021,655.13	2,021,654.89	0.24	1.43	2,021,655.35	2,021,654.93
14	1	13	2,162,210.93	2,162,210.55	0.38	78.74	2,162,211.64	2,162,210.83
21	4	14	2,195,016.77	2,195,016.51	0.26	2.59	2,195,017.08	2,195,016.56
17	2	14	2,269,165.02	2,269,164.74	0.28	12.89	2,269,165.53	2,269,164.85
22	5	14	2,283,136.20	2,283,135.93	0.27	1.68	2,283,136.50	2,283,135.98
19	3	14	2,314,681.15	2,314,680.87	0.28	5.20	2,314,681.56	2,314,680.94
20	4	13	2,511,706.15	2,511,705.79	0.35	3.09	2,511,706.51	2,511,705.86
21	5	13	2,591,917.72	2,591,917.36	0.35	1.99	2,591,918.06	2,591,917.42
13	1	12	2,680,153.40	2,680,152.85	0.55	103.20	2,680,154.40	2,680,153.24
18	3	13	2,692,408.16	2,692,407.78	0.38	6.39	2,692,408.67	2,692,407.88
16	2	13	2,695,644.24	2,695,643.85	0.39	16.25	2,695,644.90	2,695,644.00
19	4	13	2,892,577.75	2,892,577.36	0.40	3.71	2,892,578.26	2,892,577.44
20	5	12	2,959,246.58	2,959,246.10	0.48	2.36	2,959,246.97	2,959,246.17
17	3	12	3,157,212.35	3,157,211.81	0.54	7.88	3,157,212.97	3,157,211.94
15	2	12	3,236,221.68	3,236,221.12	0.56	20.70	3,236,222.55	3,236,221.33
18	4	12	3,354,812.53	3,354,811.98	0.55	4.54	3,354,813.13	3,354,812.09
12	1	11	3,377,765.49	3,377,764.65	0.83	137.95	3,377,766.92	3,377,765.23
19	5	12	3,399,753.41	3,399,752.87	0.54	2.85	3,399,753.96	3,399,752.97
16	3	11	3,735,775.72	3,735,774.96	0.76	9.78	3,735,776.48	3,735,775.13
17	4	11	3,921,442.17	3,921,441.40	0.77	5.57	3,921,442.90	3,921,441.55
14	2	11	3,931,822.28	3,931,821.45	0.82	26.68	3,931,823.43	3,931,821.75
18	5	11	3,932,709.21	3,932,708.47	0.75	3.45	3,932,709.86	3,932,708.60

(continued)

Table B.3. (continued)

n	Δn	j_{Imax}	ν_{Imax}	ν_{Ryd}	ν_{diff}	I_{max}	$\bar{\nu}$	ν_{min}
11	1	10	4,340,289.31	4,340,288.00	1.31	188.61	4,340,291.43	4,340,288.88
15	3	11	4,465,255.62	4,465,254.75	0.87	12.48	4,465,256.80	4,465,254.98
17	5	11	4,583,846.46	4,583,845.61	0.86	4.22	4,583,847.40	4,583,845.78
16	4	11	4,623,822.91	4,623,822.03	0.88	6.92	4,623,823.99	4,623,822.23
13	2	11	4,842,364.32	4,842,363.40	0.92	34.79	4,842,366.18	4,842,363.82
16	5	10	5,388,052.85	5,388,051.62	1.23	5.24	5,388,053.96	5,388,051.85
14	3	10	5,398,432.97	5,398,431.67	1.29	16.06	5,398,434.48	5,398,432.01
15	4	10	5,505,387.14	5,505,385.86	1.28	8.75	5,505,388.47	5,505,386.13
10	1	9	5,706,537.32	5,706,535.18	2.14	264.77	5,706,540.56	5,706,536.57
12	2	10	6,057,918.92	6,057,917.50	1.42	47.28	6,057,921.52	6,057,918.12
15	5	10	6,393,434.36	6,393,432.93	1.43	6.51	6,393,436.04	6,393,433.25
13	3	9	6,611,976.29	6,611,974.30	1.99	20.81	6,611,978.23	6,611,974.79
14	4	9	6,627,467.22	6,627,465.30	1.92	11.09	6,627,468.85	6,627,465.69
14	5	9	7,667,598.54	7,667,596.41	2.13	8.35	7,667,600.62	7,667,596.86
9	1	8	7,712,716.43	7,712,712.74	3.69	383.47	7,712,721.58	7,712,715.04
11	2	9	7,718,054.93	7,718,052.65	2.28	65.47	7,718,058.65	7,718,053.60
13	4	9	8,078,586.79	8,078,584.52	2.27	14.48	8,078,589.45	8,078,585.08
12	3	9	8,220,130.39	8,220,128.05	2.33	27.90	8,220,133.67	8,220,128.77
13	5	8	9,307,621.43	9,307,618.15	3.28	10.71	9,307,623.95	9,307,618.79
12	4	8	9,989,742.54	9,989,738.95	3.59	19.17	9,989,745.96	9,989,739.78
10	2	8	10,046,827.01	10,046,823.18	3.83	92.59	10,046,832.44	10,046,824.68
11	3	8	10,398,209.30	10,398,205.50	3.80	38.21	10,398,213.79	10,398,206.59
8	1	7	10,782,582.17	10,782,575.38	6.79	576.60	10,782,590.70	10,782,579.40
12	5	8	11,456,353.12	11,456,349.17	3.95	14.10	11,456,357.36	11,456,350.11
11	4	8	12,560,420.36	12,560,416.05	4.30	25.49	12,560,426.28	12,560,417.31
9	2	7	13,419,254.74	13,419,247.92	6.82	134.03	13,419,262.83	13,419,250.41
10	3	7	13,424,594.35	13,424,587.83	6.52	52.96	13,424,600.46	13,424,589.56
11	5	7	14,330,033.44	14,330,026.95	6.49	18.93	14,330,038.82	14,330,028.38
7	1	6	15,727,299.41	15,727,285.82	13.60	907.61	15,727,314.27	15,727,293.37
10	4	7	16,104,748.00	16,104,740.68	7.33	35.91	16,104,756.11	16,104,742.67
9	3	7	17,759,543.88	17,759,535.92	7.97	76.22	17,759,555.96	17,759,538.79
10	5	7	18,266,959.22	18,266,951.23	7.99	25.63	18,266,968.98	18,266,953.49
8	2	7	18,495,296.16	18,495,288.12	8.04	199.25	18,495,313.39	18,495,292.49
9	4	6	21,137,313.80	21,137,300.57	13.23	50.99	21,137,324.77	21,137,303.87
9	5	6	23,817,467.74	23,817,453.42	14.32	36.78	23,817,480.98	23,817,457.13
8	3	6	24,201,838.41	24,201,823.30	15.12	115.89	24,201,856.68	24,201,828.34
6	1	5	24,231,700.42	24,231,670.00	30.42	1,512.56	24,231,727.71	24,231,685.55
7	2	6	26,509,877.29	26,509,861.20	16.09	327.01	26,509,906.82	26,509,869.43
8	4	6	28,542,128.09	28,542,111.30	16.80	74.36	28,542,151.07	28,542,117.06

Figure B.1 illustrates the difference between the frequencies calculated by the Rydberg and Dirac equations for the H19α line at 888 GHz ($\lambda = 338\,\mu$m). All fine-structure components lying within this spectral window have been plotted. At left, a broken line marks the frequency calculated from the Rydberg equation (A.5). The vertical lines mark the fine-structure components calculated from the Dirac equation (B.1). The strongest components correspond to the case $j' - j = +1$ and the weakest correspond to $j' - j = +1$. The $j' - j = 0$ components – although plotted – generally lie within the line width of the X-axis and are too weak to be visible. The weighted-intensity

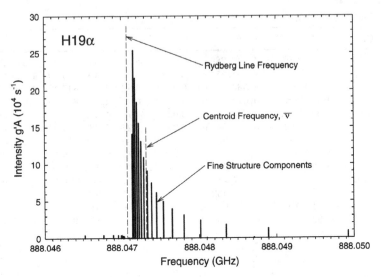

Fig. B.1 The fine-structure components of the H19α recombination line plotted against frequency. *Broken lines* mark the frequency calculated by the classical Rydberg equation and the frequency–intensity weighted mean – the centroid – of *all* fine-structure components. Strong components with frequencies greater than the Rydberg frequency are $j' - j = +1$. The weak lower-frequency components seen on either side of the Rydberg frequency come principally from $j' - j = -1$. The $j' - j = 0$ components are too weak to be visible although this plot includes them. Data from Watson (2002)

mean of all of these is marked by another broken line labeled "centroid frequency." The effect of the weighting is to shift the line to higher frequencies. The shift between the classical Rydberg frequency and the centroid of all the fine-structure components is about 250 kHz – a very small amount. Note also that the dispersion of the fine-structure components creates a line width. For the H19α line, this full width at half-intensity is about 300 kHz. Of course, the composite line profile is highly asymmetrical and cannot be easily compared to a Gaussian.

Section B.3 lists the FORTRAN code used for these calculations.

B.3 FORTRAN Code for Fine-Structure Frequencies

To facilitate calculations of the frequencies of the fine-structure lines of hydrogen, we have included FORTRAN code written by Watson (2002). This code was not intended for publication but, with the inserted comments, can be read with clarity. It produces a listing of upper principal quantum number n', upper j' value, lower principal quantum number n, lower j value, frequency in MHz (or wave number in cm^{-1}), and intensity gA in units of s^{-1}.

As discussed earlier in Sect. B.2, this code does not include corrections for relativistic effects.

```
      PROGRAM HFINE

C  FORTRAN code to calculate Hydrogen fine-structure components
C  Relativistic effects on intensities are not included
C  Uses Dirac formula for frequencies but not for intensities
C  Written by James K.G. Watson
C  Steacie Institute for Molecular Sciences
C  National Research Council of Canada
C  Ottawa
C
      IMPLICIT DOUBLE PRECISION (A-H,P-Z), INTEGER(I-N), CHARACTER*2(O)
      PARAMETER(MAXSIZ=10000)
      DIMENSION NUP(MAXSIZ),NLO(MAXSIZ),FJUP(MAXSIZ),
     +   FJLO(MAXSIZ),W(MAXSIZ),
     +   SS(MAXSIZ),INDEX(MAXSIZ),GA(MAXSIZ)

      OPEN(UNIT=5,FILE='HFINE.IN',STATUS='OLD')
C
C  INPUTS ARE (1) ISPEC = 1 FOR NORMAL OUTPUT IN A TABLE, 2
C                         AND 3 FOR SPECIAL GRAPHICS OPTIONS
C             (2) IUNIT = 1 FOR WAVENUMBER IN 1/cm AND 2 FOR MHz
C             (3) NU = UPPER PRINCIPAL QUANTUM NUMBER
C             (4) NL = LOWER PRINCIPAL QUANTUM NUMBER
C
C  Read the run parameters from the input file
C
      READ(5,5001) ISPEC,IUNIT,NU,NL
 5001 FORMAT(4I5)
C
C  Establish appropriate physical constants
C
      RINF=109737.31534D0
      IF (IUNIT.EQ.2) RINF=RINF*2.99792458D4
      ALF=7.29735308D-3
      FME=9.1093897D0/1.6605402D4
      FMH=1.007825035
      FAC=2.D0*RINF*(1.D0-FME/FMH)/(ALF*ALF)
      RH=RINF*(1.D0-FME/FMH)
C
C  Initialize variables
C
      LNCT=0
      FNL=DFLOAT(NL)
      SSUM=0.D0
      WSUM=0.D0
      SMAX=0.D0
      GAMAX=0.D0
C
C  Begin principal loop
C
      DO 99 JL2=1,(2*NL-1),2
      FJL=0.5D0*JL2
      FKL=FJL+0.5D0
      TEMP=FNL-FKL+DSQRT((FKL*FKL-ALF*ALF))
      TEMP=(ALF/TEMP)**2
      TEMPP=DSQRT((1+TEMP))
      ELO=-FAC*TEMP/(TEMPP*(1.D0+TEMPP))
      ELOO=-RINF*(1.D0-FME/FMH)/(FNL**2)
      ELO=ELO+ELOO*(FME/FMH)*(ALF/(2.D0*FNL))**2
      INTJ=INT(FJL)
      DO 99 IDJP2=1,3
      S=0.D0
      FJU=FJL+DFLOAT((IDJP2-2))
      IF (FJU.LT.0.D0) GO TO 99
```

```
          FKU=FJU+0.5D0
           DO 98 IDJL=1,2
            LL=INTJ+IDJL-1
            IF (LL.EQ.NL) GO TO 98
            FLL=DFLOAT(LL)
            DO 97 IDLP2=1,3,2
             LU=LL+IDLP2-2
             FLU=DFLOAT(LU)
C
             FNU=DFLOAT(NU)
             IF (LU.LT.0) GO TO 97
             IF (LU.GT.(NU-1)) GO TO 97
             TEMP=FNU-FKU+DSQRT((FKU*FKU-ALF*ALF))
             TEMP=(ALF/TEMP)**2
             TEMPP=DSQRT((1+TEMP))
             EUP=-FAC*TEMP/(TEMPP*(1.D0+TEMPP))
             EUPO=-RINF*(1.D0-FME/FMH)/(FNU**2)
             EUP=EUP+EUPO*(FME/FMH)*(ALF/(2.D0*FNU))**2
             FREQ=EUP-EL0
C
C Calculate line strength
             S=S+0.5D0*(2.D0*FJU+1.D0)*(2.D0*FJL+1.D0)
     &        *(FLU+FLL+1.D0)*(SIXJ(FJU,1.D0,FJL,FLL,0.5D0,FLU)
     &        *GORD(NU,LU,NL,LL))**2
C   Alternative output format
C     WRITE(6,6001) NU,LU,FJU,(RH+EUP),NL,LL,FJL,(RH+EL0),FREQ,S
C
  97     CONTINUE
  98     CONTINUE
         IF (ISPEC.NE.1) GO TO 101
C Standard unsorted list output
C     WRITE(6,6004) NU,FJU,(RH+EUP),NL,FJL,(RH+EL0),FREQ,S
 101 SSUM=SSUM+S
     WSUM=WSUM+S*FREQ
     IF (IUNIT.EQ.1) A=S*2.0261273D-6*(FREQ**3)
     IF (IUNIT.EQ.2) A=S*2.0261273D-6*((FREQ/2.99792458D4)**3)
     IF (A.LT.1.D-6) GO TO 99
     LNCT=LNCT+1
     NUP(LNCT)=NU
     FJUP(LNCT)=FJU
     NLO(LNCT)=NL
     FJLO(LNCT)=FJL
     W(LNCT)=FREQ
     SS(LNCT)=S
     GA(LNCT)=A
     IF (A.GT.GAMAX) GAMAX=A
  99 CONTINUE
     FNBAR=0.5D0*DFLOAT((NL+NU))
     IF (ISPEC.NE.1) GO TO 100
     WRITE(6,6003) NU,NL,(WSUM/SSUM),(EUPO-EL00)
6003 FORMAT(2I3,3F20.6)
 100 CONTINUE
6001 FORMAT(2I3,F4.1,1P,D15.6,3X,0P,2I3,F4.1,1P,D16.6,0P,2F20.6)
6004 FORMAT(I3,F5.1,1P,D15.6,3X,0P,I3,F5.1,1P,D15.6,0P,2F20.6)
C
C Sort output before printing
     CALL ASC2(LNCT,10000,INDEX,W)
C
C Output calculations (LNCT = line count)
C
     DO 200 I=1,LNCT
     J=INDEX(I)
     IF (ISPEC.EQ.2) WRITE(6,6005) W(I),0.D0
     IF (ISPEC.EQ.2) WRITE(6,6005) W(I),SS(J)
     IF (ISPEC.EQ.3) WRITE(6,6006) W(I),SS(J)
     IF (ISPEC.EQ.2) WRITE(6,6005) W(I),0.D0
     IF ((ISPEC.NE.1).AND.(ISPEC.NE.4)) GO TO 200
```

```
C
C  Optional filtering for threshold line strength GA
C      IF (GA(J).LT.(0.1*GAMAX)) GO TO 200
       WRITE(6,6002) NUP(J),FJUP(J),NLO(J),FJLO(J),W(I),GA(J)
   200 CONTINUE
  6002 FORMAT(I3,F5.1,3X,I3,F5.1,F15.3,F15.0)
  6005 FORMAT(2F15.6)
  6006 FORMAT(F11.3,F9.0)
C
       STOP
       END
C
C
       DOUBLE PRECISION FUNCTION gord(N1,L1,N2,L2)
       IMPLICIT DOUBLE PRECISION (A-H,O-Z), INTEGER(I-N)
C  Calculate radial matrix integrals
       GORD=0.D0
      IF (IABS(L1-L2).NE.1) GO TO 200
C
       IF (L2.GT.L1) GO TO 10
       N=N1
       L=L1
       NP=N2
       GO TO 20
    10 N=N2
       L=L2
       NP=N1
C
    20 CONTINUE
       FN=DFLOAT(N)
       FNP=DFLOAT(NP)
       FL=DFLOAT(L)
       IF (N1.NE.N2) GO TO 30
       GORD=-1.5D0*FN*DSQRT((FN*FN-FL*FL))
       GO TO 200
    30 X=-(FN-FNP)**2/(4.D0*FN*FNP)
C
       MU=MIN(N-L-1,NP-L)
       M=N+NP-2*L-1-MU
       MUP=MIN(N-L+1,NP-L)
       MP=N+NP-2*L+1-MUP
       X1=1.D0
       IF ((M-MU-1).NE.0) X1=(FN-FNP)**(M-MU-1)
       X2=1.D0
       IF (MU.NE.0) X2=(-4.D0*FN*FNP)**MU
       X3=1.D0
       IF ((MP-MUP-1).NE.0) X3=(FN-FNP)**(MP-MUP-1)
       X4=1.D0
       IF (MUP.NE.0) X4=(-4.D0*FN*FNP)**MUP
       TERM1=X2*X1*FAC(M)
      &       *HYPERG(MU,(2*L+MU-1),(M-MU+1),X)
      &          /(FAC((2*L-1+MU))*FAC((M-MU)))
       TERM2=X4*X3*FAC(MP)
      &       *HYPERG(MUP,(2*L+MUP-1),(MP-MUP+1),X)
      &          /(FAC((2*L-1+MUP))*FAC((MP-MUP))*(N+NP)**2)
C
       X5=1.D0
       IF (NP.NE.L) X5=(-1.D0)**(NP-L)
       TERM1=X5*(4.D0*FN*FNP)**(L+1)/(4.D0*(FN+FNP)**(N+NP))
      &    *DSQRT((FAC((N+L))*FAC((NP+L-1))/(FAC((N-L-1)*FAC((NP-L)))))
      &    *TERM1
       TERM2=X5*(4.D0*FN*FNP)**(L+1)/(4.D0*(FN+FNP)**(N+NP))
      &    *DSQRT((FAC((N+L))*FAC((NP+L-1))/(FAC((N-L-1))*FAC((NP-L)))))
      &    *TERM2
       GORD=TERM1-TERM2
C Print statement for radial matrix integrals
C      WRITE(6,6265) N1,L1,N2,L2,TERM1,TERM2,GORD
```

```
 6265 FORMAT(4I3,1P,5D12.3)
  200 RETURN
C
C     The hyperg(m,n,k,x) function is 2F1(-m,-n;k;x)
C
      DOUBLE PRECISION FUNCTION hyperg(m,n,k,x)
      IMPLICIT DOUBLE PRECISION (A-H,O-Z), INTEGER(I-N)
C
      TERM=1.D0
      IF (IM.EQ.0) GO TO 200
      DO 100 I=1,IM
      A1=DFLOAT((M+1-I)*(N+1-I))
      A2=DFLOAT((K-1+I)*I)
      TERM=TERM*X*A1/A2
      HYPERG=HYPERG+TERM
  100 CONTINUE
  200 RETURN
      END
C
C     Program to calculate N!
C
      DOUBLE PRECISION FUNCTION FAC(N)
      IMPLICIT DOUBLE PRECISION (A-H,O-Z), INTEGER(I-N)
      FAC=1.D0
      IF (N.EQ.0) GO TO 9000
      IF (N.GT.0) GO TO 8000
      FAC=0.D0
      GO TO 9000
 8000 X=0.D0
      DO 8001 NNN=1,N
      X=X+1.D0
 8001 FAC=FAC*X
 9000 RETURN
      END
C
C   Evaluate 6J array See Richard Zare's book on angular momentum
C
      DOUBLE PRECISION FUNCTION SIXJ(X1,X2,X3,X4,X5,X6)
C
      IMPLICIT DOUBLE PRECISION (A-H,O-Z), INTEGER(I-N)
      DIMENSION FL(400)
      COMMON/FACLOG/FL
C
      SIXJ=0.D0
C
      AA1=(-X1+X2+X3)*(X1-X2+X3)*(X1+X2-X3)
      IF (AA1.LT.0.D0) GO TO 1000
      AA2=(-X1+X5+X6)*(X1-X5+X6)*(X1+X5-X6)
      IF (AA2.LT.0.D0) GO TO 1000
      AA3=(-X4+X2+X6)*(X4-X2+X6)*(X4+X2-X6)
      IF (AA3.LT.0.D0) GO TO 1000
      AA4=(-X4+X5+X3)*(X4-X5+X3)*(X4+X5-X3)
      IF (AA4.LT.0.D0) GO TO 1000
C
      NL=400
      FL(1)=0.D0
      X=1.D0
      DO 100 I=2,NL
      X=X+1.D0
  100 FL(I)=FL(I-1)+DLOG(X)
C
      I1=IDINT((X1+X2+X3))
      FRAC=X1+X2+X3-DFLOAT(I1)
      IF (DABS(FRAC).GT.1.D-9) GO TO 900
      I2=IDINT((X1+X5+X6))
      FRAC=X1+X5+X6-DFLOAT(I2)
      IF (DABS(FRAC).GT.1.D-9) GO TO 900
```

```
      I3=IDINT((X4+X2+X6))
      FRAC=X4+X2+X6-DFLOAT(I3)
      IF (DABS(FRAC).GT.1.D-9) GO TO 900
      I4=IDINT((X4+X5+X3))
      FRAC=X4+X5+X3-DFLOAT(I4)
      IF (DABS(FRAC).GT.1.D-9) GO TO 900
C
      K1=IDINT((X2+X3+X5+X6))
      K2=IDINT((X1+X3+X4+X6))
      K3=IDINT((X1+X2+X4+X5))
C
      KMIN=MAX0(I1,I2,I3,I4)
      KMAX=MIN0(K1,K2,K3)
C
      IF (KMIN.GT.KMAX) GO TO 1000
C
      I2X1=IDINT((2.D0*X1+0.1D0))
      I2X2=IDINT((2.D0*X2+0.1D0))
      I2X3=IDINT((2.D0*X3+0.1D0))
      I2X4=IDINT((2.D0*X4+0.1D0))
      I2X5=IDINT((2.D0*X5+0.1D0))
      I2X6=IDINT((2.D0*X6+0.1D0))
C
C     WRITE(6,6211) X1,X2,X3,X4,X5,X6
C6211 FORMAT(6F5.1)
C     WRITE(6,6212) I2X1,I2X2,I2X3,I2X4,I2X5,I2X6
C6212 FORMAT(6I5)
C
      EXPON=0.5D0*(FACL(I1-I2X1)+FACL(I1-I2X2)+FACL(I1-I2X3)
     +     +FACL(I2-I2X1)+FACL(I2-I2X5)+FACL(I2-I2X6)
     +     +FACL(I3-I2X4)+FACL(I3-I2X2)+FACL(I3-I2X6)
     +     +FACL(I4-I2X4)+FACL(I4-I2X5)+FACL(I4-I2X3)
     +     -FACL(I1+1)-FACL(I2+1)-FACL(I3+1)-FACL(I4+1))
     +     +FACL(KMIN+1)
     +     -FACL(KMIN-I1)-FACL(KMIN-I2)-FACL(KMIN-I3)-FACL(KMIN-I4)
     +     -FACL(K1-KMIN)-FACL(K2-KMIN)-FACL(K3-KMIN))
      TERM=DFLOAT((-1)**KMIN)*DEXP(EXPON)
C
C     WRITE(6,6222) KMIN,TERM
C6222 FORMAT(I5,F20.9)
C
      SIXJ=TERM
C
      IF (KMAX.EQ.KMIN) GO TO 1000
C
      SUM=1.D0
      TERM=1.D0
      DO 200 K=(KMIN+1),KMAX
      XK=DFLOAT((K+1)*(K1-K+1)*(K2-K+1)*(K3-K+1))
      YK=DFLOAT((K-I1)*(K-I2)*(K-I3)*(K-I4))
      TERM=-TERM*XK/YK
      SUM=SUM+TERM
  200 CONTINUE
C
      SIXJ=SIXJ*SUM
      GO TO 1000
C
C 900 CONTINUE
  900 WRITE(6,6111)
 6111 FORMAT('    Sum of J values in a triangle of 6j noninteger ')
      WRITE(6,6112) X1,X2,X3,X4,X5,X6
 6112 FORMAT('    J values: ', 6F10.3)
C
 1000 RETURN
      END
C
C
```

```
C
      DOUBLE PRECISION FUNCTION FACL(I)
C
      IMPLICIT DOUBLE PRECISION (A-H,O-Z), INTEGER(I-N)
      DIMENSION FL(400)
      COMMON/FACLOG/FL
C
      IF (I.LE.0) FACL=0.D0
      IF (I.GE.1) FACL=FL(I)
C
      RETURN
      END
C
C
      SUBROUTINE ASC2(N,NDIM,INDEX,WAV)
C
C     Subroutine to arrange calculated spectrum in order of wave number.
C
      IMPLICIT DOUBLE PRECISION (A-H,O-Z), INTEGER(I-N)
      DIMENSION INDEX(NDIM),WAV(NDIM)
      DO 50 I=1,NDIM
   50 INDEX(I)=I
      IF (N.EQ.1) RETURN
      ILIM=N-1
      DO 300 I=1,ILIM
      JLIM=I+1
      DO 300 J=JLIM,N
      IF (WAV(J).GE.WAV(I)) GO TO 300
      TEMP=WAV(I)
      WAV(I)=WAV(J)
      WAV(J)=TEMP
      ITEMP=INDEX(I)
      INDEX(I)=INDEX(J)
      INDEX(J)=ITEMP
  300 CONTINUE
 9800 RETURN
      END
```

Appendix C
Supplemental Calculations

C.1 Early Estimates of Stark Broadening

It is interesting to revisit the calculations of the effects of line broadening with regard to the detectability of RRLs as considered by van de Hulst. We try to reproduce his calculations below following the suggestions of Sullivan (1982).

The full width of a Gaussian line at half-intensity, $\Delta\nu_T$, due to thermal broadening is given by

$$\Delta\nu_T = \nu_0 \left(4\ln 2 \frac{2kT}{Mc^2}\right)^{1/2},\tag{C.1}$$

where ν_0 is the rest frequency of the line and M is the mass of the radiating atom or molecule. Working from the van de Hulst (1945) and the Inglis and Teller (1939) papers, we derived the simplified expression for the Stark width, $\Delta\nu_{vdh}$, calculated by van de Hulst to be

$$\Delta\nu_{vdh} \approx \nu_0 \left(\frac{3\times 10^6}{\nu_0}\right)^{3/5},\tag{C.2}$$

where we have used the inverted exponent suggested by Sullivan. In this case, the correct, simplified expression for the Stark width would be

$$\Delta\nu_S \approx \nu_0 \left(\frac{3\times 10^6}{\nu_0}\right)^{5/3}.\tag{C.3}$$

The generalized expression for the line-to-continuum ratio used by van de Hulst to estimate the detectability of RRLs was

$$\frac{I_L}{I_C} = \frac{\nu_0}{\Delta\nu}\frac{h\nu_0}{kT}g,\tag{C.4}$$

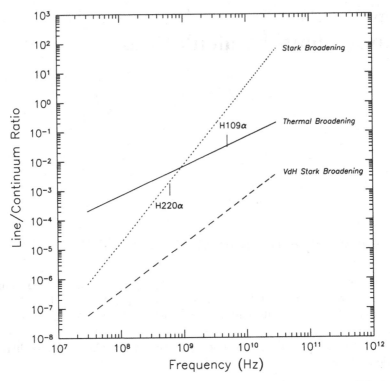

Fig. C.1 The line/continuum ratio of Hnα RRLs plotted against frequency. Calculations are based upon approximations used by van de Hulst (1945). *Solid line* assumes that line broadening is only thermal for a 10^4 K environment. *Dashed line* assumes Stark broadening estimated by the alleged "incorrect" formula with the inverted exponent of 3/5. *Dotted line*: same Stark broadening formula but with the "correct" exponent, i.e., 5/3. Also shown are observations of the H220α and H109α RRLs from the Orion nebula with appropriate *error bars*

where g is the ratio of the appropriate Gaunt factors and was taken to be ≈ 0.1. Substituting the above expressions for $\Delta \nu$ and plotting the results gives Fig. C.1.

Inspection shows that the I_L/I_C ratio is indeed low for the calculation of Stark broadening using the 3/5 exponent, in fact, lower than the thermal case by at least two orders of magnitude in the radio wavelength regime from, say, 10^7 to 10^{11} Hz. Furthermore, the VdH values of I_L/I_C are so low to constitute unrealistic detection prospects for equipment available in 1945 when the paper appeared. From the presumed calculations shown here, we can easily understand why van de Hulst rejected the possibility of detecting RRLs.

On the other hand, his approximate calculations make sense if we "correct" them by inverting the exponent to 5/3. In these calculations, Fig. C.1 shows that thermal broadening dominates the line widths at frequencies above

900 MHz. Actual observations of the H220α and H109α lines from the Orion nebula fall in the appropriate positions. The former falls below the thermal line and the latter falls on that line. Stark broadening diminishes the I_L/I_C ratio for the H220α line but has virtually no effect on the H109α line.

We conclude that Sullivan's claim may be correct. An accidental inversion of an exponent would have changed van de Hulst's conclusion of whether or not RRLs would be detectable in radio astronomy.

C.2 Refinements to the Bohr Model

While the equations derived in Sect. 1.3 describe the salient features of recombination spectra, they need additional refinements for generality. For example, the electron orbits need not be circular in a classical sense; elliptical orbits are also possible. von Sommerfeld (1916a; 1916b) extended Bohr's work by generalizing the quantization of angular momentum to radial and to azimuthal coordinates:

$$L_r = n_r \frac{h}{2\pi}; \quad n_r = 0, 1, \dots \tag{C.5}$$

and

$$L_\phi = k \frac{h}{2\pi}; \quad k = 0, 1, \dots, \tag{C.6}$$

where n_r and k are called the radial and azimuthal quantum numbers, respectively. Accordingly, the diameters of the major and minor axes of an elliptical electron orbit can be written, respectively,

$$2a = \frac{2h^2}{4\pi^2 m_R e^2} \frac{n^2}{Z} \tag{C.7}$$

and

$$2b = \frac{2h^2}{4\pi^2 m_R e^2} \frac{nk}{Z}, \tag{C.8}$$

where $n = n_r + k$. Dividing (C.7) by (C.8) shows the axial ratio $2a/2b = n/k$.

While, classically, the energy of the orbiting electron does not depend upon the ellipticity and is still given by (1.14) after the substitution of the reduced mass m_R for m, the application of relativity (von Sommerfeld, 1916a) leads to a slight modification:

$$E_{n,k} = -\frac{2\pi^2 m_R e^4}{h^2} \frac{Z^2}{n^2} \left[1 + \frac{\alpha^2 Z^2}{n} \left(\frac{1}{k} - \frac{3}{4n} \right) + \dots \right] \tag{C.9}$$

$$\approx -2.17987 \times 10^{-11} \left(1 - \frac{5.48580 \times 10^{-4}}{M_A} \right) \frac{Z^2}{n^2}$$

$$\times \left[1 + \frac{5.32514 \times 10^{-5} Z^2}{n} \left(\frac{1}{k} - \frac{3}{4n} \right) \right] \quad \text{ergs,} \qquad \text{(C.10)}$$

where M_A is the mass of the atom in amu and the calculated dimensionless "fine-structure constant" is

$$\alpha = \frac{2\pi e^2}{hc} = 7.29735337 \times 10^{-3}, \qquad \text{(C.11)}$$

compared with the currently accepted value of $7.297352533 \times 10^{-3}$ (Audi and Wapstra, 1995).

The spectral line frequencies would then result in the usual way from

$$\nu = \frac{E_{n_2} k_2 - E_{n_1} k_1}{h}. \qquad \text{(C.12)}$$

The higher-order terms of α^2 in (C.9) are usually very small and can be neglected.

Appendix D
Hydrogen Oscillator Strengths

D.1 Population of Atomic Sublevels

Excited states of the hydrogen atom can be characterized by three quantum numbers: the principal quantum number n, the orbital quantum number ℓ, and the magnetic quantum number m. Because spectral lines can originate between any of these sublevels, the oscillator strengths of the sublevels involve these quantum numbers.

The simple oscillator strengths given by (2.111) and (2.112) involve only the principal quantum numbers n. They presume an LTE distribution of atoms in the sublevels ℓ and m. It is important to know whether these distributions are adequate for cosmic RRLs, i.e., whether our simple oscillator strengths characterized only by n are sufficiently accurate to analyze these RRLs.

The number of sublevels in each principal quantum state is

$$N(n) = \frac{n^2}{2\ell + 1} N(n, \ell), \qquad \text{(D.1)}$$

where $\ell = 0, 1, \ldots, n - 1$ and $m = -\ell, -\ell + 1, \ldots, +\ell$. If the population of these states is determined by collisions, each of these states will be populated, and we can characterize them as being in LTE and confidently assume that our simple oscillator strengths are sufficient.

On the other hand, if radiative processes dominate the population of the sublevels, they may not be fully populated, the LTE approximation will not hold, and our simple oscillator strengths will be incorrect.

Appropriate calculations can resolve this situation. Pengelly and Seaton (1964) have compared the rate of collisional transitions (protons, or H$^+$s) between the sublevels of H I atoms (e.g., $n_2, \ell_2 \to n_1, \ell_1$) with the rate of radiation transitions ($n_2, \ell_2 \to n_1, \ell_2 \pm 1$) for a planetary nebula (an H II region) at 10^4 K. The critical density is where the rates are equal. Above this density, the sublevels will be fully populated and our simple approximation will be correct.

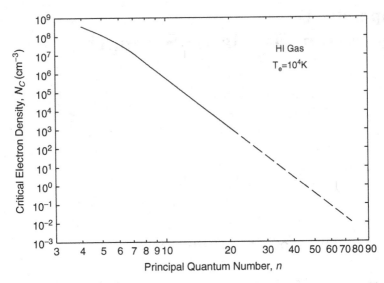

Fig. D.1 The minimum electron density required to thermalize the quantum sublevels ℓ and m in a hydrogen gas at 10^4 K. The *dashed part of the curve* indicates the extrapolation from the actual calculations. After Pengelly and Seaton (1964)

Figure D.1 shows the results for hydrogen. The situation is similar for singly ionized helium. The critical density N_c required to populate the quantum sublevels ℓ and m fully in a typical H II region has a large range depending upon the principal quantum number. For $n = 75$ corresponding to an Hα RRL with a frequency of 15.2 GHz, a critical density of only 0.01 cm^{-3} is sufficient to thermalize the sublevels. In contrast, a density of 2,000 cm^{-3} is required to thermalize the sublevels of the H20α RRL that occurs at 764 GHz in the submillimeter wavelength range.

We conclude that the sublevels of hydrogen atoms in H II regions are thermally populated throughout the radio range, even into the submillimeter range. In this regime, collisional redistribution among ℓ-states occurs faster than radiative depopulation, establishing an LTE population of the sublevels even though the principal quantum states n themselves may *not* be in LTE. The situation could be very different for H II region RRLs in the near-IR and optical ranges and for RRLs from low-density environments.

D.2 Calculation of Oscillator Strengths

Various texts on atomic theory give formulas for calculating oscillator strengths for hydrogen and for hydrogenic atoms such as singly ionized helium. In general, all of these derive from Menzel and Pekeris (1935). In particular, Goldwire (1968) discusses in detail the methodology involved in these calculations and tabulates the level-averaged absorption oscillator strengths $f(n_1, n_2)$ for a wide range of principal quantum numbers. As before, $n_2 > n_1$.

The details are as follows. Using the results of Pengelly and Seaton (1964), Goldwire calculated the level-averaged absorption oscillator strength as

$$f(n_1, n_2) \, n_1^2 = \sum_{\ell_1=1}^{n_1-1} [(2\ell_1 + 1)f(n_1, \ell_1; n_2, \ell_1 - 1)]$$

$$+ \sum_{\ell_1=0}^{n_1-1} [(2\ell_1 + 1)f(n_1, \ell_1; n_2, \ell_1 + 1)], \qquad \text{(D.2)}$$

where the oscillator strengths of the sublevels are weighted by $(2\ell_1 + 1)/n_1^2$ to account for a full population, i.e., LTE. The sublevel absorption oscillator strengths result from

$$f(n_1, \ell_1; n_2, \ell_2) = \frac{1}{2\ell_1 + 1} \sum_{m_1=-\ell_1}^{\ell_1} \sum_{m_2=-\ell_2}^{\ell_2} f(n_1, \ell_1, m_1; n_2, \ell_2, m_2) \ \text{(D.3)}$$

$$= \frac{1}{3} \left(\frac{1}{n_1^2} - \frac{1}{n_2^2} \right) \frac{\max(\ell_1, \ell_2)}{2\ell_1 + 1} \frac{1}{a_0'^2}$$

$$\times [\Re_{ZM}(n_2, \ell_2; n_1, \ell_1)]^2, \qquad \text{(D.4)}$$

where the first Bohr radius of the hydrogenic system is

$$a_0' \equiv \left(\frac{R_\infty}{R_M} \right) \frac{a_0}{Z} \qquad \text{(D.5)}$$

and a_0, M, and Z are the Bohr radius of hydrogen, the species mass, and the net electronic charge, respectively. Gordon (1929) gave equations to evaluate the radial matrix integrals of $\Re_{ZM}(n_2, \ell_2; n_1, \ell_1)$; these are also given by (4) and (5) of Goldwire (1968) or may be found in any standard text on atomic theory. In particular, Hoang-Binh (1990) has found a simple method of evaluating them given in Appendix D.3.

The formal, direct equation given by (D.2) is necessary for calculations involving alternative weightings of the degenerate sublevels. However, other expressions for oscillator strengths are available for LTE. Menzel and Pekeris (1935) give a parallel formula for the thermalized emission oscillator strength $f(n_2, n_1)$ of hydrogen in their (1.15). Perhaps, the simplest expression to use is a modified form of an expression given by Kardashev (1959), giving the thermalized absorption oscillator strength as

$$f(n_1, n_2) = \frac{2}{3} n_2^2 \frac{(4n_2 n_1)^{2n_1+2}(n_2 - n_1)^{2n_2-2n_1-4}}{(n_2 + n_1)^{2n_2+2n_1+3}} A, \qquad \text{(D.6)}$$

where the coefficient A is

$$A \equiv$$

$$\left(\frac{n_2 - 1}{n_2 - n_1 - 1} \right)^2 F^2(-n_1, -n_1, n_2 - n_1; x^{-1})$$

$$- \left(\frac{n_2}{n_2 - n_1 + 1} \right)^2 x^{-2} F^2(-n_1 + 1, -n_1 + 1, n_2 - n_1 + 2; x^{-1}). \qquad (D.7)$$

In (D.7), $F(a, b, c; z)$ is a hypergeometric function described in many mathematical reference books such as Abramowitz and Stegun (1964), and

$$x \equiv -\frac{4n_2 n_1}{(n_2 - n_1)^2}. \qquad (D.8)$$

Goldwire (1968) notes that the excellent agreement between LTE level-averaged absorption oscillator strengths derived from (D.2) and (D.6) confirms the accuracy of the latter expression.

D.3 FORTRAN Code for Evaluating Radial Matrix Integrals

FORTRAN code for calculating radial integrals for hydrogen is listed below (Hoang-Binh, 1990).

D.3.1 Radial Matrix Integrals $\mathfrak{R}_H(n_2, \ell_1 - 1; n_1, \ell_1)$

```
c    * Program AS-RFA
c    * Written by D. Hoang-Binh, 1990 Astron. Astrophys. 238:449
*-----------
c    * Hydrogen atom, Z= 1
c    * This program computes the radial integral
c    * R= R(nup, lu; nlo, lo)= ain; R**2= ain2
c    * nup= principal quantum number of upper state
c    * nlo= principal quantum number of lower state
c    * lu= orbital quantum number of upper state
c    * lo= orbital quantum number of lower state
c    * lu= lo-1
c    * SUBROUTINE FA(nlo,nup,lu,ain,ain2)
*-----------
c    * INPUT:  nlo, lo, ni, nf, inup (file='AS.r')
c    * ni= initial value of nup
c    * nf= final value of nup
c    * inup= step of increase of nup
c    * OUTPUT: R**2, f, A; for nup= ni (inup) nf (file='AS.dat')
     f= absorption oscillator strength
     A= Einstein coefficient
*-----------
     RH= 109677.576
     RKAY= RH
```

```
      iunit=5
      iread=2
      open(file='AS.dat',unit=iunit)
      open(file='AS.r',unit=iread)
*-----------
1     format('NLO,LO, NUP,LUP,R**2,f, A' //)
2     format('AS-RFA'/)
5     format(2(i4,3x), 2(1pe11.4,3x))
6     format (6(i4,3x))
7     format('nlo, lo, ni, nf, inup')
9     format('INPUT'/)
11    format(/'OUTPUT')
55    format(4(i4,3x), 4(1pe11.4,3x))
*-----------
      read (iread,6) nlo, lo, ni, nf, inup
*-----------
      write(iunit,2)
      write (iunit,9)
      write (iunit,7)
      write (iunit,6) nlo, lo, ni, nf, inup
      write(iunit,11)
      write(iunit,1)
      write(9,1)
*-----------
      lu=lo-1
      do 33 nup = ni, nf, inup
      call FA(nlo,nup,lu,ain,ain2)
*-----------
      XUP=NUP
      XLO=NLO
      EUP=RKAY-RKAY/XUP**2
      ELO=RKAY-RKAY/XLO**2
      OS=1./3.*(XUP+XLO)*(XUP-XLO)/(XUP*XLO)**2*MAX(LU,LO)/
     2 (2.*LO+1.)*AIN2
      EIN=0.66704*(2.*LO+1.)/(2.*LU+1.)*(EUP-ELO)**2*OS
*---------
      write(9,55) NLO,LO, NUP,LU, AIN2, OS, EIN
      write(iunit,55) NLO,LO, NUP,LU, AIN2, OS, EIN
33 CONTINUE
*---------
      stop
      end

      SUBROUTINE FA(n,np,lp,ain,ain2)
c     * RECURRENCE ON A

      dimension h(1100)

      l=lp+1
      B0= 0.-np+l+0.
      C0=2.*l

      B=B0
      C = C0
      cc4=1.*(n-np)
      X=-4.*N*NP/cc4**2

      h(1)=1.
      h(2)= 1.-B*X/C
      e1=0
      e2=e1
      i1=(n-l+1)

      do 4 i=2,i1
      j=i+1
      A=1.-i+0.
      H1=-A*(1.-X)*h(j-2) /(A-C)
```

```
      H2=(A*(1.-X)+(A+B*X-C))*h(j-1)/(A-C)
      h(j)= H1+H2
      if(abs(h(j)) .gt. 1.e+25) go to 30
      go to 4

30    continue
      h(j)=h(j)/1.e+25
      h(j-1)=h(j-1)/1.e+25
      h(j-2)=h(j-2)/1.e+25
      e1=e1+25.
      e2=e1

4     continue
      p1=h(i1-1)
      p2 = h(i1+1)

      cc4=1.*(n-np)
      cc4= abs(cc4)
      cc5=n+np
      FF= P1*(1.-cc4**2/cc5**2*P2/P1*10.**(e2-e1))
      alof=alog10(ABS(FF))+e1

      i2=(2*L-1)
      S1=0.

      do 6 I=1,i2
      ai=i
      S1=S1+ alog10(ai)
6     continue

      C1= - (alog10(4.)+S1)
      S=0.

      i3=(N+L)
      si3=0.

      do 7 I=1,i3
      ai=i
      si3=si3+ alog10(ai)
7     continue

      S=S+si3

      i4=(NP+L-1)
      si4=0.

      do8 I=1,i4
      ai=i
      si4=si4+ alog10(ai)
8     continue

      S=S+si4

      i5=n-l-1
      si5=0.
      if(i5 .eq. 0) go to 2

      do 9I=1,i5
      ai=i
      si5=si5+ alog10(ai)
9     continue

      S=S-si5

2     continue

      i6=np-1
```

```
      si6=0.
      if(i6 .eq. 0) go to 3

      do 12 I=1,i6
      ai=i
      si6=si6+ alog10(ai)
  12  continue

      S=S-si6

  3   continue
      C2=S/2.
      cc3=4.*n*np
      C3= (l+1.)*alog10(cc3)

      cc4=cc4
      C4= (N+NP-2.*L-2.)*alog10(cc4)

      cc5=n+np
      C5=(-N-NP)*alog10(cc5)

      C=C1+C2+C3+C4+C5

      ali =alof+C
      ain = 10.**ali
      ain2=ain**2

      return
      end

*-----------------
AS.r
0006   0001   0007   0020   0001
read (iread,6) nlo, lo, ni, nf, inup
*-----------------

*-----------------
AS.dat
AS-RFA

INPUT

nlo, lo, ni, nf, inup
   6     1     7    20     1
read (iread,6) nlo, lo, ni, nf, inup

OUTPUT
NLO,LO, NUP,LUP,R**2,f, A


    6     1     7     0    1.1811E+02   9.6717E-02   1.2644E+05
    6     1     8     0    1.6873E+01   2.2784E-02   8.1001E+04
    6     1     9     0    5.5390E+00   9.4976E-03   5.4447E+04
    6     1    10     0    2.5575E+00   5.0519E-03   3.8434E+04
    6     1    11     0    1.4192E+00   3.0770E-03   2.8203E+04
    6     1    12     0    8.8253E-01   2.0429E-03   2.1344E+04
    6     1    13     0    5.9271E-01   1.4397E-03   1.6561E+04
    6     1    14     0    4.2070E-01   1.0600E-03   1.3120E+04
    6     1    15     0    3.1129E-01   8.0704E-04   1.0577E+04
    6     1    16     0    2.3789E-01   6.3096E-04   8.6552E+03
    6     1    17     0    1.8656E-01   5.0407E-04   7.1753E+03
    6     1    18     0    1.4942E-01   4.0993E-04   6.0160E+03
    6     1    19     0    1.2180E-01   3.3844E-04   5.0950E+03
    6     1    20     0    1.0078E-01   2.8305E-04   4.3536E+03
*-----------------
```

D.3.2 Radial Matrix Integrals $\mathfrak{R}_H(n_2, \ell_1 + 1; n_1, \ell_1)$

```
c    * Program BS-RFA
c    * Written by D. Hoang-Binh, 1990 Astron. Astrophys. 238:449
*---------------
c    * Hydrogen atom, Z= 1
c    * This program computes the radial integral
c    * R= R(nup, lu; nlo, lo)= ain; R**2= ain2
c    * nup= principal quantum number of upper state
c    * nlo= principal quantum number of lower state
c    * lu= orbital quantum number of upper state
c    * lo= orbital quantum number of lower state
c    * lu=lo+1
c    * SUBROUTINE FR(nup,nlo,lo,ain,ain2)
*---------------
c    * INPUT: nlo, lo, ni, nf, inup (file='BS.r')
c    * ni= initial value of nup
c    * nf= final value of nup
c    * inup= step of increase of nup

c    * OUTPUT: R**2, f, A; for nup= ni (inup) nf (file='BS.dat')
     f= absorption oscillator strength
     A= Einstein coefficient
*-----------
     RH= 109677.576
     RKAY= RH
*-----------
     iunit=5
     iread=2
     open(file='BS.dat',unit=iunit)
     open(file='BS.r',unit=iread)
*-----------
1    format(' NLO,LO,NUP,LU,R2,f, A'/)
2    format('BS-RFA'/)
5    format(2(i4,3x), 2(1pe11.4,3x))
6    format (6(i4,3x))
7    format('nlo, lo, ni, nf, inup')
9    format('INPUT'/)
11   format(/'OUTPUT')
55   format(4(i4,3x), 4(1pe11.4,3x))
*-----------
     read (iread,6) nlo, lo, ni, nf, inup
*-----------
     write(iunit,2)
     write (iunit,9)
     write (iunit,7)
     write (iunit,6) nlo, lo, ni, nf, inup
     write(iunit,11)
     write(9,1)
     write(iunit,1)
*-----------
     lu=lo+1
     do 33 nup = ni, nf, inup
     CALL FR(nup,nlo,lo,ain,ain2)
*-----------
     XUP=NUP
     XLO=NLO
     EUP=RKAY-RKAY/XUP**2
     ELO=RKAY-RKAY/XLO**2
     OS=1./3.*(XUP+XLO)*(XUP-XLO)/(XUP*XLO)**2*MAX(LU,LO)/
    2 (2.*LO+1.)*AIN2
     EIN=0.66704*(2.*LO+1.)/(2.*LU+1.)*(EUP-ELO)**2*OS
*---------
     write(9,55) NLO,LO, NUP,LU, AIN2, OS, EIN
     write(iunit,55) NLO,LO, NUP,LU, AIN2, OS, EIN
33   CONTINUE
```

```
*----------
      stop
      end

      subroutine FR(n,np,lp,ain,ain2)
c     RECURRENCE ON B
      dimension h(1100)
      l=lp+1
      A01=-n+l+1.
      A02=A01-2.
      B0=-np+l
      C0=2.*l
      c = C0
      X=-4.*N*NP/(N-NP)**2
      B=A01
      e1=0.
      h(1)=1.
      h(2)=1.-B*X/C
      i1=(NP-L)
      if(i1 .eq. 0) go to 40
      if(i1 .eq. 1) go to 41

      do 4 I=2,i1
      j=i+1
      A=-I+1.
      H1=-A*(1.-X)*h(j-2) /(A-C)
      H2=(A*(1.-X)+(A+B*X-C))*h(j-1)/(A-C)
      h(j)= H1+H2
      if(abs(h(j)) .gt. 1.e+25) go to 30
      go to 4

30    continue
      h(j)=h(j)/1.e+25
      h(j-1)=h(j-1)/1.e+25
      e1=e1+25.

4     continue
      P1=h(i1+1)
      go to 50

40    continue
      p1=h(1)
      go to 50

41    continue
      p1=h(2)
      go to 50

50    continue
      B=A02
      E2=0.
      h(1)=1.
      h(2)=1.- B*X/C

      i1=(NP-L)
      if(i1 .eq. 0) go to 42
      if(i1 .eq. 1) go to 43

      do 5 I=2,i1
      j=i+1
      A=-I+1.
      H1=-A*(1.-X)*h(j-2)/(A-C)
      H2=(A*(1.-X)+(A+B*X-C))*h(j-1)/(A-C)
      h(j)= H1+H2
      if(abs(h(j)) .gt. 1.e+25) go to 31
      go to 5
```

```
31   continue
     h(j)=h(j)/1.e+25
     h(j-1)=h(j-1)/1.e+25
     e2=e2+25.

5    continue
     P2=h(i1+1)
     go to 51

42   continue
     p2=h(1)
     go to 51

43   continue
     p2=h(2)
     go to 51

51   continue
     cc4= n-np
     cc5=n+np
     FF= P1*(1.-cc4**2/cc5**2*P2/P1*10.**(e2-e1))
     alof=alog10(ABS(FF))+e1

c    REM CAL OF C1, C2, C3, C4, C5
     i2=(2*L-1)
     S1=0.

     do 6 I=1,i2
     ai=i
     S1=S1+ alog10(ai)
6    continue

     C1= - (alog10(4.)+S1)

     S=0.
     i3=(N+L)
     si3=0.

     do 7 I=1,i3
     ai=i
     si3=si3+ alog10(ai)
7    continue

     S=S+si3
     i4=(NP+L-1)
     si4=0.

     do8 I=1,i4
     ai=i
     si4=si4+ alog10(ai)
8    continue

     S=S+si4
     i5=n-l-1
     si5=0.

     if(i5 .eq. 0) go to 2
     do 9I=1,i5
     ai=i
     si5=si5+ alog10(ai)
9    continue

     S=S-si5
2    continue

     i6=i1
     si6=0.
```

```
      if(i6 .eq. 0)go to 3

      do 12I=1,i6
      ai=i
      si6=si6+ alog10(ai)
12    continue

      S=S-si6

3     continue
      C2=S/2.
      cc3=4.*n*np
      C3= (l+1.)*alog10(cc3)
      cc4=n-np
      cc4= abs(cc4)
      C4= (N+NP-2.*L-2.)*alog10(cc4)
      cc5=n+np
      C5=(-N-NP)*alog10(cc5)
      C=C1+C2+C3+C4+C5

      ali =alof+C
      ain =10.**ali
      ain2=ain**2
      return
      end
*--------------------

*--------------------
BS.r
0006   0000   0007   0020   0001
*--------------------

*--------------------
BS-RFA

INPUT

nlo, lo, ni, nf, inup
   6       0     7     20       1

OUTPUT
 NLO,LO,NUP,LU,R**2,f, A

   6      0      7      1    2.7419E+02    6.7356E-01    9.7844E+04
   6      0      8      1    4.2342E+01    1.7153E-01    6.7756E+04
   6      0      9      1    1.4544E+01    7.4817E-02    4.7656E+04
   6      0     10      1    6.9164E+00    4.0986E-02    3.4646E+04
   6      0     11      1    3.9173E+00    2.5480E-02    2.5949E+04
   6      0     12      1    2.4725E+00    1.7170E-02    1.9932E+04
   6      0     13      1    1.6793E+00    1.2237E-02    1.5641E+04
   6      0     14      1    1.2023E+00    9.0880E-03    1.2499E+04
   6      0     15      1    8.9579E-01    6.9673E-03    1.0146E+04
   6      0     16      1    6.8837E-01    5.4775E-03    8.3485E+03
   6      0     17      1    5.4230E-01    4.3958E-03    6.9525E+03
   6      0     18      1    4.3599E-01    3.5884E-03    5.8513E+03
   6      0     19      1    3.5653E-01    2.9720E-03    4.9712E+03
   6      0     20      1    2.9577E-01    2.4922E-03    4.2591E+03
*-----------------------
```

Appendix E
Departure Coefficients

E.1 FORTRAN Code for Calculating b_n Values

This listing is a program for calculating departure coefficients and values of β for hydrogenic atoms. Brocklehurst and Salem (1977) wrote the original version in Fortran IV for the Cambridge University (England) IBM 370/165. Walmsley (1990) modified the program to calculate departure coefficients down to level $n = 10$ when needed. Gordon converted the program to standard FORTRAN 77 at the same time.

This code does not consider collisions with detailed angular momentum levels. Consequently, departure coefficients calculated by Storey and Hummer (1995) are better for small quantum numbers and low electron densities, as described in Sect. 2.3.10.5. Where the principal quantum levels are completely degenerate, i.e., at quantum numbers appropriate to meter, decimeter, and centimeter wavelengths, and for the electron densities found in most H II regions, agreement between departure coefficients produced by the two codes is excellent.

The program elements are:

DATAIN.TXT	Input data for above (see Brocklehurst and Salem reference above for details)
BNMAIN2.FOR	Source code for main program
BNFNCTN.FOR	Source code for internal functions
BNSBRTN2.FOR	Source code for internal subroutines

Typically, modern PCs with GHz-clock rates will run the executable code in a few seconds. Calls to the subroutines will appear on the PC screen to advise of progress.

Multiple cases given in DATAIN.TXT will appear sequentially in the output file DATAOUT.TXT.

The output appears in a ASCII file on the A: drive entitled DATAOUT.TXT. This file must be renamed or deleted for each new execution because the output file specification is "NEW."

Recomputing some of the tables published by Salem and Brocklehurst (1979) generally verified the output. The results should also agree with the coefficients published by Walmsley (1990), who used the VAX Fortran version of this program.

```
        program bsubn
c
C  This program has been modified to execute in FORTRAN 77 on an
C  IBM PC or compatible, using Microsoft Fortran4.0. The original
C  JCL cards used to run it on an IBM370/165 have been commented
C  with small c's, and new code has usually been written in lower case.
C        M.A. Gordon, March 1988.
c
c
c ACXIGENERAL BN PROGRAM.  RADIO RECOMBINATION LINES FROM H REGIONS AND   ACXI0000
c1   COLD INTERSTELLAR CLOUDS: COMPUTATION OF THE BN FACTORS.             ACXI0000
c2   M. BROCKLEHURST, M. SALEM.                                          ACXI0000
cREF. IN COMP. PHYS. COMMUN. 13 (1977) 39                               ACXI0000
cJOB MS13 2355 BN PROGRAM TEST                                          ACXI0001
cROUTE PRINTER WESTCAM,POST WCAV,NOTIFY                                 ACXI0002
cMSGLEVEL=1 ACXI0003
cLIMSTORE 175K,COMP 75 SECS,PRINTER 5000                                ACXI0004
c//ONE EXEC FTG1CLG,REGG=175K,LISTC=SOURCE,MAPL=MAP    COMPILE, LOAD, GO ACXI0005
c//FORT.SYSIN DD *  SOURCE PROGRAMCARDS                                 ACXI0006
c                        WIDE TEMPERATURE RANGE BN PROGRAM              ACXI0007
c                        *********************************              ACXI0008
c                                                                       ACXI0009
C  PROGRAM FOR CALCULATION OF COEFFICIENTS OF DEPARTURE  FROM           ACXI0010
C  THERMODYNAMIC EQUILIBRIUM, BN, FOR HYDROGENIC ATOMS, AT ELECTRON     ACXI0011
C  TEMPERATURES FROM 10K TO 20 000K                                     ACXI0012
C                                                                       
c The original card deck, specified below, is now contained in a data
c file named DATAIN.TXT c                                               ACXI0013
C  INPUT CARDS -                                                        ACXI0014
C  ** ALPHANUMERIC TITLE CARD.                                          ACXI0015
C  ** NORMALLY, THE FOLLOWING 6 CARDS (SEE WRITEUP FOR EXPLANATION) -    ACXI0016
C 75  2  4                                                              ACXI0017
C 30 31 32 33 34 35 37 39 41 43 46 49 52 55 58 61 64 68 72 76 80 84 88  ACXI0018
C 929710210711211712212713213814415015616216817418018718418818222222215222 ACXI0019
C2302382462542622702792882973063153253335345355365375386397408419430441 ACXI0020
C452463474485496507                                                     ACXI0021
C 75 72 69 66                                                           ACXI0022
C                                                                       ACXI0023
C  ** ONE CARD WITH RADIATION TEMPERATURE AND EMISSION MEASURE OF       ACXI0024
C  BACKGROUND RADIATION FIELD. FORMAT 2E10.3. IF EMISSION MEASURE READ  ACXI0025
C  IS GE 10**10, IT IS TAKEN TO BE INFINITE. THIS CARD IS READ BY       ACXI0026
C                                                                       
    FUNCTION COR(N,ISW). IF THIS SUBPROGRAM IS REPLACED BY THE USER, ANY ACXI0027
C  CARDS READ BY  COR  WHEN CALLED WITH ISW=0 SHOULD BE PLACED HERE.    ACXI0028
C  (BLANK CARD = NO FIELD).                                             ACXI0029
C                                                                       ACXI0030
C  ** (ONE CARD FOR EACH CASE TO BE CALCULATED)TEMPERATURE, DENSITY,    ACXI0031
```

```
C   CASE (THIN=A - 1, THICK=B -2), PRINT CYCLE (PRINT EVERY K-TH LEVEL),   ACXI0032
C   NPLO AND NPHI, WHERE OUTPUT IS TO BE PUNCHED FROM N=NPLO TO NPHI       ACXI0033
C   (NO PUNCHED OUTPUT IF NPHI=0), ALPHANUMERIC LABEL TO BE PUNCHED IN     ACXI0034
C   COLUMNS 77-79 OF CARD OUTPUT.                                          ACXI0035
C   FORMAT 2E10.5, 4I5, 37X, A3 (I.E., LABEL IN COLS. 78-80)               ACXI0036
c
c  I'ved modified the code to accept one more parameter, ilim, to limit
c  the print out- MAG
C                                                                          ACXI0037
C   ** A BLANK CARD (WHICH ENDS EXECUTION)                                 ACXI0038
C                                                                          ACXI0039
C                                                                          ACXI0040
      character none,nbl,nskip
      character*8 icase
      COMMON /EXPDAT/ CXP(707),MAXN                                        ACXI0041
      COMMON /FITDAT/ AFIT(4,4),IVAL(4),NFIT                               ACXI0042
      COMMON /INOUT/ IREAD,IWRITE,IPUNCH,icrt                              ACXI0043
      COMMON /PARMS/ DENS,T,ITM                                           ACXI0044
      COMMON /TDEP/ TE32,TE12,CTE                                          ACXI0045
      DIMENSION CO(75), DVAL(507), IND(75,2), IPIV(75), KBOUT(507),        ACXI0046
     1 KCOUT(507), MVAL(75), SK(75,75) ,TITLE(10), VAL(507)                ACXI0047
C                                                                          ACXI0048
C   MANY MACHINES DO NOT REQUIRE THE FOLLOWING  DOUBLE PRECISION           ACXI0049
C   STATEMENT IN MAIN OR SUBPROGRAMS                                       ACXI0050
C                                                                          ACXI0051
      DOUBLE PRECISION DABS,DSQRT,DBLE,DLOG,DLOG10,DFLOAT,DEXP             ACXI0052
C                                                                          ACXI0053
      DOUBLE PRECISION AFIT,ARG,CO,COR,CTE,CX,CXP,D,DENS,DVAL,H,HH,RATIO   ACXI0054
     1,SK,T,T1,TE12,TE32,TITLE,VAL,X,XXXI                                  ACXI0055
c     DATA NONE,NBL/1HO,1H /,LPPG/45/                                      ACXI0056
      data none,nbl/'/',' '/,lppg/45/
      data icase/'(Case B)'/
c
c  Open the input and output files. IREAD, IWRITE specified by BLOCK DATA
c  If desired, you can write directly to your printer by commenting out
c  the OPEN statement for DATAOUT.TXT and resetting the IWRITE = 6 in
c  BLOCK DATA section.   Then the default of UNIT = 6 will be the printer.
c
      open(unit=IREAD,file='A:DATAIN.TXT',form='FORMATTED',access=
     1'SEQUENTIAL',status='OLD')
      open(unit=IWRITE,file='A:DATAOUT.TXT',form='FORMATTED',access=
     1'SEQUENTIAL',status='UNKNOWN')
c
      write (5,190)
190   format (//,' Begin Calculation of Departure Coefficients'/
     1' Coprocessor or Emulator version of program BSUBN2.EXE'/
     5' with Malcolm Walmsley"s extensions to n < 20'/
     2/' [See Salem & Brocklehurst, Ap.J.Supp. 39:633 (1979)'/
     3' Brocklehurst & Salem, Comp. Phys. Comm. 13:39 (1977)'/
     4' Adapted to FORTRAN 77 by M. A. Gordon, March 1988]'/)
c
c  Read the data needed to set up the calculations c
      write (5,195)
195   format (' Read Initial Parameters from DATAIN.TXT') c
      READ (IREAD,70) TITLE                                                ACXI0057
      READ (IREAD,80) IC,IR,NFIT                                          ACXI0058
      READ (IREAD,80) (MVAL(I),I=1,IC)                                     ACXI0059
```

```
          READ (IREAD,80) (IVAL(I),I=1,NFIT)                          ACXI0060
c
c  Write the headers of the output file c
          WRITE (IWRITE,90) TITLE                                     ACXI0061

          WRITE (IWRITE,100) IC,(MVAL(I),I=1,IC)                      ACXI0062
          WRITE (IWRITE,110) NFIT,(IVAL(I),I=1,NFIT)                  ACXI0063
          WRITE (IWRITE,120) IR                                       ACXI0064
          T1=0.D0                                                     ACXI0065
          MAXN=MVAL(IC)                                               ACXI0066
c
c  Reads the next data card (background radiation) and calculate
c  correction to radiative rates.  Second argument (ISW > 0) skips call
c
          H=COR(0,0)                                                  ACXI0067
c
c  Calculate and store quantities for collision rates which depend
c  only of temperature and charge for data cards to come. c
          H=COLRAT(0,0,0.D0,0.D0)                                     ACXI0068
c
c  Begin main loop of calculations
c
c10    READ (IREAD,130) T,DENS,NMIN,ICYC,NPLO,NPHI,LABEL             ACXI0069
c                                                                     ACXI0070
C  FOR MACHINES WITH THE  END=  FEATURE, USE THE FOLLOWING FORM -     ACXI0071
c10    READ (IREAD,130,END=65) T,DENS,NMIN,ICYC,NPLO,NPHI,LABEL      ACXI0072
          ilim=100
  10     read (iread,130,end=65) t,dens,nmin,icyc,nplo,nphi,ilim,label
c        IF (T.LE.0.D0.OR.DENS.LE.0.D0) STOP                         ACXI0073
         if (t.le.0.d0.or.dens.le.0.d0) go to 65
         IF (NMIN.LE.0) NMIN=2                                        ACXI0074
         if (nmin .eq. 1) icase = '(Case A)'
         if (nmin .eq. 2) icase = '(Case B)'
         write (icrt,200) T, DENS, ICASE,ilim
 200     format (1x,1P,' Read case inputs: Temperature (K) = ',G11.4/
     1   '                    Density (cm**-3) = ',G10.3/
     2   '                    Model = ',A8/
     3   '                    n Limit in Printout =',I3/)
         NPLO=MAX0(NPLO,MVAL(1))                                      ACXI0075
c  nphi is the highest quantum number in the printout
         NPHI=MIN0(NPHI,MVAL(IC))                                     ACXI0076
         ND=NPHI-NPLO+1                                               ACXI0077
         ICYC=MAX0(1,ICYC)                                            ACXI0078
         NPAGE=1                                                      ACXI0079
         NLINE=0                                                      ACXI0080
         IF (T.EQ.T1) GO TO 30                                        ACXI0081
         ITM=1                                                        ACXI0082
         IF (T.GE.1000.D0) ITM=3                                      ACXI0083
         TE12=DSQRT(T)                                                ACXI0084
         TE32=T*TE12                                                  ACXI0085
         CTE=15.778D4/T                                               ACXI0086
         DO 20 I=1,707                                                ACXI0087
         CX=0.D0                                                      ACXI0088
         ARG=CTE/DFLOAT(I**2)                                         ACXI0089
         IF (ARG.LE.165.D0) CX=DEXP(-ARG)                            ACXI0090
  20     CXP(I)=CX                                                    ACXI0091
         write (icrt,210)
```

```
210    format(1x, 'Call Colion')
       CALL COLION (0,1,T,H)                                              ACXI0092
       write (icrt,220)
220    format (6x, 'Call Radcol')
       CALL RADCOL (T,MVAL,IC,NMIN)                                       ACXI0093
 30    write(icrt,230)
230    format(11x, 'Call Reduce')
       CALL REDUCE (MVAL,IC,IR,SK)                                        ACXI0094
       write(icrt,240)
240    format(16x, 'Call RHS')
       CALL RHS (CO,MVAL,IC)                                              ACXI0095
       write(icrt,250)
250    format(21x, 'Call JMD')
       CALL JMD (SK,CO,MVAL,IC)                                           ACXI0096
       write(icrt,260)
260    format(26x, 'Call Matinv')
       CALL MATINV (SK,IC,CO,1,D,IRROR,75,IPIV,IND)                       ACXI0097
       write (icrt,270)
270    format(31x, 'Call Interp')
       CALL INTERP (MVAL,CO,VAL,DVAL,IC,2)                                ACXI0098
       J=MVAL(1)                                                         ACXI0099
       K=MVAL(IC)                                                        ACXI0100
 c  limits the printout to ilim lines
       if (ilim.ne.0) k=ilim
       IPUN=0                                                            ACXI0101
       DO 60 I=J,K                                                       ACXI0102
       RATIO=DVAL(I)/VAL(I)                                              ACXI0103
       H=T*DFLOAT(I)**3*RATIO/3.158D5                                    ACXI0104
       HH=1.D0-H                                                         ACXI0105
       XXXI=DEXP(15.778D4/(DFLOAT(I*I)*T))*VAL(I)                        ACXI0106
       X=-HH*XXXI*100.D0/T                                               ACXI0107
       IF (MOD(I,ICYC).NE.0) GO TO 50                                    ACXI0108
       NSKIP=NBL                                                         ACXI0109
 c     IF (MOD(NLINE,5).EQ.0) NSKIP=NONE                                 ACXI0110
       if (mod(nline,5).eq.0) write (IWRITE,161)
       IF (MOD(NLINE,LPPG).NE.0) GO TO 40                                ACXI0111
       write(icrt,280) npage
280    format(1x, 'Writing page ',I2,' to DATAOUT.TXT')
       WRITE (IWRITE,140) T,DENS,NMIN,icase,NPAGE                        ACXI0112
       WRITE (IWRITE,150)                                                ACXI0113
       NPAGE=NPAGE+1                                                     ACXI0114
 40    NLINE=NLINE+1                                                     ACXI0115
 c     WRITE (IWRITE,160) NSKIP,I,VAL(I),HH,DVAL(I),RATIO,H,X            ACXI0116
       write (IWRITE,160) I,VAL(I),HH,DVAL(I),RATIO,H,X
 50    IF (NPHI.EQ.0) GO TO 60                                           ACXI0117
       IF (I.LT.NPLO) GO TO 60                                           ACXI0118
       IF (I.GT.NPHI) GO TO 60                                           ACXI0119
       IPUN=IPUN+1                                                       ACXI0120
       KBOUT(IPUN)=VAL(I)*1.D4                                           ACXI0121
       KCOUT(IPUN)=DLOG10(RATIO)*1.D4                                    ACXI0122
 60    CONTINUE                                                          ACXI0123
       T1=T                                                              ACXI0124
       IF (NPHI.NE.0) CALL BCPCH (KBOUT,KCOUT,T,DENS,NMIN,LABEL,NPLO,NPHIACXI0125
      1,ND)                                                              ACXI0126
       GO TO 10                                                          ACXI0127
 65    endfile (unit=IREAD)
       close (unit=IREAD)
```

```
          endfile (unit=IWRITE)
          close (unit=IWRITE)
          STOP 'End of BSUBN'                                         ACXI0128
c                                                                     ACXI0129
 70    FORMAT (10A8)                                                  ACXI0130
 80    FORMAT (23I3)                                                  ACXI0131
c
c  Output formats modified for 80-character line
c
c 90     FORMAT(20X,10A8/20X,20(4H****)///)                           ACXI0132
  90     FORMAT (' ',10A8/24(3H***)///)
c 100   FORMAT (7H MVAL (,I3,10H VALUES) -/(1X,24I5))                 ACXI0133
 100   FORMAT (7H MVAL (,I3,10H VALUES) -/(1X,12I5)/(1X,12I5))
c 110   FORMAT (7HOIVAL (,I3,10H VALUES) -/(1X,24I5))                 ACXI0134
 110   FORMAT (7H IVAL (,I3,10H VALUES) -/(1X,12I5)/(1X,12I5))
 120   FORMAT (5H IR =,I3)                                            ACXI0135
 130   FORMAT (2G10.3,5I5,36X,A3)                                     ACXI0136
 140   FORMAT (1H1,14H TEMPERATURE =,F6.0,14H K,  DENSITY =,1PG10.3,  ACXI0137
        115HCM**-3, NMIN = ,I3,1X,A8/32X,4HPAGE,I3)                   ACXI0138
 150   FORMAT (1x,3H  N,3X,2HBN,10X,6HbsBETA,5X,6HDBN/DN,5X,
        112HD(LN(BN))/DN,2X,8H1-bsBETA,6X,4HZETA)                     ACXI0140
 160   FORMAT (1x,I3,1P,6(G12.5))                                     ACXI0141
 161   FORMAT (1X)
          END                                                        ACXI0142


          BLOCK DATA                                                  ACXI0196
          DOUBLE PRECISION DABS,DSQRT,DBLE,DLOG,DLOG10,DFLOAT,DEXP    ACXI0197
          COMMON /GAUNTS/ A1(50),A2(50),A3(50),A4(50),A5(50),A6(50),A7(50),AACXI0198
        18(50),A9(50),A10(50),A11(50),A14(50),A17(50),A20(50),A25(50),A30(5ACXI0199
        20),A40(50),A50(50),A100(50),A150(50),A225(50),A500(50),IXV(12)    ACXI0200
          COMMON /RCMB/ SV0A(33),SV0B(33),SV0C(33),SV1A(33),SV1B(33),SV1C(33ACXI0201
        1),SV2A(33),SV2B(33),SV2C(33)                                 ACXI0202
          COMMON /GAUSS/ VALUE(12)                                    ACXI0203
          COMMON /INOUT/ IREAD,IWRITE,IPUNCH,icrt                     ACXI0204
          DOUBLE PRECISION VALUE                                      ACXI0205
c                                                                     ACXI0206
C  INPUT/OUTPUT UNITS. IREAD IS CARD READER, IWRITE IS PRINTER, IPUNCH ACXI0207
C  IS CARD PUNCH.                                                     ACXI0208
          DATA IREAD,IWRITE,IPUNCH,icrt/2,3,7,6/                      ACXI0209
c                                                                     ACXI0210
C  NOTE - SOME COMPILERS REQUIRE AN IMPLIED DO LOOP WHEN THE VALUES OF ACXI0211
C  A WHOLE ARRAY ARE SET IN A DATA STATEMENT. THE FOLLOWING STATEMENTS ACXI0212
C  WILL NEEDCHANGING TO (A1(I),I=1,50), ETC. (THE ANS STANDARD FORTRAN ACXI0213
C  FORM IS EXCESSIVELY LENGTHY FOR LARGE ARRAYS.)                     ACXI0214
c                                                                     ACXI0215
C  RADIATIVE GAUNT FACTORS                                            ACXI0216
c                                                                     ACXI0217
        DATA A1/.7166,.7652,.7799,.7864,.7898,.7918,.7931,.7940,.7946,.795ACXI0218
        11,.7954,.7957,.7959,.7961,.7963,.7964,.7965,.7966,.7966,.7967,.796ACXI0219
        28,.7968,.7968,.7969,.7969,.7969,.7970,.7970,.7970,.7970,.7970,.797ACXI0220
        31,.7971,.7971,.7971,.7971,.7971,.7971,.7971,.7971,.7972,.7972,.797ACXI0221
        42,.7972,.7972,.7972,.7972,.7972,.7972/                      ACXI0222
        DATA A2/.7566,.8217,.8441,.8549,.8609,.8647,.8672,.8690,.8702,.871ACXI0223
        12,.8720,.8726,.8730,.8734,.8737,.8740,.8742,.8744,.8746,.8747,.874ACXI0224
        29,.8750,.8751,.8751,.8752,.8753,.8753,.8754,.8755,.8755,.8755,.875ACXI0225
        36,.8756,.8756,.8757,.8757,.8757,.8757,.8758,.8758,.8758,.8758,.875ACXI0226
```

```
      48,.8759,.8759,.8759,.8759,.8759,.8759,.8759/              ACXI0227
      DATA A3/.7674,.8391,.8653,.8784,.8861,.8910,.8944,.8968,.8986,.900ACXI0228
      10,.9011,.9019,.9026,.9032,.9037,.9041,.9044,.9047,.9049,.9052,.905ACXI0229
      24,.9055,.9057,.9058,.9059,.9060,.9061,.9062,.9063,.9064,.9064,.906ACXI0230
      35,.9066,.9066,.9067,.9067,.9067,.9068,.9068,.9068,.9069,.9069,.906ACXI0231
      49,.9070,.9070,.9070,.9070,.9070,.9071,.9071/              ACXI0232
      DATA A4/.7718,.8467,.8750,.8896,.8984,.9041,.9081,.9110,.9132,.914ACXI0233
      19,.9163,.9173,.9182,.9190,.9196,.9201,.9205,.9209,.9213,.9215,.921ACXI0234
      28,.9220,.9222,.9224,.9226,.9227,.9228,.9230,.9231,.9232,.9233,.923ACXI0235
      33,.9234,.9235,.9235,.9236,.9237,.9237,.9238,.9238,.9238,.9239,.923ACXI0236
      49,.9240,.9240,.9240,.9240,.9241,.9241,.9241/              ACXI0237
      DATA A5/.7741,.8507,.8804,.8960,.9055,.9118,.9162,.9195,.9220,.924ACXI0238
      10,.9255,.9268,.9278,.9287,.9294,.9300,.9306,.9310,.9314,.9318,.932ACXI0239
      21,.9324,.9326,.9329,.9331,.9332,.9334,.9335,.9337,.9338,.9339,.934ACXI0240
      30,.9341,.9342,.9343,.9344,.9344,.9345,.9345,.9346,.9347,.9347,.934ACXI0241
      48,.9348,.9348,.9349,.9349,.9349,.9350,.9350/              ACXI0242
      DATA A6/.7753,.8531,.8837,.8999,.9099,.9167,.9215,.9251,.9278,.930ACXI0243
      10,.9317,.9331,.9343,.9352,.9361,.9368,.9374,.9379,.9384,.9388,.939ACXI0244
      22,.9395,.9398,.9400,.9403,.9405,.9407,.9408,.9410,.9412,.9413,.941ACXI0245
      34,.9415,.9416,.9417,.9418,.9419,.9420,.9420,.9421,.9422,.9422,.942ACXI0246
      43,.9423,.9424,.9424,.9425,.9425,.9426,.9426/              ACXI0247
      DATA A7/.7761,.8547,.8858,.9025,.9130,.9200,.9251,.9289,.9318,.934ACXI0248
      12,.9360,.9376,.9389,.9399,.9408,.9416,.9423,.9429,.9434,.9439,.944ACXI0249
      23,.9447,.9450,.9453,.9455,.9458,.9460,.9462,.9464,.9466,.9467,.946ACXI0250
      38,.9470,.9471,.9472,.9473,.9474,.9475,.9476,.9477,.9477,.9478,.947ACXI0251
      49,.9479,.9480,.9480,.9481,.9481,.9482,.9482/              ACXI0252
      DATA A8/.7767,.8558,.8873,.9044,.9151,.9224,.9277,.9317,.9348,.937ACXI0253
      13,.9393,.9409,.9423,.9434,.9444,.9453,.9460,.9467,.9472,.9477,.948ACXI0254
      22,.9486,.9489,.9493,.9496,.9498,.9501,.9503,.9505,.9507,.9509,.951ACXI0255
      30,.9512,.9513,.9514,.9515,.9517,.9518,.9519,.9519,.9520,.9521,.952ACXI0256
      42,.9522,.9523,.9524,.9524,.9525,.9525,.9526/              ACXI0257
      DATA A9/.7771,.8565,.8884,.9058,.9167,.9242,.9297,.9338,.9370,.939ACXI0258
      16,.9417,.9434,.9449,.9461,.9472,.9481,.9489,.9496,.9502,.9507,.951ACXI0259
      22,.9517,.9520,.9524,.9527,.9530,.9533,.9535,.9537,.9539,.9541,.954ACXI0260
      33,.9545,.9546,.9548,.9549,.9550,.9551,.9552,.9553,.9554,.9555,.955ACXI0261
      46,.9557,.9557,.9558,.9559,.9559,.9560,.9561/              ACXI0262
      DATA A10/.7773,.8571,.8892,.9068,.9179,.9256,.9312,.9354,.9388,.94ACXI0263
      114,.9436,.9454,.9470,.9482,.9494,.9503,.9512,.9519,.9526,.9531,.95ACXI0264
      237,.9541,.9545,.9549,.9553,.9556,.9559,.9561,.9564,.9566,.9568,.95ACXI0265
      370,.9572,.9573,.9575,.9576,.9577,.9579,.9580,.9581,.9582,.9583,.95ACXI0266
      484,.9585,.9585,.9586,.9587,.9588,.9588,.9589/              ACXI0267
      DATA A11/.7775,.8575,.8898,.9076,.9188,.9267,.9324,.9367,.9402,.94ACXI0268
      129,.9452,.9470,.9486,.9500,.9511,.9521,.9530,.9538,.9545,.9551,.95ACXI0269
      256,.9561,.9566,.9570,.9573,.9577,.9580,.9583,.9585,.9588,.9590,.95ACXI0270
      392,.9594,.9595,.9597,.9599,.9600,.9601,.9603,.9604,.9605,.9606,.96ACXI0271
      407,.9608,.9609,.9610,.9610,.9611,.9612,.9612/              ACXI0272
      DATA A14/.7779,.8583,.8910,.9091,.9207,.9288,.9347,.9393,.9429,.94ACXI0273
      159,.9483,.9503,.9520,.9535,.9548,.9559,.9569,.9578,.9585,.9592,.95ACXI0274
      298,.9604,.9609,.9614,.9618,.9622,.9625,.9629,.9632,.9634,.9637,.96ACXI0275
      339,.9642,.9644,.9646,.9647,.9649,.9651,.9652,.9654,.9655,.9656,.96ACXI0276
      457,.9659,.9660,.9661,.9662,.9662,.9663,.9664/              ACXI0277
      DATA A17/.7781,.8587,.8916,.9099,.9217,.9300,.9361,.9408,.9446,.94ACXI0278
      176,.9502,.9523,.9541,.9557,.9571,.9582,.9593,.9602,.9611,.9618,.96ACXI0279
      225,.9631,.9637,.9642,.9646,.9651,.9655,.9658,.9662,.9665,.9668,.96ACXI0280
      370,.9673,.9675,.9677,.9679,.9681,.9683,.9685,.9686,.9688,.9689,.96ACXI0281
      491,.9692,.9693,.9694,.9695,.9696,.9697,.9698/              ACXI0282
      DATA A20/.7782,.8590,.8920,.9105,.9224,.9307,.9370,.9418,.9456,.94ACXI0283
```

```
      188,.9514,.9536,.9555,.9571,.9586,.9598,.9609,.9619,.9628,.9636,.96ACXI0284
      243,.9649,.9655,.9661,.9666,.9670,.9675,.9679,.9682,.9686,.9689,.96ACXI0285
      392,.9694,.9697,.9699,.9702,.9704,.9706,.9708,.9709,.9711,.9713,.97ACXI0286
      414,.9716,.9717,.9718,.9719,.9721,.9722,.9723/              ACXI0287
      DATA A25/.7784,.8592,.8924,.9110,.9230,.9315,.9378,.9428,.9467,.95ACXI0288
      100,.9527,.9550,.9569,.9586,.9601,.9614,.9626,.9637,.9646,.9655,.96ACXI0289
      262,.9669,.9676,.9682,.9687,.9692,.9697,.9701,.9705,.9709,.9712,.97ACXI0290
      315,.9718,.9721,.9724,.9727,.9729,.9731,.9733,.9735,.9737,.9739,.97ACXI0291
      441,.9742,.9744,.9745,.9747,.9748,.9749,.9751/              ACXI0292
      DATA A30/.7784,.8594,.8926,.9113,.9234,.9319,.9383,.9433,.9474,.95ACXI0293
      107,.9534,.9558,.9578,.9595,.9611,.9625,.9637,.9648,.9657,.9666,.96ACXI0294
      274,.9682,.9689,.9695,.9701,.9706,.9711,.9715,.9720,.9724,.9727,.97ACXI0295
      331,.9734,.9737,.9740,.9743,.9745,.9748,.9750,.9752,.9754,.9756,.97ACXI0296
      458,.9760,.9762,.9763,.9765,.9766,.9768,.9769/              ACXI0297
      DATA A40/.7785,.8595,.8928,.9116,.9237,.9324,.9389,.9440,.9480,.95ACXI0298
      114,.9542,.9567,.9587,.9606,.9622,.9636,.9649,.9660,.9670,.9680,.96ACXI0299
      288,.9696,.9703,.9710,.9716,.9722,.9727,.9732,.9737,.9741,.9745,.97ACXI0300
      349,.9752,.9756,.9759,.9762,.9765,.9768,.9770,.9773,.9775,.9777,.97ACXI0301
      479,.9781,.9783,.9785,.9787,.9788,.9790,.9791/              ACXI0302
      DATA A50/.7785,.8596,.8929,.9117,.9239,.9326,.9391,.9443,.9484,.95ACXI0303
      118,.9547,.9571,.9592,.9611,.9627,.9642,.9655,.9666,.9677,.9687,.96ACXI0304
      296,.9704,.9711,.9718,.9724,.9730,.9736,.9741,.9746,.9751,.9755,.97ACXI0305
      359,.9763,.9766,.9770,.9773,.9776,.9779,.9781,.9784,.9786,.9789,.97ACXI0306
      491,.9793,.9795,.9797,.9799,.9801,.9803,.9804/              ACXI0307
      DATA A100/.7785,.8597,.8931,.9119,.9242,.9329,.9395,.9447,.9489,.9ACXI0308
      1523,.9553,.9578,.9599,.9619,.9635,.9651,.9664,.9676,.9687,.9698,.9ACXI0309
      2707,.9716,.9724,.9731,.9738,.9744,.9750,.9756,.9761,.9766,.9771,.9ACXI0310
      3775,.9779,.9783,.9787,.9791,.9794,.9797,.9801,.9803,.9806,.9809,.9ACXI0311
      4812,.9814,.9817,.9819,.9821,.9823,.9825,.9827/              ACXI0312
      DATA A150/.7786,.8597,.8931,.9119,.9242,.9330,.9396,.9448,.9490,.9ACXI0313
      1525,.9554,.9579,.9601,.9620,.9637,.9652,.9666,.9678,.9690,.9700,.9ACXI0314
      2710,.9718,.9726,.9734,.9741,.9747,.9754,.9759,.9765,.9770,.9775,.9ACXI0315
      3779,.9783,.9787,.9791,.9795,.9799,.9802,.9805,.9808,.9811,.9814,.9ACXI0316
      4817,.9819,.9822,.9824,.9827,.9829,.9831,.9833/              ACXI0317
      DATA A225/.7786,.8597,.8931,.9120,.9243,.9330,.9396,.9448,.9490,.9ACXI0318
      1525,.9554,.9580,.9602,.9621,.9638,.9653,.9667,.9680,.9691,.9701,.9ACXI0319
      2711,.9720,.9728,.9735,.9742,.9749,.9755,.9761,.9766,.9772,.9776,.9ACXI0320
      3781,.9785,.9790,.9793,.9797,.9801,.9804,.9808,.9811,.9814,.9817,.9ACXI0321
      4819,.9822,.9825,.9827,.9829,.9832,.9834,.9836/              ACXI0322
      DATA A500/.7786,.8597,.8931,.9120,.9243,.9330,.9396,.9448,.9490,.9ACXI0323
      1525,.9555,.9580,.9602,.9621,.9639,.9654,.9668,.9680,.9692,.9702,.9ACXI0324
      2712,.9720,.9729,.9736,.9743,.9750,.9756,.9762,.9768,.9773,.9778,.9ACXI0325
      3782,.9787,.9791,.9795,.9799,.9802,.9806,.9809,.9812,.9815,.9818,.9ACXI0326
      4821,.9824,.9827,.9829,.9831,.9834,.9836,.9838/              ACXI0327
      DATA IXV/11,14,17,20,25,30,40,50,100,150,225,507/           ACXI0328
C                                                                 ACXI0329
      DATA VALUE/.4975936099985107D0,.4873642779856548D0,.46913727600136ACXI0330
      164D0,.4432077635022005D0,.4100009929869515D0,.3700620957892772D0,.ACXI0331
      2324046825968487 8D0,.2727107356944198D0,.2168967538130226D0,.157521ACXI0332
      33398480817D0,.0955594337368082D0,.0320284464313028D0/      ACXI0333
C                                                                 ACXI0334
C DATA FOR CALCULATION OF HYDROGENIC RECOMBINATION COEFFICIENTS    ACXI0335
      DATA SVOA/.06845,.07335,.07808,.08268,.08714,.09148,.09570,.09982,ACXI0336
      1.10385,.10778,.11163,.1209,.1297,.1382,.1462,.1540,.1615,.1687,.17ACXI0337
      257,.1824,.1889,.1953,.2015,.2133,.2245,.2352,.2454,.2552,.2646,.27ACXI0338
      336,.2823,.2906,.2987/                                      ACXI0339
      DATA SVOB/.3140,.3284,.3419,.3547,.3668,.3783,.3892,.3996,.4096,.4ACXI0340
```

```
         1191,.4413,.4615,.4799,.4968,.5124,.5269,.5404,.5530,.5648,.5759,.5ACXI0341
         2864,.5963,.6146,.6311,.6461,.6598,.6724,.6840,.6947,.7047,.7140,.7ACXI0342
         3226,.7384/                                                         ACXI0343
         DATA SV0C/.7524,.7649,.7761,.7862,.7955,.8039,.8117,.8188,.8254,.8ACXI0344
         1399,.8521,.8626,.8716,.8795,.8865,.8927,.8982,.9032,.9077,.9118,.9ACXI0345
         2156,.9223,.9279,.9328,.9370,.9408,.9441,.9471,.9498,.9522,.9544,2*ACXI0346
         3.9544/                                                             ACXI0347
         DATA SV1A/.00417,.00444,.00469,.00493,.00516,.00538,.00558,.00578,ACXI0348
         1.00597,.00614,.00631,.0067,.0070,.0073,.0076,.0078,.0080,.0082,.00ACXI0349
         283,.0085,.0086,.0087,.0087,.0088,.0088,.0088,.0087,.0086,.0084,.00ACXI0350
         382,.0080,.0077,.0074/                                             ACXI0351
         DATA SV1B/.0068,.0060,.0053,.0044,.0035,.0025,.0016,+.0005,-.0005,ACXI0352
         1-.0016,-.0044,-.0072,-.0101,-.0130,-.0160,-.0190,-.0220,-.0250,-.0ACXI0353
         2279,-.0308,-.0337,-.0366,-.0422,-.0478,-.0532,-.0586,-.0638,-.0689ACXI0354
         3,-.0739,-.0788,-.0835,-.0882,-.0972/                              ACXI0355
         DATA SV1C/-.1058,-.1141,-.1220,-.1295,-.1368,-.1438,-.1506,-.1571,ACXI0356
         1-.1634,-.1784,-.1923,-.2053,-.2174,-.2289,-.2397,-.2499,-.2596,-.2ACXI0357
         2689,-.2777,-.2862,-.2944,-.3099,-.3242,-.3376,-.3502,-.3622,-.3736ACXI0358
         3,-.3844,-.3948,-.4047,-.4147,2*-.4147/                            ACXI0359
         DATA SV2A/-.00120,-.00131,-.00142,-.00153,-.00164,-.00175,-.00186,ACXI0360
         1-.00197,-.00207,-.00218,-.00228,-.0025,-.0028,-.0030,-.0033,-.0035ACXI0361
         2,-.0038,-.0040,-.0043,-.0045,-.0047,-.0050,-.0052,-.0056,-.0061,-.ACXI0362
         30065,-.0070,-.0074,-.0078,-.0082,-.0086,-.0091,.0095/             ACXI0363
         DATA SV2B/-.0103,-.0110,-.0118,-.0126,-.0133,-.0141,-.0148,-.0155,ACXI0364
         1-.0162,-.0170,-.0187,-.0204,-.0221,-.0237,-.0253,-.0269,-.0284,-.0ACXI0365
         2299,-.0314,-.0329,-.0344,-.0359,-.0388,-.0416,-.0444,-.0471,-.0497ACXI0366
         3,-.0523,-.0549,-.0575,-.0600,-.0625,-.0674/                       ACXI0367
         DATA SV2C/-.0722,-.0768,-.0814,-.0859,-.0904,-.0947,-.0989,-.1031,ACXI0368
         1-.1072,-.1174,-.1272,-.1367,-.146,-.155,-.1638,-.1723,-.1807,-.189ACXI0369
         2,-.197,-.2049,-.2127,-.228,-.243,-.257,-.272,-.285,-.299,-.312,-.3ACXI0370
         325,-.337,3*-.35/                                                  ACXI0371
         END


         FUNCTION BK (N,NDASH,IS)                                           ACXI0374
C                                                                           ACXI0375
C   CALLS APPROPRIATE ROUTINES FOR CALCULATION OF ATOMIC DATA FOR           ACXI0376
C   ARRAY SK (IN MAIN)                                                      ACXI0377
C                                                                           ACXI0378
C   N=INITIAL LEVEL, NDASH=FINAL LEVEL, IS=SUBSCRIPT WHICH IDENTIFIES       ACXI0379
C   VALUE OF N IN CONDENSED MATRIX                                          ACXI0380
C                                                                           ACXI0381
         COMMON /EXPDAT/ CXP(707),MAXN                                      ACXI0382
         COMMON /PARMS/ DENS,T,ITM                                          ACXI0383
         COMMON /RCRATS/ RADTOT(75),COLTOT(75)                              ACXI0384
         COMMON /TDEP/ TE32,TE12,CTE                                        ACXI0385
         DOUBLE PRECISION DABS,DSQRT,DBLE,DLOG,DLOG10,DFLOAT,DEXP           ACXI0386
         DOUBLE PRECISION A,AN,ANDASH,BK,C,COLTOT,COR,CTE,CX,CXP,DENS,G,RADACXI0387
        1TOT,RT,T,TE12,TE32,TEMP                                           ACXI0388
         IF (N-NDASH) 20,10,50                                             ACXI0389
C                                                                           ACXI0390
C   NDASH=N                                                                 ACXI0391
   10    CALL COLION (N,1,T,RT)                                             ACXI0392
         BK=-RADTOT(IS)-(COLTOT(IS)+RT)*DENS                                ACXI0393
         IF (N.LE.20) RETURN                                                ACXI0394
         BK=BK+COR(N,3)                                                     ACXI0395
         RETURN                                                            ACXI0396
C                                                                           ACXI0397
```

```
C  NDASH GT N                                                     ACXI0398
   20    CALL RAD (A,N,NDASH,G)                                    ACXI0399
         C=COLRAT(N,NDASH,T,TE12)                                  ACXI0400
         AN=N                                                      ACXI0401
         ANDASH=NDASH                                              ACXI0402
         IF (NDASH.GT.707) GO TO 30                                ACXI0403
         CX=CXP(NDASH)                                             ACXI0404
         IF (CX.LT.1.D-30) GO TO 30                                ACXI0405
         CXN=CXP(N)                                                ACXI0406
         IF (CXN.LT.1.D-30) GO TO 30                               ACXI0407
         TEMP=CXN/CX                                               ACXI0408
         GO TO 40                                                  ACXI0409
   30    TEMP=DEXP(-CTE*(1.D0/AN**2-1.D0/ANDASH**2))               ACXI0410
   40    TEMP=(ANDASH/AN)**2*TEMP                                  ACXI0411
         BK=(A+DENS*C)*TEMP                                        ACXI0412
         IF (N.LE.20) RETURN                                       ACXI0413
         IF (NDASH.NE.N+1) RETURN                                  ACXI0414
         BK=BK+COR(N,1)                                            ACXI0415
         RETURN                                                    ACXI0416
C                                                                  ACXI0417
C  NDASH LT N                                                      ACXI0418
   50    C=COLRAT(NDASH,N,T,TE12)                                  ACXI0419
         BK=C*DENS                                                 ACXI0420
         IF (NDASH.NE.N-1) RETURN                                  ACXI0421
         IF (N.LE.20) RETURN                                       ACXI0422
         BK=BK+COR(N,2)                                            ACXI0423
         RETURN                                                    ACXI0424
         END                                                       ACXI0425

         FUNCTION CAPPA (T,TL,T32,F)                               ACXI0427
C                                                                  ACXI0428
C  COMPUTES FREE-FREE ABSORPTION COEFFICIENT  CAPPA, GIVEN ELECTRON  ACXI0429
C  TEMPERATURE (T), LOG(T) (TL), T**1.5 (T32), AND FREQUENCY IN GHZ (F)  ACXI0430
C                                                                  ACXI0431
         V=0.6529+.6666667*ALOG10(F)-TL                            ACXI0432
         IF (V.GT.-2.6) GO TO 10                                   ACXI0433
         ALIV=-1.1249*V+0.3788                                     ACXI0434
         GO TO 30                                                  ACXI0435
   10    IF (V.GT.-0.25) GO TO 20                                  ACXI0436
         ALIV=-1.2326*V+0.0987                                     ACXI0437
         GO TO 30                                                  ACXI0438
   20    ALIV=-1.0842*V+0.1359                                     ACXI0439
   30    CAPPA=4.646*EXPM1(.047993*F/T)*EXP(2.302585*ALIV)/(F**2.33333*T32) ACXI0440
         RETURN                                                    ACXI0441
         END                                                       ACXI0442

         FUNCTION COLGL (N,NDASH,T,TE12)                           ACXI0444
C                                                                  ACXI0445
C  CALCULATES COLLISION RATES FROM LEVEL  N  TO HIGHER LEVEL  NDASH AT  ACXI0446
C  ELECTRON TEMPERATURE T. TE12 IS SQRT(T). USES GAUSS-LAGUERRE   ACXI0447
C  INTEGRATION OF CROSS-SECTIONS (FUNCTION CROSS) OVER MAXWELL    ACXI0448
C  DISTRIBUTION. FUNCTION COLGL IS USED FOR VALUES OUTSIDE THE REGION  ACXI0449
C  OF VALIDITY OF FUNCTION COLRAT.                                ACXI0450
C                                                                  ACXI0451
         DOUBLE PRECISION DABS,DSQRT,DBLE,DLOG,DLOG10,DFLOAT,DEXP  ACXI0452
         DIMENSION XGL(10), WGL(10)                                ACXI0453
         DOUBLE PRECISION T,TE12                                   ACXI0454
```

```
C     DATA NGL/10/                                                  ACXI0455
C     DATA XGL/.1377935,.7294545,1.808343,3.401434,5.552496,8.330153, ACXI0456
C    A 11.84379,16.27926,21.99659,29.92070/                        ACXI0457
C     DATA WGL/.3084411,.4011199,.2180683,6.208746E-2,9.501517E-3, ACXI0458
C    A 7.530084E-4,2.825923E-5,4.249314E-7,1.839565E-9,9.911827E-13/ ACXI0459
      DATA XGL/.4157746,2.294280,6.289945,7*0./                    ACXI0460
      DATA WGL/.7110930,.2785177,1.038926E-2,7*0./                 ACXI0461
      DATA NGL/3/                                                  ACXI0462
      BETA=1.58D5/T                                                ACXI0463
      EN2=N*N                                                      ACXI0464
      END2=NDASH*NDASH                                             ACXI0465
      DE=1./EN2-1./END2                                            ACXI0466
      COLGL=0.                                                     ACXI0467
      DO 10 I=1,NGL                                                ACXI0468
      E=XGL(I)/BETA+DE                                             ACXI0469
      COLGL=COLGL+WGL(I)*CROSS(N,NDASH,E)*E*BETA                   ACXI0470
   10 CONTINUE                                                     ACXI0471
      COLGL=COLGL*6.21241E5*SNGL(TE12)*(EN2/END2)                  ACXI0472
      RETURN                                                       ACXI0473
      END                                                          ACXI0474


      FUNCTION COLRAT (N,NP,T,TE12)                                ACXI0510
C                                                                  ACXI0511
C CALCULATES RATE OF COLLISIONS FROM LEVEL  N  TO HIGHER LEVEL  NP ACXI0512
C AT ELECTRON TEMPERATURE  T.  TE12 IS SQRT(T). SETS RATE=0 FOR  NP-N ACXI0513
C GT 40, BUT THIS IS EASILY MODIFIED.                             ACXI0514
C                                                                  ACXI0515
C THIS FUNCTION MUST BE INITIALIZED BY BEING CALLED WITH N=0 BEFORE ACXI0516
C ANY COLLISION RATES ARE COMPUTED.                               ACXI0517
C THEORY: GEE, PERCIVAL, LODGE AND RICHARDS, MNRAS 175, 209-215 (1976) ACXI0518
C                                                                  ACXI0519
C RANGE OF VALIDITY OF GPLR RATES IS   10**6/N**2 LT T LLT 3*10**9 ACXI0520
C OUTSIDE THIS RANGE, NUMERICAL INTEGRATION OF THE GPLR CROSS-SECTIONS ACXI0521
C IS RESORTED TO.THESE CROSS SECTIONS ARE VALID DOWN TO ENERGIES OF ACXI0522
C 4/N**2 RYDBERGS; THE CROSS-SECTION FORMULA CAN BE USED AT LOWER  ACXI0523
C ENERGIES FOR BN CALCULATIONS, THE INACCURACY IN THE CROSS-SECTIONS ACXI0524
C HAVING LITTLE EFFECT ON THE BN'S.                               ACXI0525
C                                                                  ACXI0526
      COMMON /COLINF/ AL18S4(506),S23TRM(506)                     ACXI0527
      DOUBLE PRECISION DABS,DSQRT,DBLE,DLOG,DLOG10,DFLOAT,DEXP     ACXI0528
      DOUBLE PRECISION DRT,T,TE12                                 ACXI0529
      REAL L,J1,J2,J3,J4                                          ACXI0530
      IF (N.LE.0) GO TO 60                                        ACXI0531
      COLRAT=0.                                                   ACXI0532
      IS=NP-N                                                     ACXI0533
      IF (IS.GT.40) RETURN                                        ACXI0534
      S=IS                                                        ACXI0535
      EN2=N*N                                                     ACXI0536
      IF (SNGL(T).LT.1.E6/EN2) GO TO 50                           ACXI0537
      EN=N                                                        ACXI0538
      ENP=NP                                                      ACXI0539
      IPOW=1+IS+IS                                                ACXI0540
      POW=IPOW                                                    ACXI0541
      ENNP=N*NP                                                   ACXI0542
      BETA=1.58D5/T                                               ACXI0543
      BETA1=1.4*SQRT(ENNP)                                        ACXI0544
      BETRT=BETA1/BETA                                            ACXI0545
```

```
          BETSUM=BETA1+BETA                                          ACXI0546
          F1=0.2*S/ENNP                                              ACXI0547
          IF (F1.GT.0.02) GO TO 10                                   ACXI0548
          F1=1.-POW*F1                                               ACXI0549
          GO TO 20                                                   ACXI0550
  10      F1=(1.-F1)**IPOW                                           ACXI0551
  20      A=(2.666667/S)*(ENP/(S*EN))**3*S23TRM(IS)*F1               ACXI0552
          L=0.85/BETA                                                ACXI0553
          L=ALOG((1.+0.53*L*L*ENNP)/(1.+0.4*L))                      ACXI0554
          J1=1.333333*A*L*BETRT/BETSUM                               ACXI0555
          DRT=DSQRT(2.D0-DFLOAT(N*N)/DFLOAT(NP*NP))                  ACXI0556
          F1=0.3*S/ENNP                                              ACXI0557
          IF (F1.GT.0.02) GO TO 30                                   ACXI0558
          F1=1.-POW*F1                                               ACXI0559
          GO TO 40                                                   ACXI0560
  30      F1=(1.-F1)**IPOW                                           ACXI0561
  40      J2=0.                                                      ACXI0562
          ARG=BETA/BETA1                                             ACXI0563
          IF (ARG.LE.150.) J2=1.777778*F1*(ENP*SNGL(DRT+1.D0)/((EN+ENP)*S))*ACXI0564
         1*3*EXP(-ARG)/(BETA/(1.-AL18S4(IS)))                        ACXI0565
          XI=2./(EN2*SNGL(DRT-1.D0))                                 ACXI0566
          Z=0.75*XI*BETSUM                                           ACXI0567
          EXPZ=0.                                                    ACXI0568
          IF (Z.LE.150.) EXPZ=EXP(-Z)                                ACXI0569
          J4=2./(Z*(2.+Z*(1.+EXPZ)))                                 ACXI0570
          J3=0.25*(EN2*XI/ENP)**3*J4*ALOG(1.+0.5*BETA*XI)/BETSUM     ACXI0571
          COLRAT=EN2*EN2*(J1+J2+J3)/SNGL(TE12)**3                    ACXI0572
          RETURN                                                     ACXI0573
C                                                                    ACXI0574
C  GPLR COLLISION RATE FORMULAS INVALID AT LOW TEMPERATURES. INTEGRATE ACXI0575
C  CROSS SECTIONS.                                                   ACXI0576
C                                                                    ACXI0577
  50      COLRAT=COLGL(N,NP,T,TE12)                                  ACXI0578
          RETURN                                                     ACXI0579
C                                                                    ACXI0580
C  INITIALIZE                                                        ACXI0581
C                                                                    ACXI0582
  60      DO 70 I=1,506                                              ACXI0583
          Z=I                                                        ACXI0584
          AL18S4(I)=ALOG(18.*Z)/(4.*Z)                               ACXI0585
          S23TRM(I)=(0.184-0.04*Z**(-.6666667))                      ACXI0586
  70      CONTINUE                                                   ACXI0587
          COLRAT=0.                                                  ACXI0588
          RETURN                                                     ACXI0589
          END                                                        ACXI0590

          FUNCTION COR (N,ISW)                                       ACXI0592
C                                                                    ACXI0593
C  CORRECTION TO RATE OF POPULATION OF LEVEL N DUE TO RADIATION FIELD. ACXI0594
C  SEE WRITEUP. THIS SUBPROGRAM MAY BE REPLACED BY THE USER IF OTHER  ACXI0595
C  THAN A DILUTE BLACKBODY RADIATION FIELD IS REQUIRED.              ACXI0596
C                                                                    ACXI0597
C  THE MAIN PROGRAM CALLS COR ONCE WITH ISW=0 BEFORE STARTING THE    ACXI0598
C  CALCULATIONS. ANY INITIALIZATION (INCLUDING THE READING OF DATA IF ACXI0599
C  REQUIRED) SHOULD BE CARRIED OUT DURING THIS FIRST CALL.           ACXI0600
C                                                                    ACXI0601
C  COMPUTES TERMS OF                                                 ACXI0602
```

```
C   D(NU)RHO(NU)(N(N+1)B(N+1,N)+N(N-1)B(N-1,N)-N(N)(B(N,N-1)+B(N,N+1))) ACXI0603
C   WITH REMOVED FACTOR                                                  ACXI0604
C   (C/4PI)(H*H/2PI M KT)**3/2 N*N NE NI EXP(CHI1/N*N KT)                ACXI0605
C                                                                        ACXI0606
        COMMON /EXPDAT/ CXP(707),MAXN                                    ACXI0607
        COMMON /INOUT/ IREAD,IWRITE,IPUNCH,icrt                          ACXI0608
        DIMENSION DILT(508), DX(508)                                     ACXI0609
        DOUBLE PRECISION DABS,DSQRT,DBLE,DLOG,DLOG10,DFLOAT,DEXP         ACXI0610
        DOUBLE PRECISION A,AM,AP,COR,CXP,EX,G                            ACXI0611
        LOGICAL NOFLD                                                    ACXI0612
        DATA DILT/508*0.5/,NOFLD/.FALSE./                               ACXI0613
        IF (ISW.GT.0) GO TO 40                                          ACXI0614
        READ (IREAD,90) TBCK,EBCK                                       ACXI0615
        IF (EBCK.EQ.0. .OR. TBCK.EQ.0.) NOFLD=.TRUE.                    ACXI0616
        IF (.NOT.NOFLD) GO TO 10                                        ACXI0617
        WRITE (IWRITE,100)                                              ACXI0618
        write (icrt,111)
 111    format (' No background info detected.')
        GO TO 80                                                        ACXI0619
  10    WRITE (IWRITE,110) TBCK,EBCK                                    ACXI0620
        write (icrt,112)
 112    format (' Background info detected.  This part of the code',
       1' has not be checked.')
        C15=15.778E4/TBCK                                               ACXI0621
        MAXP=MAXN+1                                                     ACXI0622
        DO 20 I=1,MAXP                                                  ACXI0623
        AI=I*(I+1)                                                      ACXI0624
        ARG=C15*FLOAT(2*I+1)/AI**2                                      ACXI0625
        DX(I)=EXPM1(ARG)                                                ACXI0626
  20    CONTINUE                                                        ACXI0627
        IF (EBCK.GE.0.9999E10) GO TO 80                                 ACXI0628
        TBCK32=TBCK*SQRT(TBCK)                                          ACXI0629
        TL=ALOG10(TBCK)                                                 ACXI0630
        DO 30 I=1,MAXP                                                  ACXI0631
        FREQG=6.58E6/FLOAT(I)**3                                        ACXI0632
        TOW=EBCK*CAPPA(TBCK,TL,TBCK32,FREQG)                            ACXI0633
        IF (TOW.LE.20.) DILT(I)=-0.5*EXPM1(-TOW)                        ACXI0634
  30    CONTINUE                                                        ACXI0635
        GO TO 80                                                        ACXI0636
  40    IF (NOFLD) GO TO 80                                             ACXI0637
        IF (N.GT.MAXN) WRITE (IWRITE,120) N                             ACXI0638
        GO TO (50,60,70), ISW                                           ACXI0639
C   N(N+1)B(N+1,N)                                                      ACXI0640
  50    CALL RAD (A,N,N+1,G)                                            ACXI0641
        EX=DX(N)                                                        ACXI0642
        COR=(DFLOAT((N+1)**2)/DFLOAT(N*N))*DILT(N)*A/(EX*(EX+1.D0))     ACXI0643
        RETURN                                                          ACXI0644
C   N(N-1)B(N-1,N)                                                      ACXI0645
  60    CALL RAD (A,N-1,N,G)                                            ACXI0646
        EX=DX(N-1)                                                      ACXI0647
        COR=DILT(N)*A/(EX/(EX+1.D0))                                    ACXI0648
        RETURN                                                          ACXI0649
C   N(N)(B(N,N-1) + B(N,N+1))                                           ACXI0650
  70    CALL RAD (AM,N-1,N,G)                                           ACXI0651
        CALL RAD (AP,N,N+1,EX)                                          ACXI0652
        EX=DX(N-1)                                                      ACXI0653
        G=DX(N)                                                         ACXI0654
```

```
      COR=-DILT(N)*(AM/EX+(DFLOAT((N+1)**2)/DFLOAT(N*N))*AP/G)      ACXI0655
      RETURN                                                        ACXI0656
  80  COR=0.D0                                                      ACXI0657
      RETURN                                                        ACXI0658
C                                                                   ACXI0659
  90  FORMAT (2G10.3)                                               ACXI0660
 100  FORMAT (/21H NO BACKGROUND FIELD.//)                          ACXI0661
 110  FORMAT (/32H RADIATION FIELD - TEMPERATURE =,1P,G12.5,        ACXI0662
     121HK, EMISSION MEASURE =,G12.5//)                             ACXI0663
 120  FORMAT (/32H *** COR CALLED WITH N TOO LARGE,I6/)             ACXI0664
      END                                                           ACXI0665


      FUNCTION CROSS (N,NP,E)                                       ACXI0668
C                                                                   ACXI0669
C  COMPUTES CROSS SECTION FOR TRANSITION FROM LEVEL  N TO HIGHER LEVEL ACXI0670
C  NP DUE TO COLLISION WITH ELECTRON OF ENERGY  E.                 ACXI0671
C                                                                   ACXI0672
C  THE FORMULA IS VALID FOR ENERGIES IN THE RANGE 4/N**2 LT E LLT 137**2 ACXI0673
C  THIS SUBPROGRAM DOES NOT CHECK THAT  E IS WITHIN THIS RANGE.     ACXI0674
C                                                                   ACXI0675
C  THEORY: GEE, PERCIVAL, LODGE AND RICHARDS, MNRAS 175, 209-215 (1976) ACXI0676
C                                                                   ACXI0677
      COMMON /COLINF/ AL18S4(506),S23TRM(506)                       ACXI0678
      DOUBLE PRECISION DABS,DSQRT,DBLE,DLOG,DLOG10,DFLOAT,DEXP       ACXI0679
      DOUBLE PRECISION DRT                                          ACXI0680
      REAL L                                                        ACXI0681
      C2(X,Y)=X*X*ALOG(1.+.6666667*X)/(Y+Y+1.5*X)                   ACXI0682
      EN=N                                                          ACXI0683
      ENP=NP                                                        ACXI0684
      IS=NP-N                                                       ACXI0685
      IPOW=1+IS+IS                                                  ACXI0686
      POW=IPOW                                                      ACXI0687
      S=IS                                                          ACXI0688
      ENNP=N*NP                                                     ACXI0689
      EENNP=E*E*ENNP                                                ACXI0690
      EN2=N*N                                                       ACXI0691
      ENN=E*EN2                                                     ACXI0692
      D=0.2*S/ENNP                                                  ACXI0693
      IF (D.GT.0.02) GO TO 10                                       ACXI0694
      D=1.-POW*D                                                    ACXI0695
      GO TO 20                                                      ACXI0696
  10  D=(1.-D)**IPOW                                                ACXI0697
  20  A=(2.666667/S)*(ENP/(S*EN))**3*S23TRM(IS)*D                   ACXI0698
      D=0.                                                          ACXI0699
      ARG=1./EENNP                                                  ACXI0700
      IF (ARG.LT.150.) D=EXP(-ARG)                                  ACXI0701
      L=ALOG((1.+0.53*EENNP)/(1.+0.4*E))                            ACXI0702
      F=(1.-0.3*S*D/ENNP)**IPOW                                     ACXI0703
      G=0.5*(ENN/ENP)**3                                            ACXI0704
      Y=1./(1.-D*AL18S4(IS))                                        ACXI0705
      DRT=DSQRT(2.D0-DFLOAT(N*N)/DFLOAT(NP*NP))                     ACXI0706
      XP=2./(ENN*SNGL(DRT+1.D0))                                    ACXI0707
      XM=2./(ENN*SNGL(DRT-1.D0))                                    ACXI0708
      H=C2(XM,Y)-C2(XP,Y)                                           ACXI0709
      CROSS=8.797016E-17*(EN2*EN2/E)*(A*D*L+F*G*H)                  ACXI0710
      RETURN                                                        ACXI0711
      END                                                           ACXI0712
```

```
      FUNCTION DMJ1 (K,KM)                                      ACXIO714
      COMMON /FITDAT/ AFIT(4,4),IVAL(4),NFIT                    ACXIO715
      COMMON /HIGHER/ STORE1(224,4),STORE2(224),VAL(24),B,STORE3(224,5),ACXIO716
     1RTVAL(24),LIMIT                                           ACXIO717
      DOUBLE PRECISION DABS,DSQRT,DBLE,DLOG,DLOG10,DFLOAT,DEXP  ACXIO718
      DOUBLE PRECISION AFIT,B,DMJ1,RTVAL,STORE1,STORE2,STORE3,VAL  ACXIO719
      DMJ1=0.D0                                                 ACXIO720
      DO 10 J=1,NFIT                                            ACXIO721
   10 DMJ1=DMJ1+AFIT(J,KM)*STORE1(K,J)                          ACXIO722
      RETURN                                                    ACXIO723
      END                                                       ACXIO724


      FUNCTION DMJ2 (K)                                         ACXIO726
      COMMON /FITDAT/ AFIT(4,4),IVAL(4),NFIT                    ACXIO727
      COMMON /HIGHER/ STORE1(224,4),STORE2(224),VAL(24),B,STORE3(224,5),ACXIO728
     1RTVAL(24),LIMIT                                           ACXIO729
      DOUBLE PRECISION DABS,DSQRT,DBLE,DLOG,DLOG10,DFLOAT,DEXP  ACXIO730
      DOUBLE PRECISION AFIT,B,DMJ2,RTVAL,STORE1,STORE2,STORE3,SUM,TOT,VAACXIO731
     1L                                                        ACXIO732
      DMJ2=0.D0                                                 ACXIO733
      TOT=0.D0                                                  ACXIO734
      DO 20 J=1,NFIT                                            ACXIO735
      SUM=0.D0                                                  ACXIO736
      DO 10 I=1,NFIT                                            ACXIO737
   10 SUM=SUM+AFIT(J,I)                                         ACXIO738
   20 TOT=TOT+SUM*STORE1(K,J)                                   ACXIO739
      DMJ2=1.D0-TOT                                             ACXIO740
      RETURN                                                    ACXIO741
      END                                                       ACXIO742


      FUNCTION DPHI (IQ,LG,ITAU,N)                              ACXIO744
C                                                               ACXIO745
C  LAGRANGIAN INTERPOLATION                                     ACXIO746
C                                                               ACXIO747
      DIMENSION IQ(8)                                           ACXIO748
      DOUBLE PRECISION DABS,DSQRT,DBLE,DLOG,DLOG10,DFLOAT,DEXP  ACXIO749
      DOUBLE PRECISION A,DPHI                                   ACXIO750
      DPHI=0.D0                                                 ACXIO751
      DO 20 M=1,LG                                              ACXIO752
      IF (M.EQ.ITAU) GO TO 20                                   ACXIO753
      A=1.D0                                                    ACXIO754
      DO 10 L=1,LG                                              ACXIO755
      IF (L.EQ.ITAU.OR.L.EQ.M) GO TO 10                         ACXIO756
      A=A*DFLOAT(N-IQ(L))                                       ACXIO757
   10 CONTINUE                                                  ACXIO758
      DPHI=DPHI+A                                               ACXIO759
   20 CONTINUE                                                  ACXIO760
      RETURN                                                    ACXIO761
      END                                                       ACXIO762


      FUNCTION EXPM1 (X)                                        ACXIO765
C                                                               ACXIO766
C  COMPUTES  EXP(X)-1  ACCURATELY, EVEN FOR SMALL  X            ACXIO767
C                                                               ACXIO768
      IF (ABS(X).LE.0.1) GO TO 10                               ACXIO769
      EXPM1=EXP(X)-1.                                           ACXIO770
      RETURN                                                    ACXIO771
```

```
10    EXPM1=X*(.9999997+X*(0.5+X*(.1667708+X*.4166667E-1)))         ACXI0772
      RETURN                                                        ACXI0773
      END                                                           ACXI0774


      FUNCTION PHI (IQ,LG,ITAU,N)                                   ACXI1007
C                                                                   ACXI1008
C  LAGRANGIAN INTERPOLATION                                         ACXI1009
C                                                                   ACXI1010
      DIMENSION IQ(8)                                               ACXI1011
      DOUBLE PRECISION DABS,DSQRT,DBLE,DLOG,DLOG10,DFLOAT,DEXP       ACXI1012
      DOUBLE PRECISION PHI                                          ACXI1013
      PHI=1.D0                                                      ACXI1014
      DO 10 L=1,LG                                                  ACXI1015
      IF (L.EQ.ITAU) GO TO 10                                       ACXI1016
      PHI=PHI*DFLOAT(N-IQ(L))                                       ACXI1017
10    CONTINUE                                                      ACXI1018
      RETURN                                                        ACXI1019
      END                                                           ACXI1020


      FUNCTION POL (K,KM)                                           ACXI1022
      COMMON /HIGHER/ STORE1(224,4),STORE2(224),VAL(24),B,STORE3(224,5),ACXI1023
     1RTVAL(24),LIMIT                                               ACXI1024
      DOUBLE PRECISION DABS,DSQRT,DBLE,DLOG,DLOG10,DFLOAT,DEXP       ACXI1025
      DOUBLE PRECISION B,POL,RTVAL,STORE1,STORE2,STORE3,VAL          ACXI1026
      IND=K+LIMIT                                                   ACXI1027
      POL=RTVAL(K)*STORE3(IND,KM)*STORE2(IND)                       ACXI1028
      RETURN                                                        ACXI1029
      END                                                           ACXI1030


      SUBROUTINE BCPCH (KB,KC,T,DENS,NMIN,LABEL,NPLO,NPHI,NDIM)      ACXI0144
C                                                                   ACXI0145
C  PUNCHES BN AND CN VALUES IN STANDARD FORMAT                      ACXI0146
C                                                                   ACXI0147
      COMMON /INOUT/ IREAD,IWRITE,IPUNCH,icrt                       ACXI0148
      DIMENSION KB(NDIM), KC(NDIM), K1(19), K2(12)                  ACXI0149
      DOUBLE PRECISION DENL,DENS,T,TL                               ACXI0150
      EQUIVALENCE (K1(1),K2(1))                                     ACXI0151
      DOUBLE PRECISION DABS,DSQRT,DBLE,DLOG,DLOG10,DFLOAT,DEXP       ACXI0152
      CHARACTER ALPHA(35)                                           ACXI0153
C SOME COMPILERS REQUIRE  (ALPHA(I),I=1,35)  IN DATA STATEMENT      ACXI0154
      DATA ALPHA /'1','2','3','4','5','6','7','8','9','A','B',       ACXI0155
     1'C','D','E','F','G','H','I','J','K','L','M','N','O','P','Q','R',  ACXI0156
     2'S','T','U','V','W','X','Y','Z'/                              ACXI0157
      N=NPHI-NPLO+1                                                 ACXI0158
      TL=DLOG10(T)                                                  ACXI0159
      DENL=DLOG10(DENS)                                             ACXI0160
      WRITE (IPUNCH,50) TL,DENL,NMIN,NPLO,NPHI,LABEL,ALPHA(1)       ACXI0161
      NB=N/19                                                       ACXI0162
      IF (19*NB.NE.N) NB=NB+1                                       ACXI0163
      NC=N/12                                                       ACXI0164
      IF (12*NC.NE.N) NC=NC+1                                       ACXI0165
      IPOINT=0                                                      ACXI0166
      DO 20 I=1,NB                                                  ACXI0167
      DO 10 J=1,19                                                  ACXI0168
      IPOINT=IPOINT+1                                               ACXI0169
      K=0                                                           ACXI0170
      IF (IPOINT.LE.N) K=KB(IPOINT)                                 ACXI0171
```

```
          K1(J)=K                                                ACXI0172
   10     CONTINUE                                               ACXI0173
          WRITE (IPUNCH,60) K1,LABEL,ALPHA(I+1)                  ACXI0174
   20     CONTINUE                                               ACXI0175
          IPOINT=0                                               ACXI0176
          DO 40 I=1,NC                                           ACXI0177
          DO 30 J=1,12                                           ACXI0178
          IPOINT=IPOINT+1                                        ACXI0179
          K=0                                                    ACXI0180
          IF (IPOINT.LE.N) K=KC(IPOINT)                          ACXI0181
          K2(J)=K                                                ACXI0182
   30     CONTINUE                                               ACXI0183
          NBI=NB+I+1                                             ACXI0184
          WRITE (IPUNCH,70) K2,LABEL,ALPHA(NBI)                  ACXI0185
   40     CONTINUE                                               ACXI0186
          WRITE (IWRITE,80) IPUNCH                               ACXI0187
          RETURN                                                 ACXI0188
   C                                                             ACXI0189
   50     FORMAT (2F10.6,3I5,41X,A3,A1)                          ACXI0190
   60     FORMAT (19I4,A3,A1)                                    ACXI0191
   70     FORMAT (12I6,4X,A3,A1)                                 ACXI0192
   80     FORMAT (////27H *** OUTPUT PUNCHED ON UNIT,I3)         ACXI0193
          END                                                    ACXI0194


          SUBROUTINE COLION (N,IONZ,T,QI)                        ACXI0476
   C                                                             ACXI0477
   C   COMPUTES COLLISIONAL IONIZATION RATE, QI, FROM LEVEL N  FOR IONS  ACXI0478
   C   OF EFFECTIVE CHARGE  IONZ  AT ELECTRON TEMPERATURE  T.   ACXI0479
   C                                                             ACXI0480
   C   WHEN CALLED WITH  N=0, COLION COMPUTES AND STORES QUANTITIES WHICH ACXI0481
   C   DEPEND ONLY UPON TEMPERATURE AND EFFECTIVE CHARGE. IT IS ASSUMED   ACXI0482
   C   THAT  T   AND  IONZ REMAIN CONSTANT UNTIL THE NEXT CALL WITH  N=0. ACXI0483
   C                                                             ACXI0484
          COMMON /TDEP/ TE32,TE12,CTE                            ACXI0485
          DIMENSION EXPX(507)                                    ACXI0486
          DOUBLE PRECISION DABS,DSQRT,DBLE,DLOG,DLOG10,DFLOAT,DEXP  ACXI0487
          DOUBLE PRECISION CTE,QI,T,TE12,TE32                    ACXI0488
          QI=0.D0                                                ACXI0489
          IF (N.NE.0) GO TO 20                                   ACXI0490
   C  INITIALIZE                                                 ACXI0491
   c Malcolm Walmsley has modified this program to consider collisional
   c ionization from level 10.  The original program considered levels only
   c down to n = 20.
          CONS=DFLOAT(IONZ*IONZ)*CTE                             ACXI0492
          nlow=9
          nlow1=nlow+1
   c      DO 10 I=21,507                                         ACXI0493
          do 10 i=nlow1,507
   10     EXPX(I)=EXP(-CONS/FLOAT(I*I))                          ACXI0494
          RETURN                                                 ACXI0495
   c20    IF (N.LE.20) RETURN                                    ACXI0496
   20     if (n.le.507) return
          X=CONS/DFLOAT(N*N)                                     ACXI0497
          DXP=EXPX(N)                                            ACXI0498
          IF (N.GT.507) DXP=EXP(-X)                              ACXI0499
          IF (X.LE.1.) GO TO 30                                  ACXI0500
          E1=((.250621+X*(2.334733+X))/(1.681534+X*(3.330657+X)))*DXP/X  ACXI0501
```

```
        GO TO 40                                                  ACXI0502
   30   E1=-.57721566+X*(.9999193+X*(-.24991055+X*(.5519968E-1+X*(-.976004ACXI0503
        1E-2+X*.107857E-2))))-ALOG(X)                             ACXI0504
   40   EI=DXP*(1.666667-.6666667*X)/X+E1*(.6666667*X-1.)-0.5*E1*E1/DXP    ACXI0505
        QI=5.444089*EI/SNGL(TE32)                                 ACXI0506
        RETURN                                                    ACXI0507
        END                                                       ACXI0508


        SUBROUTINE HELPME (AID,KM)                                ACXI0776
        COMMON /HIGHER/ STORE1(224,4),STORE2(224),VAL(24),B,STORE3(224,5),ACXI0777
        1RTVAL(24),LIMIT                                          ACXI0778
        DOUBLE PRECISION DABS,DSQRT,DBLE,DLOG,DLOG10,DFLOAT,DEXP   ACXI0779
        DOUBLE PRECISION AID,B,C,POL,RTVAL,STORE1,STORE2,STORE3,SUM,VAL,Y ACXI0780
        SUM=0.D0                                                  ACXI0781
        DO 10 K=1,LIMIT                                           ACXI0782
        C=STORE2(K)*STORE3(K,KM)                                  ACXI0783
   10   SUM=SUM+C                                                 ACXI0784
        SUM=SUM-0.5D0*C                                           ACXI0785
        Y= .6170614899993600D-2*(POL( 1,KM)+POL( 2,KM))           ACXI0786
        Y=Y+.1426569431446683D-1*(POL( 3,KM)+POL( 4,KM))          ACXI0787
        Y=Y+.2213871940870990D-1*(POL( 5,KM)+POL( 6,KM))          ACXI0788
        Y=Y+.2964929245771839D-1*(POL( 7,KM)+POL( 8,KM))          ACXI0789
        Y=Y+.3667324070554015D-1*(POL( 9,KM)+POL(10,KM))          ACXI0790
        Y=Y+.4309508076597664D-1*(POL(11,KM)+POL(12,KM))          ACXI0791
        Y=Y+.4880932605205694D-1*(POL(13,KM)+POL(14,KM))          ACXI0792
        Y=Y+.5372213505798282D-1*(POL(15,KM)+POL(16,KM))          ACXI0793
        Y=Y+.5775283402686280D-1*(POL(17,KM)+POL(18,KM))          ACXI0794
        Y=Y+.6083523646390170D-1*(POL(19,KM)+POL(20,KM))          ACXI0795
        Y=Y+.6291872817341415D-1*(POL(21,KM)+POL(22,KM))          ACXI0796
        Y=Y+.6396909767337608D-1*(POL(23,KM)+POL(24,KM))          ACXI0797
        AID=SUM+B*Y                                               ACXI0798
        RETURN                                                    ACXI0799
        END                                                       ACXI0800


        SUBROUTINE INTERP (M,CO,VAL,DVAL,IC,IR)                   ACXI0802
   C                                                              ACXI0803
   C  COMPUTES SOLUTIONS AT ALL  N  FROM THOSE OBTAINED AT THE CONDENSED  ACXI0804
   C  POINTS                                                      ACXI0805
   C                                                              ACXI0806
        DIMENSION CO(75), DVAL(507), IQ(8), M(75), VAL(507)       ACXI0807
        DOUBLE PRECISION DABS,DSQRT,DBLE,DLOG,DLOG10,DFLOAT,DEXP   ACXI0808
        DOUBLE PRECISION CO,COT,DFL,DPHI,DVAL,FL,PHI,PHITAU,VAL    ACXI0809
   C                                                              ACXI0810
        LG=2*(IR+1)                                               ACXI0811
        IB=IC-IR                                                  ACXI0812
        IBB=IB-1                                                  ACXI0813
        ICC=IC-1                                                  ACXI0814
   C                                                              ACXI0815
        K=M(IC)                                                   ACXI0816
        DO 10 I=1,K                                               ACXI0817
        VAL(I)=0.D0                                               ACXI0818
        DVAL(I)=0.D0                                              ACXI0819
   10   CONTINUE                                                  ACXI0820
   C                                                              ACXI0821
        DO 20 IT=1,IC                                             ACXI0822
        IA=IT                                                     ACXI0823
        MIT=M(IT)                                                 ACXI0824
```

```
        VAL(MIT)=CO(IT)                                          ACXI0825
        IF ((M(IT+1)-MIT).GT.1) GO TO 30                         ACXI0826
20      CONTINUE                                                 ACXI0827
C                                                                ACXI0828
30      IF (IA.EQ.IC) GO TO 90                                   ACXI0829
        IF (IA.EQ.IB) GO TO 60                                   ACXI0830
C                                                                ACXI0831
        DO 50 IT=IA,IBB                                          ACXI0832
        N1=M(IT)+1                                               ACXI0833
        N2=M(IT+1)                                               ACXI0834
        ITR1=IT-IR-1                                             ACXI0835
        DO 40 ITAU=1,LG                                          ACXI0836
        IND=ITR1+ITAU                                            ACXI0837
40      IQ(ITAU)=M(IND)                                          ACXI0838
        DO 50 ITAU=1,LG                                          ACXI0839
        PHITAU=1.D0/PHI(IQ,LG,ITAU,IQ(ITAU))                     ACXI0840
        DO 50 N=N1,N2                                            ACXI0841
        FL=PHI(IQ,LG,ITAU,N)*PHITAU                              ACXI0842
        DFL=DPHI(IQ,LG,ITAU,N)*PHITAU                            ACXI0843
        IND=ITR1+ITAU                                            ACXI0844
        COT=CO(IND)                                              ACXI0845
        VAL(N)=VAL(N)+FL*COT                                     ACXI0846
50      DVAL(N)=DVAL(N)+DFL*COT                                  ACXI0847
60      IF (IR.EQ.0) GO TO 90                                    ACXI0848
C                                                                ACXI0849
        ICLG=IC-LG                                               ACXI0850
        DO 70 ITAU=1,LG                                          ACXI0851
        IND=ICLG+ITAU                                            ACXI0852
70      IQ(ITAU)=M(IND)                                          ACXI0853
        DO 80 ITAU=1,LG                                          ACXI0854
        PHITAU=1.D0/PHI(IQ,LG,ITAU,IQ(ITAU))                     ACXI0855
        DO 80 IT=IB,ICC                                          ACXI0856
        N1=M(IT)+1                                               ACXI0857
        N2=M(IT+1)                                               ACXI0858
        DO 80 N=N1,N2                                            ACXI0859
        FL=PHI(IQ,LG,ITAU,N)*PHITAU                              ACXI0860
        DFL=DPHI(IQ,LG,ITAU,N)*PHITAU                            ACXI0861
        IND=ICLG+ITAU                                            ACXI0862
        COT=CO(IND)                                              ACXI0863
        VAL(N)=VAL(N)+FL*COT                                     ACXI0864
80      DVAL(N)=DVAL(N)+DFL*COT                                  ACXI0865
C                                                                ACXI0866
90      RETURN                                                   ACXI0867
        END                                                      ACXI0868

        SUBROUTINE JMD (SK,CO,MVAL,IC)                           ACXI0870
        COMMON /FITDAT/ AFIT(4,4),IVAL(4),NFIT                   ACXI0871
        COMMON /GAUSS/ VALUE(12)                                 ACXI0872
        COMMON /HIGHER/ STORE1(224,4),STORE2(224),VAL(24),B,STORE3(224,5),ACXI0873
       1RTVAL(24),LIMIT                                          ACXI0874
        COMMON /PARMS/ DENS,T,ITM                                ACXI0875
        DIMENSION AZ(4), CO(75), IND(4,2), IPIV(4), MVAL(75), SK(75,75)  ACXI0876
        DOUBLE PRECISION DABS,DSQRT,DBLE,DLOG,DLOG10,DFLOAT,DEXP  ACXI0877
        DOUBLE PRECISION A,AC,AFIT,AID,AJ,AK,AKK,AZ,B,BK,CO,D,DENS,DMJ1,DMACXI0878
       1J2,RTVAL,SK,SOS,STORE1,STORE2,STORE3,T,VAL,VALUE         ACXI0879
        SOS(I,A)=DSQRT(-A)**(2*I+ITM)/DLOG(-A)                   ACXI0880
        LIMIT=200                                                ACXI0881
```

```
        NG=24                                                      ACXI0882
        DO 10 J=1,NFIT                                             ACXI0883
        K=IVAL(J)                                                  ACXI0884
        AJ=-1.D0/DFLOAT(MVAL(K))**2                                ACXI0885
        DO 10 I=1,NFIT                                             ACXI0886
10      AFIT(J,I)=SOS(I,AJ)                                        ACXI0887
        CALL MATINV (AFIT,NFIT,AZ,0,D,IRROR,4,IPIV,IND)            ACXI0888
        B=1.D0/DFLOAT(MVAL(IC)+LIMIT)**2                           ACXI0889
        A=-0.5D0*B                                                 ACXI0890
        NH=NG/2                                                    ACXI0891
        DO 20 K=1,NH                                               ACXI0892
        VAL(2*K-1)=A+VALUE(K)*B                                    ACXI0893
20      VAL(2*K)=A-VALUE(K)*B                                      ACXI0894
        DO 30 K=1,LIMIT                                            ACXI0895
        A=MVAL(IC)+K                                               ACXI0896
        AC=-1.D0/A**2                                              ACXI0897
        DO 30 J=1,NFIT                                             ACXI0898
30      STORE1(K,J)=SOS(J,AC)                                      ACXI0899
        DO 40 K=1,NG                                               ACXI0900
        DO 40 J=1,NFIT                                             ACXI0901
        INP=K+LIMIT                                                ACXI0902
40      STORE1(INP,J)=SOS(J,VAL(K))                               ACXI0903
        KK=LIMIT+NG                                                ACXI0904
        DO 60 K=1,KK                                               ACXI0905
        DO 50 J=1,NFIT                                             ACXI0906
50      STORE3(K,J)=DMJ1(K,J)                                      ACXI0907
60      STORE3(K,NFIT+1)=DMJ2(K)                                   ACXI0908
        DO 100 J=1,IC                                              ACXI0909
        I=MVAL(J)                                                  ACXI0910
        DO 70 K=1,LIMIT                                            ACXI0911
        KK=MVAL(IC)+K                                              ACXI0912
        AKK=KK                                                     ACXI0913
70      STORE2(K)=BK(I,KK,0)                                       ACXI0914
        DO 80 K=1,NG                                               ACXI0915
        AK=DSQRT(-1.D0/VAL(K))                                     ACXI0916
        RTVAL(K)=AK**3*0.5D0                                       ACXI0917
        KK=AK                                                      ACXI0918
        INP=K+LIMIT                                                ACXI0919
        STORE2(INP)=BK(I,KK,0)                                     ACXI0920
80      CONTINUE                                                   ACXI0921
        CALL HELPME (AID,NFIT+1)                                   ACXI0922
        CO(J)=CO(J)-AID                                            ACXI0923
        DO 90 KM=1,NFIT                                            ACXI0924
        CALL HELPME (AID,KM)                                       ACXI0925
        L=IVAL(KM)                                                 ACXI0926
90      SK(J,L)=SK(J,L)+AID                                        ACXI0927
100     CONTINUE                                                   ACXI0928
        RETURN                                                     ACXI0929
        END                                                        ACXI0930

        SUBROUTINE MATINV (A,N,B,L,D,IRROR,NDA,IPIV,IND)           ACXI0932
        DIMENSION A(NDA,NDA), B(NDA), IND(NDA,2), IPIV(NDA)        ACXI0933
        DOUBLE PRECISION DABS,DSQRT,DBLE,DLOG,DLOG10,DFLOAT,DEXP   ACXI0934
        DOUBLE PRECISION A,AMAX,ATEMP,B,D                          ACXI0935
C                                                                  ACXI0936
C       SOLVES SIMULTANEOUS EQUATIONS IF L=1                       ACXI0937
C                                                                  ACXI0938
```

```
C      INVERTS MATRIX A IF L=0                               ACXI0939
C                                                            ACXI0940
C      SOLUTIONS ARE RETURNED IN B                           ACXI0941
C                                                            ACXI0942
       M=IABS(L)                                             ACXI0943
       D=1.DO                                                ACXI0944
       DO 10 I=1,N                                           ACXI0945
  10   IPIV(I)=0                                             ACXI0946
       DO 190 I=1,N                                          ACXI0947
       AMAX=0.DO                                             ACXI0948
       DO 60 J=1,N                                           ACXI0949
       IF (IPIV(J)) 70,20,60                                 ACXI0950
  20   DO 50 K=1,N                                           ACXI0951
C      IF (DABS(A(J,K)).LT.1.D-50) A(J,K)=0.DO               ACXI0952
       if (dabs(a(j,k)).lt.1.d-38) a(j,k)=0.d0
       IF (IPIV(K)-1) 30,50,70                               ACXI0953
  30   IF (DABS(A(J,K))-AMAX) 50,50,40                       ACXI0954
  40   IROW=J                                                ACXI0955
       ICOL=K                                                ACXI0956
       AMAX=DABS(A(J,K))                                     ACXI0957
  50   CONTINUE                                              ACXI0958
  60   CONTINUE                                              ACXI0959
       IPIV(ICOL)=IPIV(ICOL)+1                               ACXI0960
C      IF (AMAX-1.D-50) 70,70,80                             ACXI0961
       if (amax-1.d-38) 70,70,80
  70   IRROR=1                                               ACXI0962
       RETURN                                                ACXI0963
  80   IF (IROW-ICOL) 90,120,90                              ACXI0964
  90   D=-D                                                  ACXI0965
       DO 100 K=1,N                                          ACXI0966
       AMAX=A(IROW,K)                                        ACXI0967
       A(IROW,K)=A(ICOL,K)                                   ACXI0968
 100   A(ICOL,K)=AMAX                                        ACXI0969
       IF (M) 120,120,110                                    ACXI0970
 110   AMAX=B(IROW)                                          ACXI0971
       B(IROW)=B(ICOL)                                       ACXI0972
       B(ICOL)=AMAX                                          ACXI0973
 120   IND(I,1)=IROW                                         ACXI0974
       IND(I,2)=ICOL                                         ACXI0975
       AMAX=A(ICOL,ICOL)                                     ACXI0976
       A(ICOL,ICOL)=1.DO                                     ACXI0977
       DO 130 K=1,N                                          ACXI0978
 130   A(ICOL,K)=A(ICOL,K)/AMAX                              ACXI0979
       IF (M) 150,150,140                                    ACXI0980
 140   B(ICOL)=B(ICOL)/AMAX                                  ACXI0981
 150   DO 190 J=1,N                                          ACXI0982
       IF (J-ICOL) 160,190,160                               ACXI0983
 160   AMAX=A(J,ICOL)                                        ACXI0984
       A(J,ICOL)=0.DO                                        ACXI0985
       DO 170 K=1,N                                          ACXI0986
       ATEMP=A(ICOL,K)*AMAX                                  ACXI0987
 170   A(J,K)=A(J,K)-ATEMP                                   ACXI0988
       IF (M) 190,190,180                                    ACXI0989
 180   B(J)=B(J)-B(ICOL)*AMAX                                ACXI0990
 190   CONTINUE                                              ACXI0991
       IF (L) 200,200,240                                    ACXI0992
 200   DO 230 I=1,N                                          ACXI0993
```

```
        J=N+1-I                                              ACXI0994
        IF (IND(J,1)-IND(J,2)) 210,230,210                   ACXI0995
210     IROW=IND(J,1)                                        ACXI0996
        ICOL=IND(J,2)                                        ACXI0997
        DO 220 K=1,N                                         ACXI0998
        AMAX=A(K,IROW)                                       ACXI0999
        A(K,IROW)=A(K,ICOL)                                  ACXI1000
220     A(K,ICOL)=AMAX                                       ACXI1001
230     CONTINUE                                             ACXI1002
240     IRROR=0                                              ACXI1003
        RETURN                                               ACXI1004
        END                                                  ACXI1005


        SUBROUTINE RAD (A,N,NDASH,G)                         ACXI1032
C                                                            ACXI1033
C COMPUTES EINSTEIN RADIATIVE RATES A AND GAUNT FACTORS G FOR ACXI1034
C TRANSITIONS FROM  NDASH  TO  N                             ACXI1035
C                                                            ACXI1036
        COMMON /GAUNTS/ GAUNT(50,22),IXV(12)                 ACXI1037
        DOUBLE PRECISION DABS,DSQRT,DBLE,DLOG,DLOG10,DFLOAT,DEXP ACXI1038
        DOUBLE PRECISION A,ALF2,EN,EN2,END2,G,TERM1,TERM2,TERM3 ACXI1039
C ARRAY GAUNT IS NOT DOUBLE PRECISION                        ACXI1040
        EN=N                                                 ACXI1041
        EN2=N*N                                              ACXI1042
        END2=NDASH*NDASH                                     ACXI1043
        NDMN=NDASH-N                                         ACXI1044
        IF (NDMN.GT.50) GO TO 30                             ACXI1045
        IF (N.LE.10) GO TO 20                                ACXI1046
        DO 10 J=2,12                                         ACXI1047
        IVJ=IXV(J)                                           ACXI1048
        IVJ1=IXV(J-1)                                        ACXI1049
        IF (N.GT.IVJ) GO TO 10                               ACXI1050
        P=1./FLOAT(IVJ-IVJ1)                                 ACXI1051
        Q=FLOAT(IVJ-N)*P                                     ACXI1052
        P=FLOAT(N-IVJ1)*P                                    ACXI1053
        G=Q*GAUNT(NDMN,J+9)+GAUNT(NDMN,J+10)*P               ACXI1054
        GO TO 40                                             ACXI1055
10      CONTINUE                                             ACXI1056
20      G=GAUNT(NDMN,N)                                      ACXI1057
        GO TO 40                                             ACXI1058
30      ALF2=(EN2/END2)                                      ACXI1059
        TERM1=(1.D0-ALF2)                                    ACXI1060
        TERM1=(TERM1*EN)**.666666666666666D0                 ACXI1061
        TERM2=0.1728D0*(1.D0+ALF2)/TERM1                     ACXI1062
        TERM3=0.0496D0*(1.D0-1.3333333333333D0*ALF2+ALF2*ALF2)/TERM1**2 ACXI1063
        G=1.D0-TERM2-TERM3                                   ACXI1064
40      A=G*15.7457D9/(EN*(END2-EN2)*DFLOAT(NDASH)*END2)     ACXI1065
        RETURN                                               ACXI1066
        END                                                  ACXI1067


        SUBROUTINE RADCOL (T,MVAL,IC,NMIN)                   ACXI1069
        COMMON /EXPDAT/ CXP(707),MAXN                        ACXI1070
        COMMON /RCRATS/ RADTOT(75),COLTOT(75)                ACXI1071
        COMMON /TDEP/ TE32,TE12,CTE                          ACXI1072
        DIMENSION MVAL(75)                                   ACXI1073
        DOUBLE PRECISION DABS,DSQRT,DBLE,DLOG,DLOG10,DFLOAT,DEXP ACXI1074
        DOUBLE PRECISION A,AKK,AN,C,COLTOT,CTE,CX,CXP,G,Q,RADTOT,T,TE12,TEACXI1075
```

```
        132,TOT                                                   ACXI1076
C                                                                 ACXI1077
C       RADIATIVE CASCADE COEFFICIENTS                           ACXI1078
C                                                                 ACXI1079
C       SUM FROM N TO ALL LEVELS DOWN TO NMIN (=1 OR 2)          ACXI1080
C                                                                 ACXI1081
        DO 20 M=1,IC                                              ACXI1082
        J=MVAL(M)                                                 ACXI1083
        TOT=0.D0                                                  ACXI1084
        IF (J.LE.NMIN) GO TO 20                                   ACXI1085
        K=J-1                                                     ACXI1086
        DO 10 I=NMIN,K                                            ACXI1087
        CALL RAD (A,I,J,G)                                        ACXI1088
   10   TOT=TOT+A                                                 ACXI1089
   20   RADTOT(M)=TOT                                             ACXI1090
C                                                                 ACXI1091
C       COLLISIONAL RATE TOTALS ON TO LEVEL N                    ACXI1092
C                                                                 ACXI1093
C       SUMMED FROM NMIN TO INFINITY                             ACXI1094
C                                                                 ACXI1095
c Here also, Malcolm Walmsley has changed the lower principal quantum
c number from 20 to 1 (CASE A) or 2 (CASE B) for the rates.
        nlo=nmin
        DO 70 M=1,IC                                              ACXI1096
        N=MVAL(M)                                                 ACXI1097
        AN=N                                                      ACXI1098
        TOT=0.D0                                                  ACXI1099
        L=N-1                                                     ACXI1100
c       DO 30 KK=20,L                                             ACXI1101
        do 30 kk=nlo,l
        C=COLRAT(KK,N,T,TE12)                                     ACXI1102
   30   TOT=TOT+C                                                 ACXI1103
        L=N+1                                                     ACXI1104
        MAX=N+40                                                  ACXI1105
        DO 60 KK=L,MAX                                            ACXI1106
        AKK=KK                                                    ACXI1107
        C=COLRAT(N,KK,T,TE12)                                     ACXI1108
        CX=CXP(KK)                                                ACXI1109
        IF (CX.LT.1.D-30) GO TO 40                                ACXI1110
        CXN=CXP(N)                                                ACXI1111
        IF (CXN.LT.1.D-30) GO TO 40                               ACXI1112
        Q=CXN/CX                                                  ACXI1113
        GO TO 50                                                  ACXI1114
   40   Q=DEXP(15.778D4*(1.D0/AKK**2-1.D0/AN**2)/T)              ACXI1115
   50   C=C*(AKK/AN)**2*Q                                         ACXI1116
   60   TOT=TOT+C                                                 ACXI1117
   70   COLTOT(M)=TOT                                             ACXI1118
        RETURN                                                    ACXI1119
        END                                                       ACXI1120

        SUBROUTINE RECOMB (Z,ETEMP,N1,ALPHA)                     ACXI1122
C                                                                 ACXI1123
C COMPUTES RECOMBINATION COEFFICIENT, ALPHA, ONTO LEVEL N1  FOR IONS  ACXI1124
C OF EFFECTIVE CHARGE  Z  AT ELECTRON TEMPERATURE  ETEMP          ACXI1125
C                                                                 ACXI1126
        COMMON /RCMB/ SV0(99),SV1(99),SV2(99)                    ACXI1127
C  SV0, SV1, SV2 NOT DOUBLE PRECISION                            ACXI1128
```

```
      DIMENSION XV(99)                                               ACXI1129
      DOUBLE PRECISION DABS,DSQRT,DBLE,DLOG,DLOG10,DFLOAT,DEXP        ACXI1130
      DOUBLE PRECISION ALPHA,CONST,ETEMP,F,FL,P,Q,S0,S1,S2,TE,U,V,X,XVACXI1131
     1VI,XVI1,XXX,Y,Z2,Z                                             ACXI1132
      Z2=Z*Z                                                         ACXI1133
      TE=ETEMP*1.0D-4                                                ACXI1134
      CONST=15.778D0/TE                                              ACXI1135
      FL=1.D0/(CONST*Z2)**.33333333333333D0                          ACXI1136
      CONST=5.197D-14*Z2*DSQRT(CONST)                                ACXI1137
      XV(1)=.02D0                                                    ACXI1138
      DO 10 N=2,11                                                   ACXI1139
   10 XV(N)=XV(N-1)+.002D0                                           ACXI1140
      DO 20 N=12,23                                                  ACXI1141
   20 XV(N)=XV(N-1)+.005D0                                           ACXI1142
      DO 30 N=24,33                                                  ACXI1143
   30 XV(N)=XV(N-1)+.01D0                                            ACXI1144
      DO 40 N=34,65                                                  ACXI1145
   40 XV(N)=XV(N-32)*10.D0                                           ACXI1146
      DO 50 N=66,97                                                  ACXI1147
   50 XV(N)=XV(N-32)*10.D0                                           ACXI1148
      F=N1                                                           ACXI1149
      X=15.778D0*Z2/(F*F*TE)                                         ACXI1150
      IF (X.LT.0.02D0) GO TO 80                                      ACXI1151
      IF (X.GT.20.D0) GO TO 70                                       ACXI1152
      DO 60 I=2,99                                                   ACXI1153
      XVI=XV(I)                                                      ACXI1154
      IF (X.GT.XVI) GO TO 60                                         ACXI1155
      IM1=I-1                                                        ACXI1156
      XVI1=XV(IM1)                                                   ACXI1157
      P=1.D0/(XVI-XVI1)                                              ACXI1158
      Q=(XVI-X)*P                                                    ACXI1159
      P=(X-XVI1)*P                                                   ACXI1160
      S0=P*SV0(I)+Q*SV0(IM1)                                         ACXI1161
      S1=P*SV1(I)+Q*SV1(IM1)                                         ACXI1162
      S2=P*SV2(I)+Q*SV2(IM1)                                         ACXI1163
      GO TO 90                                                       ACXI1164
   60 CONTINUE                                                       ACXI1165
   70 U=1.D0/X                                                       ACXI1166
      V=U/3.D0                                                       ACXI1167
      XXX=X**.333333333333333D0                                      ACXI1168
      S0=1.D0-U*(1.D0-2.D0*U*(1.D0-3.D0*U*(1.D0-4.D0*U)))            ACXI1169
      S1=-.1728D0*XXX*(1.D0-V*(8.D0-V*(70.D0-V*(800.D0-V*11440.D0))))ACXI1170
      S2=-.0496D0*XXX**2*(1.D0-V*(3.D0-V*(32.D0-V*448.D0)))          ACXI1171
      GO TO 90                                                       ACXI1172
   80 S0=X*DEXP(X)*(-DLOG(X)-.5772D0+X)                              ACXI1173
      XXX=X**.333333333333333D0                                      ACXI1174
      S1=.4629D0*X*(1.D0+4.D0*X)-1.0368D0*XXX**4*(1.D0+1.875D0*X)    ACXI1175
      S2=-.0672D0*X*(1.D0+3.D0*X)+.1488D0*XXX**5*(1.D0+1.8D0*X)      ACXI1176
   90 Y=(S0+FL*(S1+FL*S2))/F                                         ACXI1177
      ALPHA=CONST*Y                                                  ACXI1178
      RETURN                                                         ACXI1179
      END                                                            ACXI1180

      SUBROUTINE REDUCE (M,IC,IR,SK)                                 ACXI1182
C                                                                    ACXI1183
C     GIVEN A SET OF INTEGERS                                        ACXI1184
C         M(IT),IT=1,IC,SUCH THAT-                                   ACXI1185
```

```
C          1) M(IT+1)=M(IT)+1  FOR IT.LE.IA              ACXI1186
C          WHERE IA.GE.1  AND                            ACXI1187
C          2) (M(IT+1) - M(IT)).GT.1 FOR IT.GE.IA,       ACXI1188
C      AND GIVEN A FUNCTION SUBPROGRAM  BK               ACXI1189
C          WHICH CALCULATES THE                          ACXI1190
C          ELEMENTS OF A LARGE                           ACXI1191
C          M(IC)*M(IC) MATRIX,                           ACXI1192
C      THIS SUBROUTINE USES LAGRANGE                     ACXI1193
C          INTERPOLATION OF ORDER                        ACXI1194
C          2*(IR+1) TO CALCULATE A                       ACXI1195
C          SMALLER IC*IC MATRIX SK                       ACXI1196
C      REQUIRES A FUNCTION SUBPROGRAM                    ACXI1197
C          PHI                                           ACXI1198
C      IR MUST BE .LE. (IA-1)                            ACXI1199
C                                                        ACXI1200
       DIMENSION IQ(8), M(75), SK(75,75)                 ACXI1201
       DIMENSION STORE1(8), STORE2(8)                    ACXI1202
       DOUBLE PRECISION DABS,DSQRT,DBLE,DLOG,DLOG10,DFLOAT,DEXP   ACXI1203
       DOUBLE PRECISION BK,DUCKIT,FL,PHI,PHITAU,SK,STORE1,STORE2  ACXI1204
C                                                        ACXI1205
       LG=2*(IR+1)                                       ACXI1206
       IB=IC-IR                                          ACXI1207
       IBB=IB-1                                          ACXI1208
       ICC=IC-1                                          ACXI1209
C                                                        ACXI1210
       DO 10 IS=1,IC                                     ACXI1211
       DO 10 IT=1,IC                                     ACXI1212
   10  SK(IS,IT)=BK(M(IS),M(IT),IS)                      ACXI1213
C                                                        ACXI1214
       DO 20 IT=1,IC                                     ACXI1215
       IA=IT                                             ACXI1216
       IF ((M(IT+1)-M(IT)).GT.1) GO TO 30                ACXI1217
   20  CONTINUE                                          ACXI1218
C                                                        ACXI1219
   30  IF (IA.EQ.IC) GO TO 110                           ACXI1220
       IF (IA.EQ.IB) GO TO 80                            ACXI1221
C                                                        ACXI1222
       DO 70 IT=IA,IBB                                   ACXI1223
       N1=M(IT)+1                                        ACXI1224
       N2=M(IT+1)-1                                      ACXI1225
       DO 40 ITAU=1,LG                                   ACXI1226
       IND=IT-IR-1+ITAU                                  ACXI1227
   40  IQ(ITAU)=M(IND)                                   ACXI1228
       DO 50 ITAU=1,LG                                   ACXI1229
   50  STORE1(ITAU)=PHI(IQ,LG,ITAU,IQ(ITAU))             ACXI1230
       DO 70 N=N1,N2                                     ACXI1231
       DO 60 ITAU=1,LG                                   ACXI1232
   60  STORE2(ITAU)=PHI(IQ,LG,ITAU,N)                    ACXI1233
       DO 70 IS=1,IC                                     ACXI1234
       DUCKIT=BK(M(IS),N,IS)                             ACXI1235
       DO 70 ITAU=1,LG                                   ACXI1236
       FL=STORE2(ITAU)/STORE1(ITAU)                      ACXI1237
       IND=IT-IR-1+ITAU                                  ACXI1238
   70  SK(IS,IND)=SK(IS,IND)+DUCKIT*FL                   ACXI1239
C                                                        ACXI1240
   80  IF (IR.EQ.0) GO TO 110                            ACXI1241
C                                                        ACXI1242
```

```
      DO 90 ITAU=1,LG                                              ACXI1243
      IND=IC-LG+ITAU                                               ACXI1244
 90   IQ(ITAU)=M(IND)                                              ACXI1245
      DO 100 ITAU=1,LG                                             ACXI1246
      PHITAU=1.D0/PHI(IQ,LG,ITAU,IQ(ITAU))                         ACXI1247
      DO 100 IT=IB,ICC                                             ACXI1248
      N1=M(IT)+1                                                   ACXI1249
      N2=M(IT+1)-1                                                 ACXI1250
      DO 100 N=N1,N2                                               ACXI1251
      FL=PHI(IQ,LG,ITAU,N)*PHITAU                                  ACXI1252
      DO 100 IS=1,IC                                               ACXI1253
      IND=IC-LG+ITAU                                               ACXI1254
 100  SK(IS,IND)=SK(IS,IND)+BK(M(IS),N,IS)*FL                      ACXI1255
C                                                                  ACXI1256
 110  RETURN                                                       ACXI1257
      END                                                          ACXI1258

      SUBROUTINE RHS (CO,MVAL,IC)                                  ACXI1260
C
ACXI1261
C COMPUTES THE RIGHT HAND SIDE OF EQUATIONS (2.7) OF BROCKLEHURST, ACXI1262
C MNRAS 148, 417 (1970).                                          ACXI1263
C                                                                  ACXI1264
      COMMON /EXPDAT/ CXP(707),MAXN                                ACXI1265
      COMMON /PARMS/ DENS,T,ITM                                    ACXI1266
      COMMON /TDEP/ TE32,TE12,CTE                                  ACXI1267
      DIMENSION MVAL(75), CO(75)                                   ACXI1268
      DOUBLE PRECISION DABS,DSQRT,DBLE,DLOG,DLOG10,DFLOAT,DEXP     ACXI1269
      DOUBLE PRECISION ALFA,CO,CTE,CXP,DENS,RT,T,TE12,TE32         ACXI1270
      DO 10 I=1,IC                                                 ACXI1271
      J=MVAL(I)                                                    ACXI1272
      CALL COLION (J,1,T,RT)                                       ACXI1273
      CALL RECOMB (1.D0,T,J,ALFA)                                  ACXI1274
      CO(I)=-ALFA*CXP(J)*TE32*0.24146879D16/DFLOAT(J*J)-RT*DENS    ACXI1275
 10   CONTINUE                                                     ACXI1276
      RETURN                                                       ACXI1277
      END                                                          ACXI1278
```

```
C-----------------------------------File: DATAIN.TXT---------------------
C Pivot Points for calculation
 75  2  4                                                         ACXI0017
    5  6  7  8  9 10 11 12 13 14 15 17 19 21 23 26 29 32 35 38 41 44 48
   52 56 60 64 68 72 77 82 87 92 97102107112118124130136142148154160167
  174181188195202210218226234242250259268277286295305315325335345355366
  377388399410421432
   75 72 69 66                                                    ACXI0022
  +0.000E+00+0.000E+00 +4.000E+03+1.000E+060000020000100001000000099
  +6.000E+03+1.000E+060000020000100001000000099
```

```
C-----------------------------------File:DATAOUT.TXT---------------------
C Pivot Points from DATAIN.TXT
 ***********************************************************************
```

```
  MVAL (75 VALUES) -
    5    6    7    8    9   10   11   12   13   14   15   17
```

```
 19    21    23    26    29    32    35    38    41    44    48    52
 56    60    64    68    72    77    82    87  . 92    97   102   107
112   118   124   130   136   142   148   154   160   167   174   181
188   195   202   210   218   226   234   242   250   259   268   277
286   295   305   315   325   335   345   355   366   377   388   399
410   421   432
IVAL (4 VALUES) -
 75    72    69    66
IR = 2


NO BACKGROUND FIELD.
```

```
1 TEMPERATURE = 4000. K,   DENSITY = 1.000E+06CM**-3, NMIN =   2
(Case B)
```

PAGE 1

N	BN	bsBETA	DBN/DN	D(LN(BN))/DN	1-bsBETA	ZETA
5	6.55035E-02	1.0000	.00000	.00000	.00000	-7.93295E-03
6	.11023	1.0000	.00000	.00000	.00000	-8.24342E-03
7	.15474	1.0000	.00000	.00000	.00000	-8.65293E-03
8	.19628	1.0000	.00000	.00000	.00000	-9.08830E-03
9	.23408	1.0000	.00000	.00000	.00000	-9.52348E-03
10	.26830	1.0000	.00000	.00000	.00000	-9.95098E-03
11	.29930	1.0000	.00000	.00000	.00000	-1.03663E-02
12	.32752	1.0000	.00000	.00000	.00000	-1.07681E-02
13	.35354	1.0000	.00000	.00000	.00000	-1.11620E-02
14	.37873	1.0000	.00000	.00000	.00000	-1.15789E-02
15	.40072	1.0000	.00000	.00000	.00000	-1.19375E-02
16	.42355	-1.8973	2.36537E-02	5.58460E-02	2.8973	2.34374E-02
17	.44784	-2.4351	2.47209E-02	5.52007E-02	3.4351	3.12501E-02
18	.47186	-2.8289	2.44580E-02	5.18332E-02	3.8289	3.76914E-02
19	.49712	-3.5877	2.62511E-02	5.28065E-02	4.5877	4.97360E-02
20	.52490	-4.6127	2.90746E-02	5.53902E-02	5.6127	6.68038E-02
21	.55530	-5.6955	3.16957E-02	5.70787E-02	6.6955	8.64650E-02
22	.58829	-6.8003	3.40242E-02	5.78357E-02	7.8003	.10851
23	.62304	-7.7302	3.52945E-02	5.66491E-02	8.7302	.12973
24	.65839	-8.4010	3.53487E-02	5.36899E-02	9.4010	.14808
25	.69340	-8.8514	3.45154E-02	4.97770E-02	9.8514	.16343
26	.72711	-9.0309	3.27622E-02	4.50581E-02	10.031	.17403
27	.75844	-8.8901	3.00873E-02	3.96700E-02	9.8901	.17794
28	.78718	-8.6640	2.73595E-02	3.47564E-02	9.6640	.17930
29	.81311	-8.3084	2.45010E-02	3.01323E-02	9.3084	.17700
30	.83617	-7.8567	2.16548E-02	2.58975E-02	8.8567	.17160
31	.85648	-7.3654	1.89875E-02	2.21693E-02	8.3654	.16431
32	.87423	-6.8625	1.65611E-02	1.89436E-02	7.8625	.15588
33	.88975	-6.3968	1.44585E-02	1.62500E-02	7.3968	.14754
34	.90323	-5.9100	1.25369E-02	1.38801E-02	6.9100	.13808
35	.91491	-5.4497	1.08659E-02	1.18764E-02	6.4497	.12873
36	.92507	-5.0404	9.45546E-03	1.02213E-02	6.0404	.12017
37	.93388	-4.6362	8.20399E-03	8.78480E-03	5.6362	.11141

38	.94154	-4.2611	7.12709E-03	7.56964E-03	5.2611	.10308
39	.94820	-3.9202	6.20930E-03	6.54853E-03	4.9202	9.53702E-02
40	.95400	-3.5991	5.41246E-03	5.67345E-03	4.5991	8.79815E-02
41	.95906	-3.3023	4.72659E-03	4.92836E-03	4.3023	8.10577E-02
42	.96348	-3.0278	4.13539E-03	4.29212E-03	4.0278	7.45800E-02
43	.96736	-2.7755	3.62663E-03	3.74901E-03	3.7755	6.85689E-02
44	.97076	-2.5425	3.18728E-03	3.28329E-03	3.5425	6.29751E-02
45	.97375	-2.3236	2.80393E-03	2.87952E-03	3.3236	5.76772E-02
46	.97638	-2.1253	2.47507E-03	2.53494E-03	3.1253	5.28533E-02
47	.97871	-1.9425	2.18990E-03	2.23753E-03	2.9425	4.83840E-02
48	.98078	-1.7731	1.94159E-03	1.97965E-03	2.7731	4.42252E-02
49	.98260	-1.6110	1.72169E-03	1.75217E-03	2.6110	4.02307E-02

1 TEMPERATURE = 4000. K, DENSITY = 1.000E+06CM**-3, NMIN =   2

(Case B)

PAGE 2

N	BN	bsBETA	DBN/DN	D(LN(BN))/DN	1-bsBETA	ZETA
50	.98423	-1.4662	1.53306E-03	1.55763E-03	2.4662	3.66498E-02
51	.98568	-1.3320	1.36807E-03	1.38795E-03	2.3320	3.33252E-02
52	.98697	-1.2070	1.22307E-03	1.23922E-03	2.2070	3.02200E-02
53	.98813	-1.0862	1.09319E-03	1.10632E-03	2.0862	2.72122E-02
54	.98916	-.97829	9.81137E-04	9.91886E-04	1.9783	2.45218E-02
55	.99009	-.87787	8.82279E-04	8.91106E-04	1.8779	2.20145E-02
56	.99093	-.78383	7.94668E-04	8.01940E-04	1.7838	1.96639E-02
57	.99168	-.69278	7.15652E-04	7.21652E-04	1.6928	1.73853E-02
58	.99237	-.61081	6.46821E-04	6.51797E-04	1.6108	1.53324E-02
59	.99298	-.53419	5.85621E-04	5.89761E-04	1.5342	1.34122E-02
60	.99354	-.46221	5.30999E-04	5.34452E-04	1.4622	1.16071E-02
61	.99404	-.39279	4.81564E-04	4.84449E-04	1.3928	9.86531E-03
62	.99450	-.32948	4.37991E-04	4.40412E-04	1.3295	8.27622E-03
63	.99492	-.27000	3.98956E-04	4.00992E-04	1.2700	6.78291E-03
64	.99530	-.21388	3.63868E-04	3.65585E-04	1.2139	5.37343E-03
65	.99565	-.15975	3.31958E-04	3.33409E-04	1.1598	4.01370E-03
66	.99597	-.10991	3.03567E-04	3.04796E-04	1.1099	2.76166E-03
67	.99626	-6.28759E-02	2.77960E-04	2.79004E-04	1.0629	1.57984E-03
68	.99652	-1.83102E-02	2.54797E-04	2.55685E-04	1.0183	4.60073E-04
69	.99677	2.47899E-02	2.33613E-04	2.34371E-04	.97521	-6.22883E-04
70	.99699	6.46838E-02	2.14639E-04	2.15286E-04	.93532	-1.62526E-03
71	.99720	.10245	1.97432E-04	1.97987E-04	.89755	-2.57408E-03
72	.99739	.13830	1.81792E-04	1.82268E-04	.86170	-3.47486E-03
73	.99756	.17318	1.67392E-04	1.67801E-04	.82682	-4.35106E-03
74	.99772	.20552	1.54437E-04	1.54790E-04	.79448	-5.16325E-03
75	.99787	.23618	1.42638E-04	1.42942E-04	.76382	-5.93339E-03
76	.99801	.26533	1.31868E-04	1.32131E-04	.73467	-6.66532E-03
77	.99814	.29311	1.22018E-04	1.22246E-04	.70689	-7.36279E-03
78	.99825	.32023	1.12894E-04	1.13091E-04	.67977	-8.04382E-03
79	.99836	.34539	1.04650E-04	1.04822E-04	.65461	-8.67533E-03
80	.99846	.36934	9.70984E-05	9.72479E-05	.63066	-9.27620E-03
81	.99856	.39217	9.01680E-05	9.02983E-05	.60783	-9.84911E-03

82	.99864	.41400	8.37956E-05 8.39095E-05	.58600	-1.03967E-02
83	.99872	.43534	7.78671E-05 7.79666E-05	.56466	-1.09319E-02
84	.99880	.45521	7.24808E-05 7.25680E-05	.54479	-1.14303E-02
85	.99887	.47417	6.75224E-05 6.75989E-05	.52583	-1.19057E-02
86	.99893	.49230	6.29507E-05 6.30179E-05	.50770	-1.23601E-02
87	.99899	.50966	5.87287E-05 5.87878E-05	.49034	-1.27953E-02
88	.99905	.52661	5.47911E-05 5.48431E-05	.47339	-1.32200E-02
89	.99910	.54249	5.11905E-05 5.12364E-05	.45751	-1.36179E-02
90	.99915	.55769	4.78616E-05 4.79021E-05	.44231	-1.39984E-02
91	.99920	.57224	4.47798E-05 4.48157E-05	.42776	-1.43628E-02
92	.99924	.58620	4.19229E-05 4.19546E-05	.41380	-1.47123E-02
93	.99928	.59979	3.92532E-05 3.92813E-05	.40021	-1.50526E-02
94	.99932	.61262	3.67973E-05 3.68223E-05	.38738	-1.53735E-02

1 TEMPERATURE = 4000. K, DENSITY = 1.000E+06CM**-3, NMIN =   2
(Case B)

PAGE   3

N	BN	bsBETA	DBN/DN	D(LN(BN))/DN	1-bsBETA	ZETA
95	.99936	.62490	3.45180E-05	3.45402E-05	.37510	-1.56809E-02
96	.99939	.63670	3.24000E-05	3.24197E-05	.36330	-1.59759E-02
97	.99942	.64803	3.04296E-05	3.04472E-05	.35197	-1.62593E-02
98	.99945	.65905	2.85845E-05	2.86002E-05	.34095	-1.65349E-02
99	.99948	.66949	2.68785E-05	2.68924E-05	.33051	-1.67960E-02

1 TEMPERATURE = 6000. K, DENSITY = 1.000E+06CM**-3, NMIN =   2
(Case B)

PAGE   1

N	BN	bsBETA	DBN/DN	D(LN(BN))/DN	1-bsBETA	ZETA
5	.14001	1.0000	.00000	.00000	.00000	-6.68090E-03
6	.19784	1.0000	.00000	.00000	.00000	-6.84550E-03
7	.24882	1.0000	.00000	.00000	.00000	-7.09266E-03
8	.29287	1.0000	.00000	.00000	.00000	-7.36148E-03
9	.33108	1.0000	.00000	.00000	.00000	-7.63444E-03
10	.36435	1.0000	.00000	.00000	.00000	-7.89897E-03
11	.39387	1.0000	.00000	.00000	.00000	-8.15795E-03
12	.41992	1.0000	.00000	.00000	.00000	-8.40096E-03
13	.44398	1.0000	.00000	.00000	.00000	-8.64554E-03
14	.46613	1.0000	.00000	.00000	.00000	-8.88430E-03
15	.48708	1.0000	.00000	.00000	.00000	-9.12443E-03
16	.50742	-2.1042	2.02406E-02	3.98894E-02	3.1042	1.97207E-02
17	.52777	-2.6388	2.05739E-02	3.89827E-02	3.6388	2.54225E-02
18	.54874	-3.3472	2.15287E-02	3.92330E-02	4.3472	3.32005E-02
19	.57105	-4.2895	2.31787E-02	4.05899E-02	5.2895	4.39103E-02
20	.59529	-5.4528	2.52727E-02	4.24543E-02	6.4528	5.77766E-02
21	.62160	-6.7344	2.73239E-02	4.39573E-02	7.7344	7.40556E-02
22	.64992	-8.0594	2.91042E-02	4.47809E-02	9.0594	9.21745E-02
23	.67955	-9.2067	3.00044E-02	4.41533E-02	10.207	.10959
24	.70953	-10.071	2.99080E-02	4.21520E-02	11.071	.12466
25	.73910	-10.690	2.91033E-02	3.93768E-02	11.690	.13734
26	.76749	-10.995	2.75689E-02	3.59208E-02	11.995	.14622
27	.79385	-10.923	2.53107E-02	3.18833E-02	11.923	.14983

28	.81803	-10.737	2.30205E-02 2.81413E-02	11.737	.15138
29	.83986	-10.387	2.06393E-02 2.45746E-02	11.387	.15002
30	.85931	-9.9155	1.82850E-02 2.12786E-02	10.916	.14622
31	.87648	-9.3788	1.60718E-02 1.83368E-02	10.379	.14081
32	.89152	-8.8152	1.40553E-02 1.57655E-02	9.8152	.13439
33	.90471	-8.2831	1.23005E-02 1.35961E-02	9.2831	.12795
34	.91619	-7.7176	1.06956E-02 1.16740E-02	8.7176	.12056
35	.92617	-7.1752	9.29497E-03 1.00359E-02	8.1752	.11316
36	.93487	-6.6863	8.10633E-03 8.67108E-03	7.6863	.10632
37	.94244	-6.1994	7.05023E-03 7.48086E-03	7.1994	9.92640E-02
38	.94902	-5.7433	6.13848E-03 6.46823E-03	6.7433	9.25119E-02
39	.95476	-5.3255	5.35869E-03 5.61260E-03	6.3255	8.62214E-02
40	.95977	-4.9293	4.68003E-03 4.87619E-03	5.9293	8.01559E-02
41	.96415	-4.5607	4.09438E-03 4.24661E-03	5.5607	7.44430E-02
42	.96799	-4.2184	3.58857E-03 3.70725E-03	5.2184	6.90783E-02
43	.97135	-3.9018	3.15203E-03 3.24499E-03	4.9018	6.40723E-02
44	.97431	-3.6083	2.77423E-03 2.84737E-03	4.6083	5.93949E-02
45	.97692	-3.3317	2.44422E-03 2.50198E-03	4.3317	5.49559E-02
46	.97921	-3.0799	2.16031E-03 2.20616E-03	4.0799	5.08933E-02
47	.98125	-2.8469	1.91364E-03 1.95021E-03	3.8469	4.71165E-02
48	.98305	-2.6304	1.69851E-03 1.72780E-03	3.6304	4.35918E-02
49	.98465	-2.4231	1.50793E-03 1.53143E-03	3.4231	4.02037E-02

1 TEMPERATURE = 6000. K, DENSITY = 1.000E+06CM**-3, NMIN =   2
(Case B)

PAGE 2

N	BN	bsBETA	DBN/DN	D(LN(BN))/DN	1-bsBETA	ZETA
50	.98608	-2.2369	1.34398E-03	1.36296E-03	3.2369	3.71517E-02
51	.98735	-2.0641	1.20037E-03	1.21576E-03	3.0641	3.43107E-02
52	.98848	-1.9026	1.07401E-03	1.08652E-03	2.9026	3.16512E-02
53	.98950	-1.7465	9.60794E-04	9.70992E-04	2.7465	2.90738E-02
54	.99041	-1.6066	8.62911E-04	8.71268E-04	2.6066	2.67598E-02
55	.99123	-1.4761	7.76463E-04	7.83335E-04	2.4761	2.45994E-02
56	.99196	-1.3538	6.99788E-04	7.05456E-04	2.3538	2.25708E-02
57	.99263	-1.2354	6.30641E-04	6.35324E-04	2.2354	2.06046E-02
58	.99323	-1.1285	5.70299E-04	5.74187E-04	2.1285	1.88278E-02
59	.99377	-1.0285	5.16606E-04	5.19844E-04	2.0285	1.71636E-02
60	.99426	-.93439	4.68654E-04	4.71358E-04	1.9344	1.55973E-02
61	.99471	-.84365	4.25251E-04	4.27513E-04	1.8436	1.40856E-02
62	.99512	-.76074	3.86950E-04	3.88649E-04	1.7607	1.27037E-02
63	.99548	-.68280	3.52619E-04	3.54218E-04	1.6828	1.14039E-02
64	.99582	-.60920	3.21745E-04	3.23095E-04	1.6092	1.01760E-02
65	.99613	-.53821	2.93664E-04	2.94805E-04	1.5382	8.99118E-03
66	.99641	-.47277	2.68660E-04	2.69628E-04	1.4728	7.89879E-03
67	.99667	-.41098	2.46098E-04	2.46921E-04	1.4110	6.86695E-03
68	.99690	-.35241	2.25682E-04	2.26383E-04	1.3524	5.88876E-03
69	.99712	-.29578	2.07010E-04	2.07608E-04	1.2958	4.94263E-03
70	.99732	-.24331	1.90274E-04	1.90785E-04	1.2433	4.06600E-03
71	.99750	-.19362	1.75091E-04	1.75530E-04	1.1936	3.23575E-03

72	.99767	-.14643	1.61286E-04	1.61663E-04	1.1464	2.44713E-03
73	.99782	-.10052	1.48574E-04	1.48898E-04	1.1005	1.67992E-03
74	.99797	-5.79291E-02	1.37132E-04	1.37411E-04	1.0579	9.68160E-04
75	.99810	-1.75243E-02	1.26705E-04	1.26947E-04	1.0175	2.92881E-04
76	.99822	2.08957E-02	1.17186E-04	1.17395E-04	.97910	-3.49228E-04
77	.99833	5.75251E-02	1.08476E-04	1.08657E-04	.94247	-9.61407E-04
78	.99844	9.33029E-02	1.00406E-04	1.00563E-04	.90670	-1.55934E-03
79	.99853	.12651	9.31113E-05	9.32480E-05	.87349	-2.11422E-03
80	.99862	.15812	8.64261E-05	8.65453E-05	.84188	-2.64246E-03
81	.99871	.18827	8.02885E-05	8.03926E-05	.81173	'-3.14641E-03
82	.99878	.21711	7.46431E-05	7.47340E-05	.78289	-3.62831E-03
83	.99886	.24532	6.93896E-05	6.94691E-05	.75468	-4.09954E-03
84	.99892	.27160	6.46141E-05	6.46838E-05	.72840	-4.53860E-03
85	.99898	.29668	6.02163E-05	6.02775E-05	.70332	-4.95771E-03
86	.99904	.32068	5.61599E-05	5.62137E-05	.67932	-5.35851E-03
87	.99910	.34367	5.24125E-05	5.24598E-05	.65633	-5.74254E-03
88	.99915	.36611	4.89166E-05	4.89583E-05	.63389	-6.11738E-03
89	.99920	.38716	4.57183E-05	4.57551E-05	.61284	-6.46887E-03
90	.99924	.40730	4.27603E-05	4.27928E-05	.59270	-6.80516E-03
91	.99928	.42659	4.00209E-05	4.00497E-05	.57341	-7.12738E-03
92	.99932	.44511	3.74805E-05	3.75060E-05	.55489	-7.43660E-03
93	.99936	.46315	3.51061E-05	3.51287E-05	.53685	-7.73774E-03
94	.99939	.48017	3.29208E-05	3.29409E-05	.51983	-8.02185E-03

1 TEMPERATURE = 6000. K, DENSITY = 1.000E+06CM**-3, NMIN =   2
(Case B)

PAGE 3

N	BN	bsBETA	DBN/DN	D(LN(BN))/DN	1-bsBETA	ZETA
95	.99942	.49649	3.08919E-05	3.09098E-05	.50351	-8.29422E-03
96	.99945	.51216	2.90060E-05	2.90219E-05	.48784	-8.55566E-03
97	.99948	.52722	2.72510E-05	2.72652E-05	.47278	-8.80694E-03
98	.99951	.54186	2.56072E-05	2.56199E-05	.45814	-9.05135E-03
99	.99953	.55575	2.40867E-05	2.40980E-05	.44425	-9.28306E-03

Appendix F
Observational Units

Radio telescopes use "antenna temperature," or T_A, as units of intensity. Gordon et al. (1992) describe the derivation of these units, relate them to the even more peculiar units of T_A^* and T_R^* used in millimeter wave astronomy, and relate all of these to the units of physics suitable for physical analyses. Below, we quote sections from that paper.

F.1 What Radio Telescopes Measure

The definition of "spectral flux density" from a source of specific intensity I_ν is

$$S_\nu \equiv \int_{\text{source}} I_\nu \, d\Omega, \tag{F.1}$$

and, in the radio range when $h\nu \ll kT$, we use the Rayleigh–Jeans approximation[1] for the specific intensity to obtain

$$S_\nu = \frac{2k}{\lambda^2} \int_{\text{source}} T(\theta, \phi, \nu) \, d\Omega, \tag{F.2}$$

where $T(\theta, \phi, \nu)$ is the equivalent temperature of a black body that radiates I_ν at the frequency ν in the direction (θ, ϕ). It parameterizes the specific intensity. For extragalactic molecular lines, astronomers report a somewhat different quantity, F, to characterize the flux density received in the line:

---

[1] Actually, $h\nu \approx kT$ for many observations in the millimeter wave range. In these cases, consider T to be an *effective radiation temperature*, i.e., a surrogate for a more complex expression (see (2) of Ulich and Haas (1976)). Because of the calibration techniques used with millimeter wave telescopes, expressions involving T still prove to be useful although the parameter may no longer be the radiation temperature.

$$F \equiv \int_{\text{line}} S_\nu \, d\nu \tag{F.3}$$

$$= \frac{2k}{\bar{\lambda}^2} \int_{\text{source}} \int_{\text{line}} T(\theta, \phi, \nu) \, d\nu \, d\Omega \tag{F.4}$$

a quantity that is the integral of spectral flux density over the width of the line. The parameter $\bar{\lambda}$ is the observed wavelength at the center of the spectral line.

In general, $T(\theta, \phi, \nu)$ is not a measured quantity. It varies within the telescope beam and cannot be observed directly. Furthermore, the units used by observers to report F for spectral lines are often in a telescope-dependent form that cannot be easily compared with observations made with other telescopes.

Kutner and Ulich (1981) have considered this problem in some detail. They concentrated upon correction of spectral observations for wide-angle scattering, stray radiation, and atmospheric extinction. In this chapter, we extend their work by considering the coupling of the antenna to sources of angular extent less than the beam so as to derive equations for reporting telescope-independent quantities.

F.2 How Radio Telescopes Measure

We consider below two circumstances. The first case deals with sources smaller than the main beam, i.e., where the source size can range from as small as a point to as large as the distance between the first "nulls" of the main beam. Most observations of spectral lines from external galaxies fall into this category. Therefore, this category is the principal subject of this chapter. The second case deals with objects of an angular size larger than the main beam, such as observations of galactic molecular clouds with millimeter wave telescopes of intermediate to large diameters (≥ 10 m).

F.2.1 Sources Smaller Than the Beam Size

In the commonly used "on–off" observing technique, we measure the direct product[2] of the telescope response and the source distribution over each point, (θ, ϕ), within the solid angle of the source, Ω_S. In this case, we require Ω_S to be smaller than the solid angle of the main beam, Ω_B. For simplification, we omit the "subscript" ν in T and in S although most quantities are functions of ν. Here, the measured antenna temperature T_A of a source with a

---

[2] Observations made with a moving beam involve a convolution of beam and source rather than a direct product.

brightness temperature distribution $T_R(\theta, \phi)$ observed with an antenna with a normalized beam $f(\theta, \phi)$ is (see, e.g., Baars (1973))

$$T_A = \frac{\eta_R}{\Omega_A} \int_{\text{source}} T_R(\theta, \phi) f(\theta, \phi) \, d\Omega, \tag{F.5}$$

where η_R is the radiation efficiency of the antenna accounting for ohmic losses, and $\Omega_A = \int_{4\pi} f(\theta, \phi) \, d\Omega$ is the solid angle of the antenna pattern. Normally, η_R is close to 1 for a well-designed telescope surface. Using the relationship

$$\frac{\eta_R}{\Omega_A} = \frac{G}{4\pi} = \frac{A}{\lambda^2}, \tag{F.6}$$

where G is the antenna gain, λ is the wavelength of the observations, and $A \equiv \eta_A \pi (D/2)^2$ is the effective area of the antenna of diameter D with an *aperture efficiency*, η_A, we obtain

$$T_A = \frac{A}{\lambda^2} T_R \int_{\text{source}} \psi(\theta, \phi) f(\theta, \phi) \, d\Omega, \tag{F.7}$$

where we have introduced the normalized source brightness distribution function, $\psi(\theta, \phi)$. The parameter T_R is the source brightness temperature at the position $(\theta, \phi) = (0, 0)$.

The substitution of (F.2) into (F.7) yields (Baars, 1973)

$$T_A = \frac{SA}{2k} \frac{1}{\Omega_S} \int_{\text{source}} \psi(\theta, \phi) f(\theta, \phi) \, d\Omega \tag{F.8}$$

$$= \frac{SA}{2k} \frac{\Omega_\Sigma}{\Omega_S}, \tag{F.9}$$

where we have defined the source solid angle

$$\Omega_S \equiv \int_{\text{source}} \psi(\theta, \phi) \, d\Omega \tag{F.10}$$

and the beam-weighted source solid angle

$$\Omega_\Sigma \equiv \int_{\text{source}} \psi(\theta, \phi) f(\theta, \phi) \, d\Omega. \tag{F.11}$$

The factor $K \equiv \Omega_S / \Omega_\Sigma$ corrects the measured antenna temperature for the weighting of the source distribution by the large antenna beam. Therefore, (F.9) gives the spectral flux density of a source smaller than the beam as

$$S = \frac{2k}{A} K T_A, \tag{F.12}$$

and the flux density received in a spectral line, given by (F.3), becomes

$$F = \frac{8k}{\pi D^2} \frac{K}{\eta_A} \int_{\text{line}} T_A \, d\nu, \qquad \Omega_B > \Omega_S, \tag{F.13}$$

in terms of observational units.

For sources with Gaussian or disk distributions, the correction factor K can be written explicitly as

$$K = \begin{cases} 1 + x^2 & \text{Gaussian source} \\ \frac{x^2}{1 - \exp(-x^2)} & \text{disk source} \quad x \leq 1, \end{cases} \tag{F.14}$$

where the quantity x is defined by

$$x = \begin{cases} \theta_S/\theta_B & \text{Gaussian source} \\ \sqrt{\ln 2} \, \theta_D/\theta_B & \text{disk source} \end{cases} \tag{F.15}$$

where θ_S and θ_B are the widths of the source and beam at half-intensity, respectively, and θ_D is the angular diameter of the disk source. Table F.1 tabulates K as a function of source size for both a Gaussian and disk source.

Note that the basic characteristic of the antenna required to evaluate (F.13) is the aperture efficiency η_A that can normally be accurately determined[3] from the observation of a point source ($K = 1$) or a small source of known size and brightness distribution such as a planet where K may be determined from Table F.1. As long as the source is smaller than the beam, there is no need to invoke the *beam efficiency*, defined as

$$\eta_B \equiv \frac{1}{\Omega_A} \int_{\text{mainbeam}} f(\theta, \phi) \, d\Omega = \Omega_B/\Omega_A, \tag{F.16}$$

which is more difficult to determine since the entire main beam shape must be measured.

Table F.1 Correction factor K

	K	
x	Gaussian	Disk
0.0	1.000	1.000
0.1	1.010	1.005
0.2	1.040	1.020
0.3	1.090	1.046
0.4	1.160	1.082
0.5	1.250	1.130
0.6	1.360	1.191
0.7	1.490	1.265
0.8	1.640	1.354
0.9	1.810	1.459
1.0	2.000	1.582

[3] The effect of an error pattern from an imperfect reflector is, in this case, only to decrease η_A – precisely the quantity that we measure by observations of a point source of known flux density.

From (F.6) and (F.16), we obtain

$$\eta_B = \frac{\pi \eta_A D^2 \Omega_B}{4 \eta_R \lambda^2}. \qquad (F.17)$$

Combining (F.12) and (F.17) and putting $\eta_R = 1$, we find

$$S = \frac{2k}{\lambda^2} \frac{1}{\eta_B} T_A K \Omega_B, \qquad (F.18)$$

which for a uniformly bright source that just fills the main beam ($\Omega_\Sigma = \Omega_B$) is reduced to the well-known relationship

$$S = \frac{1}{\eta_B} \frac{2k}{\lambda^2} T_A \Omega_S. \qquad (F.19)$$

If the beam shape is known, one can convert the antenna efficiency η_A into beam efficiency η_B using (F.17). For example, the solid angle of a symmetrical Gaussian beam is

$$\Omega_B = 1.133 \, \theta_B^2, \qquad (F.20)$$

where the full width at half-flux density is given by

$$\theta_B = F_t \frac{\lambda}{D}. \qquad (F.21)$$

For a quadratic illumination function, Table F.2 gives the taper factor F_t as a function of the edge taper.

Assuming a feed with a 12-db taper – a common illumination for parabolic reflectors used in radio astronomy – and substituting (F.20) and (F.21) into (F.17), we find

$$\eta_B = 1.2 \frac{\eta_A}{\eta_R}, \qquad (F.22)$$

where, in many cases, the radiation efficiency $\eta_R \approx 1$.

Table F.2 Taper factor F_t

Taper (db)	F_t
0	1.020
-8	1.115
-10	1.135
-12	1.155
-14	1.170
-16	1.186
-18	1.198
-20	1.208
-22	1.218
-24	1.227
$-\infty$	1.267

We now arrive at a telescope-independent expression for the line flux density F of a spectral line observed from a source of angular size less than the beam. Evaluating the factors in (F.13), we find

$$\frac{F}{[\mathrm{W\,m^{-2}}]} = 3.515 \times 10^{-23} \left(\frac{D}{[\mathrm{m}]}\right)^{-2} \frac{K}{\eta_A} \times$$

$$\int_{\mathrm{line}} \frac{T_A}{[\mathrm{K}]} \frac{d\nu}{[\mathrm{Hz}]}, \qquad \Omega_B > \Omega_S. \qquad (\mathrm{F.23})$$

To convert F to $(\mathrm{ergs\,s^{-1}\,cm^{-2}})$ or to $(\mathrm{Jy\,Hz})$, multiply by 10^3 or 10^{26}, respectively.

Note that F is independent of the telescope size. (F.9) shows the measured T_A to be proportional to $\eta_A D^2/K$, thus precisely canceling out the factor $K/(\eta_A D^2)$ in (F.23).

Sanders et al. (1991) have also considered this problem, but their (A6) and (A8) do not quite follow standard antenna theory. Furthermore, the denominator of (A11) is missing a factor of $2\ln 2 \approx 1.4$. Therefore, the numerical results in their resulting equations (A12)–(A15) that relate the observations to astrophysical quantities need to be multiplied by this factor.

F.2.2 Sources Larger Than the Beam Size

This case is more difficult. The random imperfections of most telescopes give rise to an error beam that is many times wider than the main diffraction beam of the telescope. Even weak radiation entering the error beam can contribute significantly to the resulting spectrum because of the large solid angle of the error beam. The coupling of the overall beam to the source region is often too complex to be corrected by a simple mathematical procedure. This situation is encountered when observing giant molecular clouds in our galaxy with a large millimeter wave antenna.

To obtain accurate measurements of the line flux density F from such extended sources, one needs a detailed knowledge of the antenna pattern out to an angle at least as large as that of the source *and* a detailed knowledge of the source brightness distribution so as to calculate the coupling of the antenna and source. The large-scale antenna pattern can be difficult to measure and the source distribution is, of course, usually unknown.

As a practical approach, we suggest the use of a quantity that we call the *effective beam efficiency*, η_B',

$$\eta_B'(\Theta) \equiv \int_\Theta f(\theta, \phi)\, d\Omega/\Omega_A, \qquad (\mathrm{F.24})$$

in which the integration is extended over a solid angle Θ equal to that of the source. So, if the source size is known, η'_B will be the best representation of the coupling of the beam to the source. If we also assume that the source is uniformly bright over its angle Θ, then the measured antenna temperature relates to the effective brightness temperature of the source by

$$T_A = \eta'_B T_R. \tag{F.25}$$

Alternatively, (F.19) is valid for this case if η'_B replaces η_B.

Most observers follow procedures described by Kutner and Ulich (1981), which describe observations in terms of the parameter T_R^* that corrects observations for all telescope-dependent parameters *except* the coupling of the antenna to the source brightness distribution. In terms used by Kutner and Ulich (1981), and under the assumption of a uniformly bright source,

$$\eta'_B = \eta_c \eta_s, \tag{F.26}$$

where η_c and η_s are their "coupling" and "extended source" efficiencies, respectively. Unfortunately, η_c generally cannot be measured and can be calculated only with simplifying assumptions.

In our approach, it is possible to estimate η'_B by observing a series of sources of different sizes using the planets (a few arcseconds to an arcminute) and the moon ($\approx 30'$). Interpolation between $1'$ and $30'$ could result in large errors. Extrapolation of η'_B beyond $30'$ could be determined using the complete forward beam efficiency (over 2π sr), which may be obtained from the standard "sky tips" used to measure the atmospheric extinction. Note that the contribution from the main beam, sidelobes, and error pattern are all present in η'_B. Thus, a direct measurement of η'_B is more accurate than any theoretical calculation. We repeat that such measurements require that the source size is known and that the brightness distribution is constant over the source.

If a reasonable estimate of $\eta'_B(\Theta)$ over a source size Θ is available, one could correct the measured antenna temperature at each point into a "main beam" value by multiplying by η_B/η'_B. The mapped source could then be processed as if it had been observed with a "clean beam" of efficiency η_B.

For extragalactic sources that are only a few times larger than the main beam of a well-behaved antenna, we recommend that observers map the source at $\theta_B/2$ (Nyquist sampling) intervals with respect to θ_B given by (F.21). Each measurement of line flux density can be corrected to telescope-independent quantities using expressions given in this chapter. The total line flux density for the source would then be the sum of these measurements. Although significant errors due to beam imperfections would still exist in the sum owing to radiation entering the sidelobes and error beam, the restriction to sources only a few times larger than the beam would minimize these contributions. Although the resulting total F would still overestimate the actual line flux density, we do not know of a better, alternative procedure.

F.2.3 Antenna Temperature Scale

Filled-aperture, centimeter wave telescopes calibrated by hot and cold loads placed in front of the receiver produce spectra in units of T_A described in this chapter. We assume that such observations have been corrected for atmospheric extinction, if present.

Unlike centimeter wave telescopes, millimeter wave telescopes use the atmosphere in calibration procedures involving choppers or vanes and produce spectra in intensity units of T_A^*, T_R^*, or derivatives thereof (see especially Kutner and Ulich (1981), Guilloteau (1988), and Downes (1989)). In effect, these units presume angular sizes for the emitting region. Using the definitions given by Kutner and Ulich (1981),[4] we find

$$T_A = T_A^* \eta_l \qquad (F.27)$$

and

$$T_A = T_R^* \eta_s, \qquad (F.28)$$

to relate millimeter wave intensity units to our unit T_A. Here, η_l is the "forward beam efficiency" and η_s is the "extended source efficiency" defined to be $\eta_l \eta_{fss}$, where η_{fss} is called the "forward spillover and scattering efficiency." The efficiency η_l results from a sky tip by extrapolation of the measured antenna temperature as a function of air mass to the point where the air mass is zero. The determination of η_s is less straightforward, because it involves a choice for the size of the "diffraction" beam as described by Kutner and Ulich (1981). Usually, η_s is measured by observations of the moon.

The NRAO 12-m telescope produces spectra in intensity units of T_R^*. The temperature scale of its spectra can be converted into our units by using an efficiency η_s of ≈ 0.64 for observations from 70 to 310 GHz and ≈ 0.59 for observations from 330 to 360 GHz (Jewell, 1990).

The IRAM 30-m telescope produces spectral intensities in a variety of units depending upon what the observer enters in the command SET EFFICIENCY of the observing program OBS. Entering the "forward efficiency" (η_l) produces spectra in units of T_A^*; entering the "extended efficiency" (η_s), T_R^*; and entering the "main beam efficiency," T_{mb}.

Table F.3 lists efficiencies that obtain for the IRAM 30-m telescope at this writing that have been taken from Thum (1986), Mauersberger et al. (1989), Baars et al. (1989), and Greve (1992). Depending upon which efficiency was

---

[4] Some of these definitions are clarifications of ones originally given by Ulich and Haas (1976).

Table F.3 Efficiencies for the IRAM 30-m telescope

Frequency (GHz)	η_l	η_s	η_{mb}
80–115	0.90	0.69	0.59
140–160	0.86	0.68	0.55
210–260	0.90	0.75	0.46
330–360	0.84	0.60	0.20

entered into OBS, either (F.27) or (F.28) may be used to convert spectral intensities taken with the IRAM 30-m telescope into the general units of T_A used in this chapter. In addition, if spectra are reported in units of main beam brightness temperature, one should multiply these intensities by η_{mb} to convert them into our traditional units of T_A.

Similar procedures should apply to the temperature scales used at other millimeter wave telescopes.

References

Abramowitz, M. and Stegun, I. A.: 1964, *Handbook of Mathematical Functions*, United States Government Printing Office, Washington, DC.

Adler, D. S., Wood, D. O. S. and Goss, W. M.: 1996, *Astrophys. J.* **471**, 871.

Afflerbach, A., Churchwell, E. B., Accord, J. M., Hofner, P., Kurtz, S. and DePree, C. G.: 1996, *Astrophys. J., Suppl. Ser.* **106**, 423.

Altenhoff, W. J., Mezger, P. G., Wendker, H. and Westerhout, G.: 1960, Meßprogramme bei der Wellenlänge 11 cm am 25m-Radioteleskop Stockert. IV. Die Durchmusterung der Milchstraße und die Quellen-Durchmusterung bei 2.7 GHz, *Technical Report 59*, Veröff. Univ. Sternwarte Bonn.

Altenhoff, W. J., Strittmatter, P. A. and Wendker, H. J.: 1981, *Astron. Astrophys.* **93**, 48.

Anantharamaiah, K. R.: 1986, *J. Astrophys. Astron.* **7**, 131.

Anantharamaiah, K. R., Erickson, W. C., Payne, H. E. and Kantharia, N. G.: 1994, *Astrophys. J.* **430**, 682.

Anantharamaiah, K. R., Erickson, W. C. and Radhakrishnan, V.: 1985, *Nature* **315**, 647.

Anantharamaiah, K. R. and Goss, W. M.: 1990, *in* M. A. Gordon and R. L. Sorochenko (eds), *Radio Recombination Lines: 25 Years of Investigation*, IAU Colloquium 125, Kluwer Academic Publishers, Dordrecht, p. 267.

Anantharamaiah, K. R., Goss, W. M. and Dewdney, P. E.: 1990, *in* M. A. Gordon and R. L. Sorochenko (eds), *Radio Recombination Lines: 25 Years of Investigation*, IAU Colloquium 125, Kluwer Academic Publishers, Dordrecht, p. 123.

Anantharamaiah, K. R. and Kantharia, N. G.: 1999, *in* A. R. Taylor, T. L. Landecker and G. Joncas (eds), *New Perspective on the Interstellar Medium*, Vol. CS-167, ASP Conf. Ser., Astron. Soc. Pacific, Provo, Utah, p. 123.

Anantharamaiah, K. R., Payne, H. E. and Erickson, W. C.: 1988, *Mon. Not. R. Astron. Soc.* **235**, 151.

Anantharamaiah, K. R., Viallefond, F., Mohan, N. R., Goss, W. M. and Zhao, J.-H.: 2000, *Astrophys. J.* **537**, 613.

Anantharamaiah, K. R., Zhao, J.-H., Goss, W. M. and Viallefond, F.: 1993, *Astrophys. J.* **419**, 585.

Ariskin, V. I., Alekseev, Y. I., Gladyshev, A. S., Lekht, E. E. and Rudnitskij, G. M.: 1979, *Astron. Tsirk.* **1049**, 1.

Ariskin, V. I., Kolotovkina, S. A., Lekht, E. E., Rudnitskij, G. M. and Sorochenko, R. L.: 1982, *Astron. Zh.* **59**, 38. English translation: 1982 Sov. Astron. – AJ, 26:23.

Audi, G. and Wapstra, A. H.: 1995, *Nucl. Phys. A* **4**, 409.

Baars, J. W. M.: 1973, *IEEE Trans. Ant. Prop.* **AP-21**, 461.

Baars, J. W. M., Güsten, R. and Schultz, A.: 1989, The 345 GHz cooled Schottky receiver at the 30-m telescope, *Technical report 71*, MPIfR-Division of mm-Technology.

Baker, J. G. and Menzel, D. H.: 1938, *Astrophys. J.* **88**, 52.

Balick, B., Gammon, R. H. and Hjellming, R. M.: 1974, *Proc. Astron. Soc. Pacific* **86**, 616.

Ball, J. A., Cesarsky, D., Goldberg, L. and Lilley, A. E.: 1970, *Astrophys. J.* **162**, L25.

Balser, D. S., Bania, T. M., Rood, R. T. and Wilson, T. L.: 1999, *Astrophys. J.* **510**, 759.

Bania, T. M.: 2001, Private communication to M. A. Gordon. Figure only. The data appear in Bania et al. 2000.

Bania, T. M., Rood, R. T. and Balser, D. S.: 2000, *in* L. da Silva, M. Spite and J. R. de Medeiros (eds), *The Light Elements and Their Evolution*, Vol. Proceedings of IAU Symposium 198, Astron. Soc. Pacific, Provo, Utah, p. 214.

Bania, T. M., Rood, R. T. and Balser, D. S.: 2002, *Nature* **415**, 54.

Becker, W. and Fenkart, R.: 1963, *Z. Ap* **56**, 257.

Beigman, I. I.: 1977, *Zh. Ehksp. Teor. Fiz.* **73**, 1729. English translation: 1977 Sov. Phys. – JETP, 46:908.

Bell, M. B.: 1997, *Publ. Astron. Soc. Pac.* **109**, 609.

Bell, M. B., Avery, L. W., Seaquist, E. R. and Vallée, J. P.: 2000, *Publ. Astron. Soc. Pac.* **112**, 1236.

Bell, M. B. and Seaquist, E. R.: 1977, *Astron. Astrophys.* **56**, 461.

Bell, M. B. and Seaquist, E. R.: 1978, *Astrophys. J.* **233**, 378.

Bell, M. B., Seaquist, E. R., Mebold, U., Reif, K. and Shaver, P. A.: 1984, *Astron. Astrophys.* **130**, 1.

Berger, P. S. and Simon, M.: 1972, *Astrophys. J.* **171**, 191.

Berulis, I. I., Smirnov, G. T. and Sorochenko, R. L.: 1975, *Pis'ma Astron. Zh.* **1**, 28. English translation: 1975 Sov. Astron. Lett. 1:187.

Berulis, I. I. and Sorochenko, R. L.: 1973, *Astron. Zh.* **50**, 270. English translation: 1977 Sov. Astron. – AJ, 17:179.

Berulis, I. I. and Sorochenko, R. L.: 1983, *in* N. V. Andreenko (ed.), *Proceedings of the XV All-Union Conference of Galactic and Extragalactic Radioastronomy*, Inst. Rad. Elect., Kharkov, p. 190.

Bethe, H. A. and Salpeter, E. E.: 1957, *Quantum Mechanics of One- and Two-Electron Systems*, Academic Press, Inc., New York.

Bieging, J. H., Goss, W. M. and Wilcots, E. M.: 1991, *Astrophys. J., Suppl. Ser.* **75**, 999.

Black, J. H. and Dalgarno, A.: 1973, *Astrophys. Lett.* **15**, 79.

Blake, D. H., Crutcher, R. M. and Watson, W. D.: 1980, *Nature* **287**, 707.

Blitz, L., Fich, M. and Stark, A. A.: 1982, *Astrophys. J., Suppl. Ser.* **49**, 183.

Bochkarev, N. G.: 1988, *Astron. Nachr.* **310**, 399.

Bohr, N.: 1913, *Phil. Mag.* **26**, 1.

Bohr, N.: 1914, *Fysisk Tidsskraft* **12**, 97.

Bohr, N.: 1923, *Proc. Phys. Soc.* **35**, 275.

Boreiko, R. T. and Clark, T. A.: 1986, *Astron. Astrophys.* **157**, 353.

Bracewell, R. M.: 1965, *The Fourier Transform and Its Applications*, first edn, McGraw-Hill, New York.

Brackett, F. S.: 1922, *Astrophys. J.* **56**, 154.

Brault, J. W. and Noyes, R. W.: 1983, *Astrophys. J.* **269**, L61.

Brocklehurst, M. and Leeman, S.: 1971, *Ap. Letters* **9**, 36.

Brocklehurst, M. and Salem, M.: 1977, *Computer Physics Communications* **13**, 39.

Brocklehurst, M. and Seaton, M. J.: 1972, *Mon. Not. R. Astron. Soc.* **157**, 179.

Brown, R. L. and Gómez-González, J.: 1975, *Astrophys. J.* **200**, 598.

Brown, R. L. and Lockman, F. J.: 1975, *Astrophys. J.* **200**, L155.

Burles, S., Nollett, K. M. and Turner, M. S.: 2001, *Astrophys. J.* **552**, L1.

Burton, W. B.: 1974, *in* G. L. Verschuur and K. I. Kellermann (eds), *Galactic and Extra-Galactic Radio Astronomy*, first edn, Springer-Verlag, New York, p. 85.

Burton, W. B. and Gordon, M. A.: 1978, *Astron. Astrophys.* **63**, 7.

Cameron, A. G. W.: 1973, *Space Sci. Rev.* **15**, 121.

Cane, H. V.: 1978, *Aust. J. Phys.* **31**, 561.

Casse, J. L. and Shaver, P. A.: 1977, *Astron. Astrophys.* **61**, 805.

Caswell, J. L. and Haynes, R. F.: 1987, *Astron. Astrophys.* **171**, 261.

Cesarsky, C. J. and Cesarsky, D. A.: 1971, *Astrophys. J.* **169**, 293.

Cesarsky, D. A. and Cesarsky, C. J.: 1973, *Astrophys. J.* **184**, 83.

Chaisson, E. J.: 1975, *Astrophys. J.* **197**, L65.

Chaisson, E. J. and Rodríguez, L. F.: 1977, *Astrophys. J.* **214**, L111.

Chandrasekhar, S.: 1950, *Radiative Transfer*, Clarendon Press, Oxford.

Chang, E. S.: 1984, *J. Phys. B* **17**, L11.

Chang, E. S. and Noyes, R. W.: 1983, *Astrophys. J.* **275**, L125.

Chapman, S. and Cowling, T. G.: 1960, *The Mathematical Theory of Non-Uniform Gases*, Cambridge University Press, Cambridge.

Churchwell, E.: 1971, *Astron. Astrophys.* **15**, 90.

Churchwell, E.: 1975, *in* T. L. Wilson and D. Downes (eds), *H*II *Regions and Related Topics*, Springer-Verlag, Berlin, p. 245.

Churchwell, E. B.: 1980, *in* P. A. Shaver (ed.), *Radio Recombination Lines*, Reidel, Dordrecht, p. 225.

Churchwell, E. B., Smith, L. F., Mathis, J., Mezger, P. G. and Huchtmeier, W.: 1978, *Astron. Astrophys.* **70**, 719.

Churchwell, E. B. and Walmsley, C. M.: 1975, *Astron. Astrophys.* **38**, 451.

Churchwell, E., Mezger, P. G. and Huchtmeier, W.: 1974, *Astron. Astrophys.* **32**, 283.

Churchwell, E., Terzian, Y. and Walmsley, C. M.: 1976, *Astron. Astrophys.* **48**, 331.

Clark, B. G.: 1965, *Astrophys. J.* **142**, 1398.

Clark, T. A., Naylor, D. A. and Davis, G. R.: 2000a, *Astron. Astrophys.* **357**, 757.

Clark, T. A., Naylor, D. A. and Davis, G. R.: 2000b, *Astron. Astrophys.* **361**, L60.

Cohen, M., Bieging, J. H., Dreher, J. W. and Welch, W. J.: 1985, *Astrophys. J.* **292**, 249.

Colley, D.: 1980, *Mon. Not. R. Astron. Soc.* **193**, 495.

Cox, P., Martín-Pintado, J., Bachiller, R., Bronfman, L., Cernicharo, J., Lyman, L. A. and Roelfsema, P. R.: 1995, *Astron. Astrophys.* **295**, L39.

Davies, R. D.: 1971, *Astrophys. J.* **163**, 479.

Davis, J. T. and Vaughan, J. M.: 1963, *Astrophys. J.* **137**, 1302.

Dickel, H. R., Goss, W. M. and Condon, G. R.: 1996, *Astrophys. J.* **373**, 158.

Dieter, N. H.: 1967, *Astrophys. J.* **150**, 435.

Downes, D.: 1989, *in* D. M. Alloin and J.-M. Mariotti (eds), *Difraction-Limited Imaging with Very Large Telescopes*, Kluwer, Dordrecht, p. 53.

Downes, D., Wilson, T. L., Bieging, J. and Wink, J.: 1980, *Astron. Astrophys. Suppl. Ser.* **40**, 379.

Dravskikh, A. F. and Dravskikh, Z. V.: 1969, *Astron. Zh.* **46**, 455. English translation: 1969 Sov. Astron.-A.J. 13:360.

Dravskikh, A. F., Dravskikh, Z. V., Kolbasov, V. A., Misezhnikov, G. S., Nikulin, D. E. and Shteinshleiger, V. B.: 1965, *Dok. Akad. Nauk SSSR* **163**, 332. English translation:1966, Sov. Phys. – Dokl.10:627.

Dravskikh, Z. V.: 1994, Personal letter to M. A. Gordon.

Dravskikh, Z. V.: 1996, Email message to M. A. Gordon.

Dravskikh, Z. V. and Dravskikh, A. F.: 1964, *Astron. Tsirk.* **282**, 2.

Dravskikh, Z. V. and Dravskikh, A. F.: 1967, *Astron. Zh.* **44**, 35.

Dravskikh, Z. V., Dravskikh, A. F. and Kolbasov, V. A.: 1966, *Trans. IAU* **XIIB**, 360. Paper presented by Y. Pariijski at the XII General Assembly of the IAU in Hamburg, Germany, 1964.

Dupree, A. K.: 1968, *Astrophys. J.* **152**, L125.

Dupree, A. K.: 1974, *Astrophys. J.* **187**, 25.

Egorova, T. M. and Ryzkov, N. F.: 1960, *Izv. Glavn. Astrofiz. Obs.* **21**, 140.

Elmegreen, B. G. and Lada, C. J.: 1977, *Astrophys. J.* **214**, 725.

Elmegreen, B. G. and Morris, M.: 1979, *Astrophys. J.* **229**, 593.

Emerson, D. T.: 2002, Private communication to M. A. Gordon.

Erickson, W. C., McConnel, D. and Anantharamaiah, K. R.: 1995, *Astrophys. J.* **454**, 125.

Ershov, A. A. and Berulis, I. I.: 1989, *Pis'ma Astron. Zh.* **15**, 413. English translation: 1989 Sov. Astron. Lett. 15:413.

Ershov, A. A., Iljsov, Y. P., Lekht, E. E., Smirnov, G. T., Solodkov, V. T. and Sorochenko, R. L.: 1984, *Pis'ma Astron. Zh.* **10**, 833. English translation: 1984 Sov. Astron. Lett. 10:348.

Ershov, A. A., Lekht, E. E., Rudnitskij, G. M. and Sorochenko, R. L.: 1982, *Pis'ma Astron. Zh.* **8**, 694. English translation: 1982 Sov. Astron. Lett. 8:374.

Ershov, A. A., Lekht, E. E., Smirnov, G. T. and Sorochenko, R. L.: 1987, *Pis'ma Astron. Zh.* **13**, 19. English translation: 1987 Sov. Astron. Lett. 13:8.

Evans, N. J., Munday, L. G., Davies, J. H. and Vanden Bout, P. A.: 1987, *Astrophys. J.* **312**, 344.

Ewen, H. J. and Purcell, E. M.: 1951, *Nature* **168**, 356.

Federman, S. R., Glassgold, A. E. and Kwan, J.: 1979, *Astrophys. J.* **227**, 466.

Ferguson, E. and Shlüter, H.: 1963, *Ann. Phys. (N.Y.)* **22**, 351.

Field, G. B.: 1974, *Astrophys. J.* **187**, 453.

Field, G. B., Goldsmith, D. W. and Habing, H. J.: 1969, *Astrophys. J.* **155**, L149.

Finn, G. D. and Mugglestone, D. E.: 1965, *Mon. Not. R. Astron. Soc.* **129**, 221.

Fowler, W. A., Reeves, H. and Silk, J.: 1970, *Astrophys. J.* **162**, 49.

Frey, A., Lemke, D., Fahrbach, U. and Thum, C.: 1979, *Astron. Astrophys.* **74**, 133.

Garay, G., Gathier, R. and Rodríguez, L. F.: 1989, *Astron. Astrophys.* **266**, 101.

Garay, G., Gómez, Y., Lisano, S. and Brown, R. L.: 1998, *Astrophys. J.* **501**, 699.

Garay, G., Lizano, S., Gómez, Y. and Brown, R. L.: 1998, *Astrophys. J.* **501**, 710.

Garay, G. and Rodríguez, L. F.: 1983, *Astrophys. J.* **266**, 263.

Gardner, F. F. and McGee, R. X.: 1967, *Nature* **213**, 480.

Gaume, R. A., Wilson, T. L. and Johnston, K. J.: 1994, *Astrophys. J.* **425**, 127.

Gee, C. S., Percival, I. C., Lodge, I. C. and Richards, D.: 1976, *Mon. Not. R. Astron. Soc.* **175**, 209.

Georgelin, Y. M. and Georgelin, Y. P.: 1976, *Astron. Astrophys.* **49**, 57.

Giannani, T., Nisino, B., Lorenzetti, D., DiGiorgio, A. M., Spinoglio, L., Benedettini, M., Saraceno, P., Smith, H. A. and White, G. J.: 2000, *Astron. Astrophys.* **358**, 310.

Goldberg, L.: 1966, *Astrophys. J.* **144**, 1225.

Goldberg, L.: 1968, *in* Y. Terzian (ed.), *Interstellar Ionized Hydrogen*, W. A. Benjamin, Inc., New York, p. 373.

Goldberg, L. and Dupree, A. K.: 1967, *Nature* **215**, 41.

Goldreich, P. and Kwan, J.: 1974, *Astrophys. J.* **189**, 441.

Goldwire, Jr., H. C.: 1968, *Astrophys. J., Suppl. Ser.* **17**(152), 445.

Goldwire, Jr., H. C. and Goss, W. M.: 1967, *Astrophys. J.* **149**, 15.

Gómez, Y., Garay, G. and Lizano, S.: 1995, *Astrophys. J.* **453**, 727.

Gómez, Y., Lebron, M., Rodríguez, L. F., Garay, G., Lizano, S., Escalante, V. and Canto, J.: 1998, *Astrophys. J.* **503**, 297.

Gómez, Y., Moran, J. M., Rodríguez, L. F. and Garay, G.: 1989, *Astrophys. J.* **345**, 862.

Gordon, M. A.: 1969, *Astrophys. J.* **158**, 479.

Gordon, M. A.: 1988, *in* G. L. Verschuur and K. I. Kellermann (eds), *Galactic and Extra-Galactic Radio Astronomy*, second edn, Springer-Verlag, New York, p. 37.

Gordon, M. A.: 1989, *Astrophys. J.* **337**, 782.

Gordon, M. A.: 1992, *Astrophys. J.* **387**, 701.

Gordon, M. A.: 1994, *Astrophys. J.* **421**, 314.

Gordon, M. A.: 2003, *Astrophys. J.* **589**, 953.

Gordon, M. A., Baars, J. W. M. and Cocke, W. J.: 1992, *Astron. Astrophys.* **264**, 337.

Gordon, M. A., Brown, R. L. and Gottesman, S. T.: 1972, *Astrophys. J.* **178**, 119.

Gordon, M. A. and Burton, W. B.: 1976, *Astrophys. J.* **208**, 346.

Gordon, M. A. and Cato, T.: 1972, *Astrophys. J.* **176**, 587.

Gordon, M. A. and Churchwell, E.: 1970, *Astron. Astrophys.* **9**, 307.

Gordon, M. A. and Gottesman, S. T.: 1971, *Astrophys. J.* **168**, 361.

Gordon, M. A., Holder, B. P., Jisonna, L. J., Jorgenson, R. A. and Strelnitski, V. S.: 2001, *Astrophys. J.* **559**, 402.

Gordon, M. A. and Meeks, M. L.: 1967, *Astrophys. J.* **149**, L21.

Gordon, M. A. and Walmsley, C. M.: 1990, *Astrophys. J.* **365**, 606.

Gordon, W.: 1929, *Ann. der Phys.* **2**, 1031.

Goss, W. M., Kalberla, P. M. W. and Dickel, H. R.: 1984, *Astron. Astrophys.* **139**, 317.

Gottesman, S. T. and Gordon, M. A.: 1970, *Astrophys. J.* **162**, L93.

Graf, U. U., Eckart, A., Genzel, R., Harris, A. I., Poglitsch, A., Russell, A. P. G. and Stutzki, J.: 1993, *Astrophys. J.* **405**, 249.

Greisen, E. W.: 1973, *Astrophys. J.* **184**, 363.

Greve, A.: 1975, *Sol. Phys.* **40**, 329.

Greve, A.: 1977, *Sol. Phys.* **52**, 417.

Greve, A.: 1992, Private communication to J. W. M. Baars.

Griem, H. R.: 1960, *Astrophys. J.* **149**, 883.

Griem, H. R.: 1967, *Astrophys. J.* **148**, 547.

Griem, H. R.: 1974, *Spectral line broadening by plasmas*, Vol. 39 of *Pure and applied physics*, Academic Press, New York.

Griem, H. R.: 2005, *Astrophys. J.* **609**, L133.

Griem, H. R., Kolb, A. C. and Shen, K. Y.: 1959, *Phys. Rev.* **116**, 4.

Gudnov, V. M. and Sorochenko, R. L.: 1967, *Astron. Zh.* **44**, 1001. English translation: 1968 Sov. Astron. – AJ, 11:805.

Guilloteau, S.: 1988, Spectral Line Calibration on the 30-m and 15-m Antennas, *Technical report*, IRAM.

Gulyaev, S. A. and Nefedov, S. A.: 1989, *Astron. Nachr.* **310**(5), 403.

Gulyaev, S. A. and Sorochenko, R. L.: 1974, *Astron. Zh.* **51**, 1237. English translation: 1974 Sov. Astron. – AJ, 18:737.

Gulyaev, S. A. and Sorochenko, R. L.: 1983, The Catalogue of Radio Recombination Lines, *Technical Report 145, 146, and 168*, Lebedev. Phys. Inst.

Gulyaev, S. A., Sorochenko, R. L. and Tsivilev, A. P.: 1997, *Pis'ma Astron. Zh.* **23**, 3. English translation: 1997 Astron. Lett. 23:165.

Hamann, F. and Simon, M.: 1986, *Astrophys. J.* **311**, 909.

Hamann, F. and Simon, M.: 1988, *Astrophys. J.* **327**, 876.

Hart, L. and Pedlar, A.: 1976, *Mon. Not. R. Astron. Soc.* **176**, 547.

Hart, L. and Pedlar, A.: 1980, *Mon. Not. R. Astron. Soc.* **193**, 781.

Harvey, P. M., Thronson, H. A. and Gatley, I.: 1979, *Astrophys. J.* **231**, 115.

Heiles, C.: 1994, *Astrophys. J.* **436**, 720.

Heiles, C., Reach, W. T. and Koo, B.-C.: 1996, *Astrophys. J.* **466**, 191.

Herrmann, F., Madden, S. C., Nikola, T., Poglitsch, A., Timmermann, R., Geis, N., Townes, C. H. and Stacey, G. J.: 1997, *Astrophys. J.* **481**, 343.

Hewish, A., Bell, S. J., Pilkington, J. D. H., Scott, P. F. and Collins, R. A.: 1968, *Nature* **217**, 709.

Hill, J. K.: 1977, *Astrophys. J.* **212**, 692.

Hjellming, R. M.: 1999, *in* A. N. Cox (ed.), *Allen's Astrophysical Quantities*, fourth edn, Springer Verlag, New York.

Hjellming, R. M., Gordon, C. P. and Gordon, K. J.: 1969, *Astron. Astrophys.* **2**, 202.

Hjellming, R. M., Wade, C. M., Vandenberg, N. R. and Newell, R. T.: 1979, *Astron. J.* **84**, 1619.

Hjerting, F.: 1938, *Astrophys. J.* **88**, 508.

Hoang-Binh, D.: 1968, *Astrophys. Lett.* **2**, 231.

Hoang-Binh, D.: 1972, *Mém. Roy. Soc. Liége* **3**, 367.

Hoang-Binh, D.: 1982, *Astron. Astrophys.* **112**, L3.

Hoang-Binh, D.: 1990, *Astron. Astrophys.* **238**, 449.

Hoang-Binh, D. and Walmsley, C. M.: 1974, *Astron. Astrophys.* **35**, 49.

Hodge, P. W. and Kennicutt, R. C.: 1983, *Astrophys. J.* **267**, 563.

Hogerheijde, M. R., Jansen, D. J. and van Dishoeck, E. F.: 1995, *Astron. Astrophys.* **294**, 792.

Höglund, B. and Mezger, P. G.: 1965, *Science* **150**, 339.

Hollenbach, D.: 2002, Private communication to M. A. Gordon.

Hollenbach, D., Johnstone, D., Lizano, S. and Shu, F.: 1994, *Astrophys. J.* **428**, 654.

Howe, J. E.: 1999, Private communication to R. L. Sorochenko.

Howe, J. E., Jaffe, D. T., Genzel, R. and Stacey, G. J.: 1991, *Astrophys. J.* **373**, 158.

Hughes, M. P., Thompson, A. R. and Colvin, R. S.: 1971, *Astrophys. J., Suppl. Ser.* **23**, 323.

Humason, M. L. and Wahlquist, H. D.: 1955, *Astron. J.* **60**, 254.

Humphreys, C. J.: 1953, *J. Res. Nat. Bur. Standards* **50**.

Hunten, D. M., Roach, F. E. and Chamberlain, J. W.: 1956, *J. Atmos. Terr. Phys.* **8**, 345.

Inglis, D. R. and Teller, E.: 1939, *Astrophys. J.* **90**, 439.

Innanen, K. A.: 1973, *Astrophys. Space. Sci.* **22**, 343.

Jackson, P. D. and Kerr, F. J.: 1971, *Astrophys. J.* **168**, 723.

Jackson, P. D. and Kerr, F. J.: 1975, *Astrophys. J.* **196**, 723.

Jaffe, D. T. and Pankonin, V.: 1978, *Astrophys. J.* **226**, 869.

Jaffe, D. T., Zhou, S., Howe, I. E. and Stacey, G. J.: 1994, *Astrophys. J.* **436**, 203.

Jefferies, J. T.: 1968, *Spectral Line Formation*, Blaisdell, Waltham.

Jewell, P. R.: 1990, NRAO 12m User's Manual, *Internal report*, National Radio Astronomy Observatory.

Kantharia, N. G. and Anantharamaiah, K. R.: 2000, *in* A. P. Rao and G. Swarup (eds), *The Universe at Low Radio Frequencies*, Vol. In Press, IAU Symposium 199, Astron. Soc. Pacific, Provo, Utah, p. 337.

Kantharia, N. G., Anantharamaiah, K. R. and Payne, H. E.: 1998, *Astrophys. J.* **506**, 758.

Kardashev, N. S.: 1959, *Astron. Zh.* **36**, 838. English translation: 1960 Sov. Astron. – AJ 3:813.

Kardashev, N. S.: 2002, Email to M. A. Gordon.

Kawamura, J. and Masson, C.: 1996, *Astrophys. J.* **461**, 282.

Khersonskii, V. K. and Varshalovich, D. A.: 1980, *Astron. Zh.* **57**, 621. English translation: 1980 Sov. Astron. – AJ. 24:359.

Kinman, T. D.: 1959, *Mon. Not. R. Astron. Soc.* **119**, 559.

Kitaev, V. V., Smirnov, G. T., Sorochenko, R. L. and Lekht, E. E.: 1994, *Turkish J. Physics* **18**, 908.

Knapp, G. R., Brown, R. L., Kuiper, T. B. N. and Kakar, R. K.: 1976, *Astrophys. J.* **204**, 781.

Kogan, V. I., Lisitsa, V. S. and Sholin, G. V.: 1987, *in* B. B. Kadomtcev (ed.), *Review of Plasma Physics*, Vol. 13, Consultant Bureau, New York, London, p. 261.

Konovalenko, A. A.: 1984, *Pis'ma Astron. Zh.* **10**, 846. English translation: 1984 Sov. Astron. Lett. 10:353.

Konovalenko, A. A.: 1990, *in* M. A. Gordon and R. L. Sorochenko (eds), *Radio Recombination Lines: 25 Years of Investigation*, IAU Colloquium 125, Kluwer Academic Publishers, Dordrecht, p. 175.

Konovalenko, A. A.: 1995, Private communication to R. L. Sorochenko.

Konovalenko, A. A. and Sodin, L. G.: 1979, *Pis'ma Astron. Zh.* **5**, 663. English translation: 1979 Sov. Astron. Lett., 5:355.

Konovalenko, A. A. and Sodin, L. G.: 1980, *Nature* **283**, 360.

Konovalenko, A. A. and Sodin, L. G.: 1981, *Nature* **294**, 135.

Koo, B.-C., Heiles, C. and Reach, W. T.: 1992, *Astrophys. J.* **390**, 108.

Krügel, E. and Tenorio-Tagle, G.: 1978, *Astron. Astrophys.* **70**, 51.

Krügel, E., Thum, C., Martín-Pintado, J. and Pankonin, V.: 1982, *Astron. Astrophys. Suppl. Ser.* **48**, 345.

Kurucz, R. L.: 1979, *Astrophys. J., Suppl. Ser.* **40**, 1.

Kutner, M. L. and Ulich, B. L.: 1981, *Astrophys. J.* **250**, 341.

Kwok, S.: 1999, *The Origin and Evolution of Planetary Nebulae*, number 31 in *Cambridge Astrophysics Series*, Cambridge University Press, Cambridge.

Lada, C. J.: 1987, *in* M. Peimbert and J. Jugaku (eds), *Star Formation, Proc. IAU Symp. 115*, IAU, Reidel, Dordrecht, p. 1.

Lamers, H. J. G. and Cassinelli, J. P.: 1999, *Introduction to Stellar Winds*, Cambridge University Press, Cambridge.

Lang, K. R. and Lord, S. D.: 1976, *Mon. Not. R. Astron. Soc.* **175**, 217.

Launay, I. M. and Roueff, E.: 1977a, *J. Phys. B* **10**, 879.

Launay, I. M. and Roueff, E.: 1977b, *Astron. Astrophys.* **56**, 289.

Lekht, E. E., Smirnov, G. T. and Sorochenko, R. L.: 1989, *Pis'ma Astron. Zh.* **15**, 396. English translation: 1989 Sov. Astron. Lett. 15:171.

Lichten, S. M., Rodríguez, L. F. and Chaisson, E. J.: 1979, *Astron. Astrophys.* **229**, 524.

Lilley, A. E., Menzel, D. H., Penfield, H. and Zuckerman, B.: 1966, *Nature* **209**, 468.

Lilley, A. E. and Palmer, P.: 1968, *Astrophys. J., Suppl. Ser.* **16**(144), 143.

Lilley, A. E., Palmer, P., Penfield, H. and Zuckerman, B.: 1966, *Nature* **211**, 174.

Lin, C. C. and Shu, F. H.: 1964, *Astrophys. J.* **140**, 646.

Linblad, P. and Langebartel, R. G.: 1953, *Stokholm Obs. Ann.* **17**, 6.

Lis, D. C., Keene, J. and Schilke, P.: 1997, *in* W. B. Latter, S. J. E. Radford, P. R. Jewell, J. G. Mangum and J. Bally (eds), *CO: 25 Years of Millimeter Wave Spectroscopy*, Kluwer, Dordrecht, p. 128.

Lockman, F. J.: 1976, *Astrophys. J.* **209**, 429.

Lockman, F. J.: 1980, *in* P. A. Shaver (ed.), *Radio Recombination Lines*, Riedel, Dordrecht, p. 185.

Lockman, F. J.: 1989, *Astrophys. J., Suppl. Ser.* **71**, 469.

Lockman, F. J.: 1990, *in* M. A. Gordon and R. L. Sorochenko (eds), *Radio Recombination Lines: 25 Years of Investigation*, IAU Colloquium 125, Kluwer Academic Publishers, Dordrecht, p. 225.

Lockman, F. J. and Brown, R. L.: 1975a, *Astrophys. J.* **201**, 134.

Lockman, F. J. and Brown, R. L.: 1975b, *Astrophys. J.* **201**, 134.

Lockman, F. J., Pisano, D. J. and Howard, G. J.: 1996, *Astrophys. J.* **472**, 173.

Loren, R. B., Plambeck, R. L., Davis, J. H. and Snell, R. L.: 1981, *Astrophys. J.* **245**, 495.

Lorentz, H. A.: 1906, *Proc. Acad. Sci. Amsterdam* **8**, 591.

Luhman, M. I., Jaffe, D. T., Sternberg, A., Herrmann, F. and Poglitsch, A.: 1997, *Astrophys. J.* **482**, 298.

Makuiti, S., Shibai, H., Okuda, H., Nakagawa, T., Matsuhara, H., Hiromoto, N. and Okumura, K.: 1996, *Publ. Astron. Soc. Jpn.* **48**, L71.

Martín-Pintado, J., Bachiller, R., Thum, C. and Walmsley, C. M.: 1989, *Astron. Astrophys.* **215**, L13.

Martín-Pintado, J., Bujarrabal, V., Bachiller, R., Gómez-González, J. and Planesas, P.: 1988, *Astron. Astrophys.* **197**, L15.

Masson, C. R.: 1986, *Astrophys. J.* **302**, L30.

Mathis, J. S.: 1971, *Astrophys. J.* **167**, 261.

Mathis, J. S.: 1980, *in* P. A. Shaver (ed.), *Radio Recombination Lines*, Riedel, Dordrecht, p. 269.

Matthews, H. E., Pedlar, A. and Davies, R. D.: 1973, *Mon. Not. R. Astron. Soc.* **165**, 149.

Mauersberger, R., Guélin, M., Martín-Pintado, J., Thum, C., Hein, H. and Navarro, S.: 1989, *Astron. Astrophys. Suppl. Ser.* **79**, 217.

Mazing, M. A. and Wrubleskaja, N. A.: 1966, *Sov. Phys. - JETP* **50**, 343.

McGee, R. X. and Gardner, F. F.: 1967, *Nature* **213**, 579.

McGee, R. X. and Newton, L. M.: 1981, *Mon. Not. R. Astron. Soc.* **196**, 889.

McKee, C. F. and Ostriker, J. P.: 1977, *Astrophys. J.* **218**, 148.

Meaburn, J. and Walsh, J. R.: 1980, *Mon. Not. R. Astron. Soc.* **191**, 5.

Meeks, M. L. (ed.): 1976, *Astrophysics, Part C: Radio Observations*, Vol. 12 of *Methods of Experimental Physics*, Academic Press, New York, p. 308.

Menzel, D. H.: 1968, *Nature* **218**, 756.

Menzel, D. H.: 1969, *Astrophys. J., Suppl. Ser.* **18**(161), 221.

Menzel, D. H. and Pekeris, C. L.: 1935, *Mon. Not. R. Astron. Soc.* **96**, 77.

Merrill, P. W. and Burwell, C. G.: 1933, *Astrophys. J.* **78**, 87.

Mezger, P. G.: 1960, Meßprogramme bei der Wellenlänge 11 cm am 25m-Radioteleskop Stockert. I. Radioteleskop und Meßtechnik bei 2.7 GHz, *Technical Report 59*, Veröff. Univ. Sternwarte Bonn.

Mezger, P. G.: 1965, Letter to R. L. Sorochenko.

Mezger, P. G.: 1970, *in* W. Becker and G. Contopoulos (eds), *The Spiral Structure of Our Galaxy*, IAU Symp. 38, Reidel, Dordrecht, p. 107.

Mezger, P. G.: 1978, *Astron. Astrophys.* **70**, 565.

Mezger, P. G.: 1980, *in* P. A. Shaver (ed.), *Radio Recombination Lines*, Reidel, Dordrecht, p. 81.

Mezger, P. G.: 1992, *Blick in das kalte Weltall: Protosterne, Staubscheiben und Löcher*, Deutsche Verlag-Anstalt, Stuttgart.

Mezger, P. G.: 1994, Personal letter to M. A. Gordon.

Mezger, P. G., Chini, R., Kreysa, E., Wink, J. E. and Salter, C. J.: 1988, *Astron. Astrophys.* **191**, 44.

Mezger, P. G. and Ellis, S. A.: 1968, *Astrophys. Lett.* **1**, 159.

Mezger, P. G. and Höglund, B.: 1967, *Astrophys. J.* **147**, 579.

Mezger, P. G., Pankonin, V., Schmidt-Burgk, J., Thum, C. and Wink, J.: 1979, *Astron. Astrophys.* **80**, L3.

Mezger, P. G. and Smith, L. F.: 1975, *in* E. K. Karadze (ed.), *Proceedings of the 3rd European Astronomy Meeting*, Acad. Sci. Georgian SSR, Tbilisi, p. 369.

Mezger, P. G., Smith, L. F. and Churchwell, E.: 1974, *Astron. Astrophys.* **32**, 269.

Mihalas, D.: 1978, *Stellar Atmospheres*, second edn, W. H. Freeman, San Francisco, California.

Minaeva, L. A., Sobelman, I. I. and Sorochenko, R. L.: 1967, *Astron. Zh.* **44**, 995.

Miranda, L. F., Torrelles, J. M. and Eiroa, C.: 1995, *Astrophys. J.* **446**, L39.

Mochizuki, K. and Nakagawa, T.: 2000, *Astrophys. J.* **535**, 118.

Mohr, P. J. and Taylor, B. N.: 1999, Codata recommended values of the fundamental physical constants: 1998, *J. Phys. Chem. Ref. Data* **28**(6).

Moran, J. M.: 1994, *in* R. Genzel and A. I. Harris (eds), *The Nucleii of Normal Galaxies*, Kluwer, Dordrecht, p. 475.

Moran, J. M.: 2002, Email to M. A. Gordon.

Muller, C. A. and Oort, J. H.: 1951, *Nature* **168**, 357.

Münch, G.: 1957, *Astrophys. J.* **125**, 42.

Murcray, F. J., Goldman, A., Murcray, F. H., Bradford, C. M., Murcray, D. G., Coffey, M. T. and Mankin, W. G.: 1981, *Astrophys. J.* **247**, L97.

Nakagawa, T., Yui, Y. Y., Doi, Y., Okuda, H., Shibai, H., Mochizuki, K., Nishimura, T. and Low, F. J.: 1998, *Astrophys. J., Suppl. Ser.* **115**, 259.

Natta, A., Walmsley, C. M. and Tielens, A. G. G. M.: 1994, *Astrophys. J.* **428**, 209.

Odegard, N.: 1985, *Astrophys. J., Suppl. Ser.* **57**, 571.

O'Dell, C. R.: 1966, *Astrophys. J.* **143**, 168.

Oks, E.: 2004, *Astrophys. J.* **609**, L25.

Olnon, F. M.: 1975, *Astron. Astrophys.* **39**, 217.

Onello, J. S. and Phillips, J. A.: 1995, *Astrophys. J.* **448**, 727.

Onello, J. S., Phillips, J. A., Benaglia, P., Goss, W. M. and Terzian, Y.: 1994, *Astrophys. J.* **426**, 249.

Onello, J. S., Phillips, J. A. and Terzian, Y.: 1991, *Astrophys. J.* **383**, 693.

Oster, L.: 1961, *Rev. Mod. Phys.* **33**, 525.

Oster, L.: 1970, *Astron. Astrophys.* **9**, 318.

Osterbrock, D. E.: 1989, *Astrophysics of Gaseous Nebulae and Active Galaxies*, University Science Books, Mill Valley, California.

Osterbrock, D. E. and Flather, E.: 1959, *Astrophys. J.* **129**, 26.

Palmer, P.: 1968, *Radio Frequency Recombination Spectra of Diffuse Nebulae*, Ph.d dissertation, Harvard University.

Palmer, P.: 2001, Email message to M. A. Gordon.

Palmer, P. and Zuckerman, B.: 1966, *Nature* **209**, 1118.

Palmer, P., Zuckerman, B., Penfield, H., Lilley, A. E. and Mezger, P. G.: 1967, *Nature* **215**, 40.

Panagia, N.: 1979, *Mem. Soc. Astron. Ital.* **50**, 79.

Pankonin, V.: 1980, *in* P. A. Shaver (ed.), *Radio Recombination Lines*, Reidel, Dordrecht, p. 111.

Pankonin, V., Walmsley, C. M. and Thum, C.: 1980, *Astron. Astrophys.* **89**, 173.

Pankonin, V., Walmsley, C. M., Wilson, T. L. and Thomasson, P.: 1977, *Astron. Astrophys.* **57**, 341.

Parijskij, Y. N.: 2002, Email to M. A. Gordon.

Payne, H. E., Anantharamaiah, K. R. and Erickson, W. C.: 1989, *Astrophys. J.* **341**, 890.

Payne, H. E., Anantharamaiah, K. R. and Erickson, W. C.: 1994, *Astrophys. J.* **430**, 690.

Pedlar, A. and Davies, R. D.: 1972, *Mon. Not. R. Astron. Soc.* **159**, 129.

Pedlar, A., Davies, R. D., Hart, L. and Shaver, P. A.: 1978, *Mon. Not. R. Astron. Soc.* **182**, 473.

Peimbert, M.: 1967, *Astrophys. J.* **150**, 825.

Peimbert, M.: 1979, *in* W. B. Burton (ed.), *The Large-Scale Characteristics of the Galaxy. IAU Symp. No. 84*, Reidel, Dordrecht, p. 307.

Peimbert, M. and Goldsmith, D. W.: 1972, *Astron. Astrophys.* **19**, 398.

Peimbert, M. and Spinrad, H.: 1970, *Astrophys. J.* **159**, 809.

Peimbert, M. and Torres-Peimbert, S.: 1977, *Mon. Not. R. Astron. Soc.* **179**, 217.

Penfield, H., Palmer, P. and Zuckerman, B.: 1967, *Astrophys. J.* **148**, 25.

Pengelly, R. M. and Seaton, M. J.: 1964, *Mon. Not. R. Astron. Soc.* **127**, 165.

Petuchowski, S. J. and Bennet, C. L.: 1993, *Astrophys. J.* **405**, 591.

Pfund, A. H.: 1924, *J. Opt. Soc. Am. A* **9**, 193.

Phookun, B., Anantharamaiah, K. R. and Goss, W. M.: 1998, *Mon. Not. R. Astron. Soc.* **295**, 156.

Pikelner, S. B.: 1967, *Astron. Zh.* **44**, 915. English translation: 1968 Sov. Astron. Lett. 11:737.

Planesas, P., Martín-Pintado, J. and Serabyn, E.: 1992, *Astrophys. J.* **386**, L23.

Ponomarev, V. O.: 1994, *Pis'ma Astron. Zh.* **20**, 184. English translation: 1994 Astron. Lett. 12:155.

Ponomarev, V. O., Smith, H. A. and Strelnitski, V. S.: 1994, *Astrophys. J.* **424**, 976.

Ponomarev, V. O. and Sorochenko, R. L.: 1992, *Pis'ma Astron. Zh.* **18**, 541. English translation: 1992 Sov. Astron. Lett. 18:215.

Pottasch, S. R.: 1984, *Planetary Nebulae*, Vol. 107 of *Astrophysics and Space Science Library*, Reidel, Dordrecht.

Price, R. M.: 2002, Private communication to M. A. Gordon.

Puxley, P. J., Brand, P. W. J. L., Moore, T. J., Mountain, C. M., Nakai, N. and Yamashita, T.: 1989, *Astrophys. J.* **345**, 163.

Puxley, P. J., Brand, P. W. J. L., Moore, T. J. T., Mountain, C. M. and Nakai, N.: 1991, *Mon. Not. R. Astron. Soc.* **248**, 585.

Reber, G.: 1944, *Astrophys. J.* **100**, 279.

Reber, G. and Greenstein, J. L.: 1947, *Observatory* **67**, 15.

Recillas-Cruz, E. and Peimbert, M.: 1970, *Bol. Obs. Tonantz. Tacub.* **35**, 247.

Reifenstein, E. C., Wilson, T. L., Burke, B. F. and Altenhoff, W. J.: 1970, *Astron. Astrophys.* **4**, 357.

Reynolds, R. J.: 1984, *Astrophys. J.* **282**, 191.

Reynolds, R. J.: 1990, *in* S. Bowyet and C. Leinert (eds), *Galactic and Extragalactic Background Radiation, Proc. IAU Symp. No. 139*, Kluwer, Dordrecht, p. 157.

Reynolds, R. J.: 1991, *Astrophys. J.* **372**, L17.

Reynolds, R. J.: 1993, *in* J. P. Cassinelli and E. B. Churchwell (eds), *Massive Stars: Their Lives in the Interstellar Medium*, Vol. CS-35 of *Conference Proceedings*, Astron. Soc. Pacific, San Francisco, p. 338.

Reynolds, S. P.: 1988, *in* G. L. Verschuur and K. I. Kellermann (eds), *Galactic and Extra-Galactic Radio Astronomy*, Springer-Verlag, New York, p. 439.

Roberts, W. W.: 1972, *Astrophys. J.* **173**, 259.

Roberts, W. W.: 1975, *Vistas Astron.* **19**, 91.

Roberts, W. W.: 2001.

Robinson, B.: 2001.

Roelfsema, P. R., Goss, W. M. and Geballe, T. R.: 1988, *Astron. Astrophys.* **207**, 132.

Roelfsema, P. R., Goss, W. M. and Geballe, T. R.: 1989, *Astron. Astrophys.* **222**, 347.

Roelfsema, P. R., Goss, W. M. and Mallik, D. C. V.: 1992, *Astrophys. J.* **394**, 188.

Roelfsema, P. R., Goss, W. M., Whiteoak, J. B., Gardner, F. F. and Pankonin, V.: 1987, *Astron. Astrophys.* **175**, 219.

Russell, R. W., Melnick, G., Gull, G. E. and Harwit, M.: 1980, *Astrophys. J.* **240**, L99.

Rydberg, J. R.: 1890, *Zs. Phys. Chem. (Leipzig)* **5**(227).

Salem, M. and Brocklehurst, M.: 1979, *Astrophys. J., Suppl. Ser.* **39**, 633.

Sanders, D. B., Scoville, N. Z. and Soifer, B. T.: 1991, *Astrophys. J.* **370**, 158.

Schmidt, M.: 1965, Rotation parameters and distribution of mass in the galaxy, *in* A. Blaauw and M. Schmidt (eds), *Galactic Structure*, Vol. V of *Stars and Stellar Systems*, University of Chicago Press, Chicago, chapter 65, p. 513.

Seaquist, E. R. and Bell, M. B.: 1977, *Astron. Astrophys.* **60**, L1.

Seaquist, E. R., Carlstrom, J. E., Bryant, P. M. and Bell, M. B.: 1996, *Astrophys. J.* **465**, 691.

Seaquist, E. R., Kerton, C. R. and Bell, M. B.: 1994, *Astrophys. J.* **429**, 612.

Seaton, M. J.: 1959, *Mon. Not. R. Astron. Soc.* **119**, 90.

Seaton, M. J.: 1964, *Mon. Not. R. Astron. Soc.* **127**, 177.

Sejnowski, T. J. and Hjellming, R. M.: 1969, *Astrophys. J.* **156**, 915.

Shaver, P. A.: 1975, *Pramana* **5**, 1.

Shaver, P. A.: 1976a, *Astron. Astrophys.* **49**, 149.

Shaver, P. A.: 1976b, *Astron. Astrophys.* **49**, 1.

Shaver, P. A.: 1978, *Astron. Astrophys.* **68**, 97.

Shaver, P. A.: 1980, *Astron. Astrophys.* **91**, 279.

Shaver, P. A., Churchwell, E. and Rots, A. H.: 1977, *Astron. Astrophys.* **55**, 435.

Shaver, P. A., Churchwell, E. and Walmsley, C. M.: 1978, *Astron. Astrophys.* **64**, 1.

Shaver, P. A., McGee, R. X., Newton, L. M., Danks, A. C. and Pottasch, S. R.: 1983, *Mon. Not. R. Astron. Soc.* **204**, 53.

Shaver, P. A., Pedlar, A. and Davies, R. D.: 1976, *Mon. Not. R. Astron. Soc.* **177**, 45.

Shibai, H., Okuda, H., Nakagawa, T., Matsuhara, H., Maihara, T., Mizutani, K., Kobayashi, Y., Hiromoto, N., Nishimura, T. and Low, F. J.: 1991, *Astrophys. J.* **374**, 522.

Shimabukuro, F. J. and Wilson, W. J.: 1973, *Astrophys. J.* **183**, 1025.

Shklovsky, I. S.: 1956a, *Astron. Zh.* **33**, 315.

Shklovsky, I. S.: 1956b, *Kosmicheskoe Radioisluchenie*, Gostechizdat, Moscow. see Shklovsky (1960) for the English translation.

Shklovsky, I. S.: 1960, *Cosmic Radio Waves*, Harvard Univ. Press, Cambridge. Translated by R. B. Rodman and C. M. Varsavsky.

Silverglate, P. R.: 1984, *Astrophys. J.* **278**, 604.

Silverglate, P. R. and Terzian, Y.: 1978, *Astrophys. J.* **224**, 437.

Simonson, S. C. and Mader, G. L.: 1973, *Astron. Astrophys.* **27**, 337.

Simpson, J. P.: 1973a, *Astrophys. Space. Sci.* **20**, 187.

Simpson, J. P.: 1973b, *Proc. Ast. Soc. Pac.* **85**, 479.

Smirnov, G. T.: 1985, PhD thesis, Lebedev Physical Institute, Moscow, USSR.

Smirnov, G. T., Kitaev, V. V., Sorochenko, R. L. and Schegolev, A. F.: 1996, *Ann. Sess. of Sci. Council Astrocosmic. Ctr.*, P. N. Lebedev Phys. Inst., Moscow, p. 3.

Smirnov, G. T., Sorochenko, R. L. and Pankonin, V.: 1984, *Astron. Astrophys.* **135**, 116.

Smirnov, G. T., Sorochenko, R. L. and Walmsley, C. M.: 1995, *Astron. Astrophys.* **300**, 923.

Smith, H. A., Strelnitski, V. S., Miles, J. W., Kelley, D. M. and Lacy, J. H.: 1997, *Astron. J.* **114**, 2658.

Smith, M. G. and Weedman, D. W.: 1970, *Astrophys. J.* **160**, 65.

Sobelman, I. I.: 1992, *Atomic Spectra and Radiative Transitions*, Vol. 12 of *Springer Series on Atoms and Plasmas*, second edn, Springer, New York.

Sobelman, I. I., Vainshtein, L. A. and Yukov, E. A.: 1995, *Excitation of Atoms and Broadening of Spectral Lines*, Vol. 15 of *Springer Series on Atomic, Optical and Plasma Physics*, second edn, Springer, New York.

Sorochenko, R. L.: 1965, *Trudy Fiz. Inst., P. N. Lebedeva Akad. Nauk USSR* **28**, 90. English translation: 1966, Proc. P .N. Lebedev Phys. Inst. 28:65.

Sorochenko, R. L.: 1996, *Astron. Astrophys. Trans.* **11**, 199.

Sorochenko, R. L. and Berulis, I. I.: 1969, *Ap. Letters* **4**, 173.

Sorochenko, R. L. and Borodzich, E. V.: 1965, *Dokl. Akad. Nauk SSSR* **163**, 603. English translation: 1966, Sov. Phys. – Dokl. 10, 588.

Sorochenko, R. L. and Borodzich, E. V.: 1966, *Trans. IAU* **XIIB**, 360. Paper presented by Vitkevich at the XII General Assembly of the IAU in Hamburg, Germany, 1964.

Sorochenko, R. L., Puzanov, V. A., Salomonovich, A. E. and Steinschleiger, V. B.: 1969, *Astrophys. Lett.* **3**, 7.

Sorochenko, R. L., Rydbeck, G. and Smirnov, G. T.: 1988, *Astron. Astrophys.* **198**, 233.

Sorochenko, R. L. and Smirnov, G. T.: 1987, *Pis'ma Astron. Zh.* **13**, 191. English translation: 1987 Sov. Astron. Lett. 13:77.

Sorochenko, R. L. and Smirnov, G. T.: 1990, *in* M. A. Gordon and R. L. Sorochenko (eds), *Radio Recombination Lines: 25 Years of Investigation*, IAU Colloquium 125, Kluwer Academic Publishers, Dordrecht, p. 189.

Sorochenko, R. L., Smirnov, G. T. and Lekht, E. E.: 1991, *in* Y. N. Parijskij (ed.), *Proceedings of the XXXIII All-Union Conference of Galactic and Extragalactic Radioastronomy*, Physical-Technical Inst., Ilim, Ashkhabad, p. 111.

Sorochenko, R. L. and Tsivilev, A. P.: 2000, *Astron. Zn.* **77**, 488. English translation: 2000 Astron. Reports 44:426.

Sorochenko, R. L. and Walmsley, C. M.: 1991, *Astron. Astrophys. Trans.* **1**, 31.

Spitzer, L.: 1978, *Physical Processes in the Interstellar Medium*, Wiley, New York.

Spitzer, L. and Tomasko, M. G.: 1968, *Astrophys. J.* **152**, 971.

Stacey, G. J., Viscuso, P. J., Fuller, C. E. and Kurtz, N. T.: 1985, *Astrophys. J.* **289**, 803.

Stark, J.: 1913, *Sitzungber. Kungl. Akad. Wiss.* p. 932.

Stepkin, S. V., Konovalenko, A. A., Kanthia, N. G. and Shankar, N. U.: 2007, *Mon. Not. R. Astron. Soc.* **374**, 852.

Storey, P. J. and Hummer, D. G.: 1995, *Mon. Not. R. Astron. Soc.* **272**, 41.

Strelnitski, V. S., Haas, M. R., Smith, H. A., Erickson, E. F., Colgan, S. W. J. and Hollenbach, D. J.: 1996, *Science* **272**, 1459.

Strelnitski, V. S., Ponomarev, V. O. and Smith, H. A.: 1996, *Astrophys. J.* **470**, 1118.

Strelnitski, V. S., Smith, H. A. and Ponomarev, V. O.: 1996, *Astrophys. J.* **470**, 1134.

Strömgren, B.: 1939, *Astrophys. J.* **89**, 526.

Stutzki, J., Stacey, G. J., Harris, A. I., Jaffe, D. T. and Lugten, J. B.: 1988, *Astrophys. J.* **332**, 379.

Subramanyan, R.: 1992a, *Mon. Not. R. Astron. Soc.* **254**, 292.

Subramanyan, R.: 1992b, *Mon. Not. R. Astron. Soc.* **254**, 719.

Subramanyan, R. and Goss, W. M.: 1996, *Mon. Not. R. Astron. Soc.* **281**, 239.

Sullivan, W. T.: 1982, *Classics in Radio Astronomy*, D. Reidel, Dordrecht.

Sunyaev, R. A.: 1966, *Astron. Zh.* **43**, 1237.

Taylor, J. H. and Cordes, J. M.: 1993, *Astrophys. J.* **411**, 674.

Taylor, J. H. and Manchester, R. N.: 1977, *Astrophys. J.* **215**, 885.

Terzian, Y.: 1990, *in* M. A. Gordon and R. L. Sorochenko (eds), *Radio Recombination Lines: 25 Years of Investigation*, IAU Colloquium 125, Kluwer Academic Publishers, Dordrecht, p. 141.

Terzian, Y.: 2002, Personal Communciation to M. A. Gordon.

Terzian, Y., Higgs, L. A., MacLeod, J. M. and Doherty, L. H.: 1974, *Astron. J.* **79**, 1018.

Terzian, Y. and Parrish, A.: 1970, *Astrophys. Letters* **5**, 261.

Thum, C.: 1986, Temperature calibration of line observations at the 30-m telescope, *Internal report*, IRAM.

Thum, C., Martín-Pintado, J. and Bachiller, R.: 1992, *Astron. Astrophys.* **256**, 507.

Thum, C., Matthews, H. E., Harris, A. I., Taconi, L. J., Shuster, K. F. and Martín-Pintado, J.: 1994, *Astron. Astrophys.* **288**, L25.

Thum, C. and Morris, D.: 1999, *Astron. Astrophys.* **344**, 923.

Thum, C., Strelnitski, V. S., Martín-Pintado, J., Matthews, H. E. and Smith, H. A.: 1995, *Astron. Astrophys.* **300**, 843.

Tielens, A. G. G. M. and Hollenbach, D.: 1985, *Astrophys. J.* **291**, 722.

Towle, J. P., Feldman, P. A. and Watson, J. K. G.: 1996, *Astrophys. J., Suppl. Ser.* **107**, 747.

Troland, T. H., Crutcher, R. M. and Heiles, C.: 1985, *Astrophys. J.* **298**, 808.

Tsivilev, A. P., Ershov, A. A., Smirnov, G. T. and Sorochenko, R. L.: 1986, *Pis'ma Astron. Zh.* **12**, 848. English translation: 1986 Sov. Astron. Lett. 12:355.

Turner, B. E.: 1970, *Astrophys. Lett.* **6**, 99.

Ulich, B. L. and Haas, R. W.: 1976, *Astrophys. J., Suppl. Ser.* **30**, 247.

Vallée, J. P.: 1987a, *Astron. Astrophys.* **178**, 237.

Vallée, J. P.: 1987b, *Astrophys. J.* **317**, 693.

Vallée, J. P.: 1987c, *Astrophys. Space. Sci.* **133**, 275.

Vallée, J. P.: 1987d, *Astron. J.* **92**, 204.

Vallée, J. P.: 1989, *Astrophys. J.* **341**, 238.

van de Hulst, H. C.: 1945, *Nederladsch Tidjschrift voor Naturkunde* **11**, 230.

van der Werf, P. P., Goss, W. M. and Vanden Bout, P. A.: 1988, *Astron. Astrophys.* **201**, 311.

van der Werf, P. P., Stutzki, J., Sternberg, A. and Krabbe, A.: 1996, *Astron. Astrophys.* **313**, 633.

van Dishoeck, E. F. and Black, J. H.: 1986, *Astrophys. J., Suppl. Ser.* **62**, 109.

Vázquez, R., Torrelles, J. M., Rodríguez, L. F., Gómez, Y., López, J. A. and Miranda, L. F.: 1999, *Astrophys. J.* **515**, 633.

Vidal, C. R.: 1964, *Naturforsch.* **A19**, 1947.

Vidal, C. R.: 1965, *Proc. Seventh. Intern. Conf. on Phenomena in Ionized Gases*, p. 168.

Viner, M. R., Clarke, J. N. and Hughes, J. N.: 1976, *Astron. J.* **81**, 512.

Voigt, W.: 1913, *Phys. Zeits.* **14**, 377.

von Sommerfeld, A.: 1916a, *Ann. Phys. (Leipzig)* **50**, 385.

von Sommerfeld, A.: 1916b, *Ann. Phys. (Leipzig)* **51**, 1.

Walmsley, C. M.: 1990, *Astron. Astrophys. Suppl. Ser.* **82**, 201.

Walmsley, C. M., Churchwell, E. and Terzian, Y.: 1981, *Astron. Astrophys.* **96**, 278.

Walmsley, C. M. and Watson, W. D.: 1982, *Astrophys. J.* **260**, 317.

Waltman, W. B., Waltman, E. B., Schwartz, P. R., Johnson, K. J. and Wilson, W. J.: 1973, *Astrophys. J., Lett.* **185**, L135.

Wannier, P. G., Lichten, S. M. and Morris, M.: 1983, *Astrophys. J.* **255**, 738.

Watson, J. K. G.: 2002, Email to M. A. Gordon.

Watson, J. K. G.: 2007, *J. Phys. B* **39**, 1889.

Watson, W. D., Western, L. R. and Christensen, R. B.: 1980, *Astrophys. J.* **240**, 956.

Weedman, D. W.: 1966, *Astrophys. J.* **145**, 965.

Weinreb, S., Barrett, A. H., Meeks, M. L. and Henry, J. C.: 1963, *Nature* **200**, 829.

Weisskopf, V.: 1932, *Z. Phys.* **75**, 287.

Weisskopf, V. and Wigner, E.: 1930, *Z. Phys.* **63**, 54.

White, R. L. and Becker, R. H.: 1985, *Astrophys. J.* **297**, 677.

Wild, J. P.: 1952, *Astrophys. J.* **115**, 206.

Wilson, O. C., Münch, G., Flather, E. M. and Coffeen, M. F.: 1959, *Astrophys. J., Suppl. Ser.* **4**, 199.

Wilson, T. L., Bieging, J. H. and Wilson, W. E.: 1979, *Astron. Astrophys.* **71**, 205.

Wilson, T. L. and Filges, L.: 1990, *in* M. A. Gordon and R. L. Sorochenko (eds), *Radio Recombination Lines: 25 Years of Investigation*, IAU Colloquium 125, Kluwer Academic Publishers, Dordrecht, p. 105.

Wilson, T. L., Hoang-Binh, D., Stark, A. A. and Filges, L.: 1990, *Astron. Astrophys.* **238**, 331.

Wilson, T. L., Mauersberger, R., Muders, D., Przewodnik, A. and Olano, C. A.: 1993, *Astron. Astrophys.* **280**, 221.

Wilson, T. L., Mezger, P. G., Gardner, F. F. and Milne, D. K.: 1970, *Astron. Astrophys.* **6**, 364.

Wilson, T. L. and Pauls, T.: 1984, *Astron. Astrophys.* **138**, 225.

Wink, J., Wilson, T. L. and Bieging, J. H.: 1983, *Astron. Astrophys.* **127**, 211.

Wyrowski, F., Schilke, P., Hofner, P. and Walmsley, C. M.: 1997, *Astrophys. J.* **487**, L171.

Zelik, M. and Lada, C. J.: 1978, *Astrophys. J.* **222**, 896.

Zhao, J.-H., Anantharamaiah, K. R., Goss, W. M. and Viallefond, F.: 1996, *Astrophys. J.* **472**, 54.

Zhao, J.-H., Anantharamaiah, K. R., Goss, W. M. and Viallefond, F.: 1997, *Astrophys. J.* **482**, 186.

Zuckerman, B. and Ball, J. A.: 1974, *Astrophys. J.* **190**, 35.

Zuckerman, B. and Palmer, P.: 1968, *Astrophys. J.* **153**, L145.

Author Index

Subject Index